ELECTRONIC HEALTH RECORD

A Systems Analysis of the Medications Domain

ELECTRONIC HEALTH RECORD

A Systems Analysis of the Medications Domain

Alexander Scarlat, MD

Foreword by John Halamka, MD

CRC Press
Taylor & Francis Group
Boca Raton London New York

CRC Press is an imprint of the
Taylor & Francis Group, an **informa** business
A PRODUCTIVITY PRESS BOOK

CRC Press
Taylor & Francis Group
6000 Broken Sound Parkway NW, Suite 300
Boca Raton, FL 33487-2742

Version Date: 2011913

International Standard Book Number: 978-1-4398-7852-1 (hbk)

Library of Congress Cataloging-in-Publication Data

Scarlat, Alexander.
Electronic health record : a systems analysis of the medications domain / Alexander Scarlat.
p. ; cm.
Includes bibliographical references and index.
ISBN 978-1-4398-7852-1 (pbk. : alk. paper)
I. Title.
[DNLM: 1. Electronic Health Records--organization & administration. 2. Electronic Prescribing. 3. Medication Systems--organization & administration. 4. Systems Analysis. WX 175]

610.285--dc23 2011037684

Visit the Taylor & Francis Web site at
http://www.taylorandfrancis.com

and the CRC Press Web site at
http://www.crcpress.com

For Evemine, Adrian, Yuval, Dan, May, and Adam

And especially for my best friend and better half

Dahlia

Contents

List of Figures

Foreword

What is the ideal medication workflow?

Simply put, it is from a doctor's brain to a patient's vein without handwriting, handoffs, or hassle.

Although Beth Israel Deaconess Medical Center (BIDMC) has not used handwritten medication orders since 2001 and has an e-prescribing rate of 96%, the medication workflow is not ideal.

For 20 years, we have automated many existing processes: provider ordering, pharmacy supply chain, positive patient identification, medication administration records, and billing. However, automating an imperfect process does not make it better. Automation just makes it faster. Over the past year, we have spent 3,000 hours reexamining the entire medication workflow using a multidisciplinary Lean process. What is the best way to package and deliver medications? How should care delivery teams reconcile medications and communicate issues? How do we involve the patient?

We are now turning this idealized design into a detailed specification that we can automate.

Doing this work takes three kinds of people: medication experts with domain knowledge of the workflow, information technology (IT) experts with domain knowledge of available hardware/software, and project managers who glue the two together.

As an IT professional, I have worked on hundreds of projects. The most successful are those with business owners who understand their requirements and can communicate them to IT implementers. In one recent large project, we taught systems analysis to the stakeholders because the project bottleneck was communicating the idealized workflow. The project is now on time to go live.

Dr. Alexander Scarlat's *Electronic Health Record: A Systems Analysis of the Medications Domain* is important for two reasons. First, it provides a framework that will enable clinicians to communicate with technologists. As the pace of IT change accelerates, so does demand for workflow automation. Technology is no longer the rate-limiting step. Translating user needs into IT products and services is the biggest challenge we face. Dr. Scarlat's clear explanations empower healthcare professionals with tools that will enhance any IT project.

Second, the medication domain is the highest-priority area for healthcare stakeholders to improve quality, safety, and efficiency. Meaningful Use Stage 1 included numerous medication lists, e-prescribing, allergy, reconciliation, and decision support requirements. Meaningful Use Stage 2 will include even more medication-related activities, emphasizing the importance of automating these processes correctly.

I plan to use this book in the BIDMC medication work, which seeks to achieve zero defects, cost reductions, and patient engagement. Both clinicians and IT professionals should find the book is a valuable resource as they create the re-formed healthcare delivery system of the future, beyond Meaningful Use.

John D. Halamka, MD
Boston, Massachusetts

Preface

Why This Book?

During the years I have worked in healthcare information technology (IT), either with vendors of electronic health records (EHRs) or the hospitals implementing this technology, I have been asked many times by colleagues to recommend a practical book on the subject. The clinicians—physicians, nurses, pharmacists—are looking for a book to introduce them to the modeling techniques and engineering tools the IT department or their vendor uses, while the information technologists are interested in an introductory text on the clinical domain, something that can be assimilated without going to medical school. Realizing these two highly educated groups of professionals rarely speak the same language, they need a communication tool.

Systems analysis in general and structured systems analysis specifically offer a rigorous methodology and can create such communication tools in the form of models, diagrams, and other succinct, systematic artifacts. Systems analysis has been successfully applied in thousands of IT projects in other industries, but somehow it just never caught on in the healthcare IT domain. One can find books and articles on the vendor selection process, best practices for user training, and the psychological, sociological, and financial aspects of informatics in medicine as well as the cultural and organizational change management that accompanies the implementation of IT in clinical environments. However, to my best knowledge, very little if any has been published in the domain of healthcare informatics systems analysis.

Even when an organization does have the systems analysis related to EHRs it has developed or implemented, this reference is considered an intellectual property to be guarded accordingly, as a secret possession. Unfortunately, a software blueprint kept in a vault cannot be improved, lessons learned by others cannot be applied to or derived from it, and the professionals involved in the development or implementation of EHRs may find themselves "reinventing the wheel" solving problems others have already solved or worse—making the same mistakes others did before them.

Finding no better (published) alternative, I decided to write this book with two goals in mind:

1. Introduce the clinicians to a methodology and tools set so they can better communicate with the IT specialists developing or implementing EHRs
2. Clarify the complexity of the medications domain, a major part of an EHR—the prescription, dispensation, computerized physician order entry, and administration of drugs—for IT professionals to better communicate with their clinical colleagues

In addition, the last chapter on interoperability presents to both clinicians and informatics professionals the standards and vocabularies, rules, and regulations governing the EHR as recommended by U.S. authorities in the recently published American Recovery and Reinvestment Act (ARRA). According to ARRA, meaningful use of certified EHR systems in the United States will entitle clinicians to substantial funds from the U.S. government over the next couple of years. Thus, the need for a structured, systematic analysis of EHRs has aligned with a major, current development in the U.S. healthcare IT industry: the publication of government-backed standards for the industry to adopt to be considered a meaningful user of this technology.

What Is in This Book?

The first chapter is a short primer on structured systems analysis intended to introduce beginners to the tools used in the book: conceptual and logical data models, entity relationship diagrams, context and data flow diagrams.

The second chapter initiates the analysis of the medications domain from two perspectives: workflows and data structures. Workflows are decomposed into processes, activities, and tasks, while data structures are partitioned in the conceptual, entity-relationship, and logical data models.

The analysis continues in a more detailed fashion in Chapters 3 through 6, which respectively discuss the process to prescribe, order, dispense, and administer drugs.

The next three chapters illustrate how the system is expected to interact optimally with users: Chapter 7 is on the user interface; Chapter 8, "Clinical Decision Support," involves what the system can and should do with the information it possesses; and the output expected is discussed in Chapter 9, "Report."

The final chapter presents the standards, vocabularies, rules, and regulations, which if adopted and implemented correctly will enable various disparate EHR systems to play together in the proverbial healthcare IT sandbox.

What Is Not in This Book?

Throughout this book, we examine together the EHR medications domain using structured systems analysis methodology and tools. The system described is a vendor-, platform-, and computer language-independent generic EHR. The reader will not find any vendor or hospital, technology or specific database, application or product mentioned by name. At least initially, system analysis is blind to these issues.

The goal of the book is not to provide the complete, correct, and consistent technical specs for such an ideal, hypothetical system to be built. The book presents the tools and methods to enable and facilitate the communications between clinicians and IT so they in turn can interpret, develop, and implement such clinical-technical specifications. Thus, do not expect to find actual software code, hardware specs, testing scripts, or physical data model minutiae in the book.

In the same spirit, the pharmacological examples used are for IT demonstrative and illustrative purposes only; the examples are not intended for actual clinical or medical use.

Is This Book for You?

The book should be accessible to both clinicians and IT professionals: No initial knowledge of medicine or computer sciences is assumed or required.

Clinicians: Physicians, nurses, pharmacists, C-level administrators, and other healthcare professionals contemplating or in the midst of the implementation of EHR technology will benefit from the succinct diagrammatic representations of the medications domain: workflows and data elements.

IT: Software engineers, architects, data and process modelers, product and project managers, database administrators, analysts, testers, and consultants will find the book useful in clarifying the terms and verbiage used by the clinicians in their work, while reviewing modeling techniques used by other industries for more than four decades.

Whether in the process of requirements elicitation, analysis, design, code, test, configuration, implementation, or maintenance of EHRS, both clinicians and IT professionals can benefit from better techniques to facilitate the abstraction of complex ideas into a set of easily understood models and diagrams. It is hoped both groups should find a common ground, clear and easy-to-use methodology and tools to make their collaboration on the medications informatics domain more productive.

Each chapter ends with a set of questions that should help the reader review the material covered; answers are found in the back of the book. Students and teachers of healthcare informatics will find the book is both methodical and didactic for their academic purposes; thus, it could be used as a textbook for a course on the subject.

When applied correctly, systems analysis can accelerate development, decrease costs, and increase the quality of a software product and eventually user satisfaction. I believe systems analysis can be applied to healthcare informatics with the same successful results demonstrated in other industries.

Acknowledgments

I am thankful to many individuals who have provided useful feedback and encouragement during this journey: John Halamka, William Bria, Pini Matia, Joseph Finn, Karl Matuszewski, Philip Anderson, Dorel Abramovici, Jane Brokel, Laurette Floru, Steven Lins, Jon Patrick, Ruth Tamari, Mansnimar Singh, Scot Silverstein, Brian Berenbach, John Frownfelter, Saswat Mohanty, Ziv Mayanski, and Eyal Frank.

Many thanks to Kristine Mednansky and others from Taylor & Francis/CRC Press for helping bring the idea to the masses.

Suggestions?

This book may contain errors, and most probably can be improved. If you want to contribute suggestions, additional material, ideas, or diagrams, feel free to provide feedback at the book's Web site: http://drscarlat.wordpress.com/.

I hope you will find the book useful and enjoy it as much as I derived pleasure from writing it.

A. Scarlat, MD

About the Author

Alexander Scarlat, MD, is a physician with a degree in computer sciences and a strong background in clinical informatics. Dr. Scarlat's career spans more than two decades, during which he has worked with both vendors developing electronic health records and healthcare organizations implementing this technology. Knowledgeable and experienced in medicine and informatics, Dr. Scarlat is an efficient liaison between these two highly educated groups of professionals—clinicians and information technology—that rarely speak the same language.

Chapter 1

Short Primer on Structured Systems Analysis

Everything should be made as simple as possible but not simpler.

Albert Einstein

The first chapter is a short primer on structured systems analysis: It explains what a system is, why systems analysis is needed, and the reasons this book is based specifically on structured systems analysis.

We discuss the methodology and introduce a set of tools that will be used throughout the book. If you are familiar with structured systems analysis, you may skip to the next chapter.

What Is a System?

According to the Merriam-Webster dictionary online (1), a *system* is

> *a regularly interacting or interdependent group of items forming a unified whole ... an organized set of doctrines, ideas, or principles usually intended to explain the arrangement or working of a systematic whole ... an organized or established procedure ... harmonious arrangement or pattern.*

We are continuously interacting with natural and human-made systems all our lives. Some systems have been thoroughly analyzed (digestive system, highway system, solar system), and some like Electronic Health Record (EHR) systems and the medications domain are only sparingly documented.

Why Systems Analysis?

As no one would build a house without a set of blueprints, no one should consider building a nontrivial computer application without a systematic, structured set of plans. As seen in this book, healthcare information technology (IT) in general and software applications for prescribing, ordering, dispensing, and administering medications specifically are far from trivial systems. The rationale for writing this book is to present the medications domain in a structured, systematic manner. The goal of systems analysis is to create an efficient model of users' needs, the plan for the system under consideration (2,3). When done correctly, systems analysis:

- Creates a set of communication tools for clinicians and IT professionals
- Optimizes all stages of a system life cycle: scope definition and control, requirements elicitation, analysis, design, code writing, testing, implementation, training, and maintenance
- Defines the system in a complete, correct, and concise manner
- Emphasizes certain aspects of a system while deemphasizing other areas
- Partitions a huge thing—"the system"—into smaller work chunks
- Increases consistency between the system parts: modules and data structures
- Minimizes redundancy between components of the system
- Improves maintainability of the system

Can we achieve the same goals using only text documents? Can systems analysis be done with text and spreadsheets? This is not really possible for the following reasons:

- Text-based and spreadsheet documentation of a system tends to grow into intimidating monsters of hundreds and even thousands of pages. Nobody really likes to read this stuff. The old saying about a picture being worth a thousand words is painfully true when documenting IT systems.

- Text and spreadsheets are monolithic and thus cannot be systematically chunked or partitioned into smaller working units. IT systems are usually developed by a number of teams (database, application, user interface, testing, etc.), and the capability to work in parallel on several parts of a system can be advantageous from a productivity perspective.
- It is easier to document ambiguous statements and requirements in text and spreadsheets than in structured diagrams.
- Due to the dynamic interactions between users and developers while defining and refining the system and the ever-changing nature of the requirements, the moment a text document or spreadsheet is published, it is already outdated, irrelevant, and out of sync with reality. Maintaining text documents and spreadsheets is more difficult than maintaining a set of diagrams.
- Text documentation is more prone to suffer from redundancy issues: The same fact—be it a requirement or a solution—may be documented in more than one place. If a fact is updated on one page, there is an immediate need to update all other pages that may document this fact. Redundancy breeds inconsistency unless a significant effort is invested in keeping multiple documents in sync.

The truth is systems analysis is not optional; this is a paraphrase on *Simsion and Witt's* famous adage, *"Data modeling is not optional"* (4). While it may seem to be a self-evident point, I have seen my share of software applications being developed without a blueprint, usually with less-than-satisfactory outcomes. "Some text and spreadsheets in a folder" do not count as a blueprint, and one cannot expect become efficient in systems analysis if one uses only text and spreadsheets as the sole tools.

Why Structured Systems Analysis?

There are two main systems analysis methodologies: *Structured Systems Analysis* (*SSA*) and *Unified Modeling Language* (*UML*) (5). When comparing the two methodologies, I prefer *SSA* to *UML* since the former is easier to understand, learn, become efficient with, and explain to nonexperts. *SSA* is also less prone to misunderstandings and errors than *UML*. In his book, *Modern Structured Analysis, Ed Yourdon* listed the characteristics of SSA tools (3):

- It is graphical, with appropriate textual support.
- It allows the system to be viewed in a top-down, partitioned fashion.
- It suffers from minimal redundancy.
- It helps predict the system's behavior.
- It is transparent and easy to learn, understand, and use.

To this list, I would like to humbly add a sixth characteristic:

- It is technology and implementation independent—the modeling tools should not be limited to any particular implementation or platform technology during the analysis phase. Nor should the clinicians or IT professionals waste precious time on discussing technology-related issues at this early embryonic stage.

Processes and Data

Using the house-building analogy, as the builders need more than one kind of plan (architectural, structural, plumbing, electricity/wiring, etc.), the IT craftspeople need several modeling tools as well. If we agree that an information system is a data-processing system, it will not come as a surprise that *SSA* proceeds from two main perspectives: processes and data (3,6). A top-down analysis of system functions reveals the following concepts at an increased resolution and higher details:

System → *Workflow* → *Process* → *Activity* → *Task*

A system may have multiple workflows. Each workflow may be a collection of several processes. A process can be decomposed into several activities. An activity can be decomposed into several tasks. An activity or task can be defined in one page of graphic or text. For example,

System: Medications → *Workflow:* Prescribe → *Process:* Modify Rx [prescription] → *Activity:* Select medication → *Task:* Select dose unit

Similar top-down analysis of system data structures uncovers the following terms:

System → *Repository* → *Database* → *Table* → *Record* → *Field*

A system may use multiple repositories. Each repository may be composed of several databases. One database usually has many tables. A table has many records, and one record is composed of several fields. For example,

System: Electronic Health Record → *Repository:* Clinical Repository → *Database:* Medications → *Table:* Patient Medication → *Record:* Patient ID, Drug ID, Strength, Strength Unit, … → *Field:* Strength

The following modeling tools are used throughout the book:

- Dataflow diagram (DFD) details the functions the system is to perform and represents data in *motion.*
- Entity-relationship diagram (ERD) details data structures used by the system for storage and represents data at *rest.*

These modeling tools create diagrams that show the main components of a system and the interactions between these components. A technical specifications document (but not this book) may also have two sets of highly structured supporting documents to provide the meaning and definitions of the components and interactions: *Data Dictionary* and *Functional Primitives Specifications*.

If the system under consideration has aspects of a "real-time system," then a third tool, a *State Transition Diagram,* may be employed (3). A real-time system is defined as a hardware/software configuration that can receive high-volume and process at high-speed information coming from the environment in a way that it can control some aspects of the environment or direct a process that takes place in the environment. Real-time systems are embedded in missile guidance, automatic pilots, or mundane washing machines and elevators. Since the medications system rarely if ever exhibits characteristics of real-time systems, we do not detail *State Transition Diagrams* any further.

Dataflow Diagram

Also known as workflow diagram, bubble chart, bubble diagram, or process model, the DFD is a tool used to model data in motion—the system functions (Figure 1.1):

- The origins and destinations of data within a system
- The workflows, processes, activities, and tasks of the system
- The transformation of inputs into outputs

The components of a DFD are

1. *External entity*: a rectangle representing people, groups of users, personae, roles, other computer systems or organizations that interact with the system.
2. *Dataflow*: an arrow representing data in motion, the path for data to move from one part of the system to another, the input needed or the output generated by a component of the system.
3. *Process*: a bubble representing an activity or task performed by the system. It is something the system does: the transformation of an input into an output or the process of moving data from one place to another.
4. *Data store*: two parallel lines representing data at rest, a location for the data to be stored for future use by the system's components.

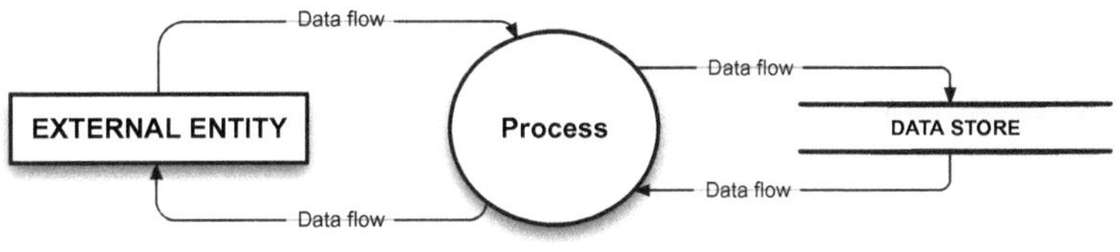

Figure 1.1 DFD symbols.

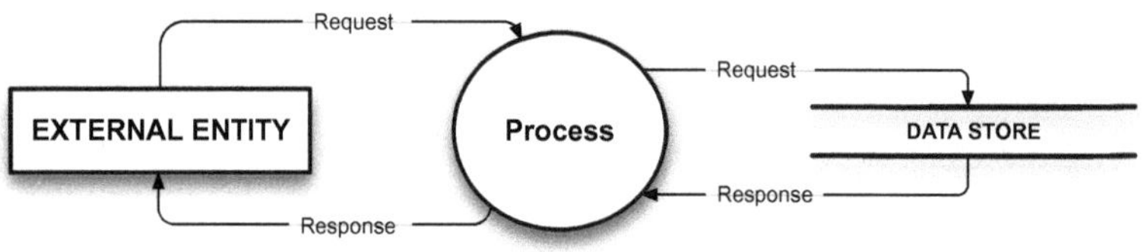

Figure 1.2 DFD request/response format.

Most of the dataflows represent a series of asynchronous messages, a request/response type of communication: A component requests information, and another component responds accordingly (Figure 1.2).

Since some of the DFDs we are to encounter can become quite complex and thus potentially cluttered, we can simplify the diagrams using dataflows with double-headed arrows—a dialog format. Usually the request name is on the top and the response name on the bottom of the dataflow arrow (Figure 1.3).

Let us examine a DFD example showing the process of prescribing a new medication for a specific condition or indication, such as hypertension (Figure 1.4). What can we learn from this DFD?

PRESCRIBER, an external entity, sends the *Patient ID* as a request for a patient's clinical information to process 1.1 *Review Patient Clinical Data*. Process 1.1 *Review Patient Clinical Data* queries the *CLINICAL REPOSITORY* data store, which responds with patient-specific information: age, weight, gender, conditions patient may suffer from, medications patient is taking, allergies, and so on. Process 1.1 displays the information returned, the query results, to *PRESCRIBER*.

PRESCRIBER sends an indication (such as hypertension) to process 1.2 *Select Drug*, which queries a different data store: *RECOMMENDED DRUG DOSE*. The query results, a list of drugs indicated for treating the condition, is returned to process 1.2, which displays the list to *PRESCRIBER*.

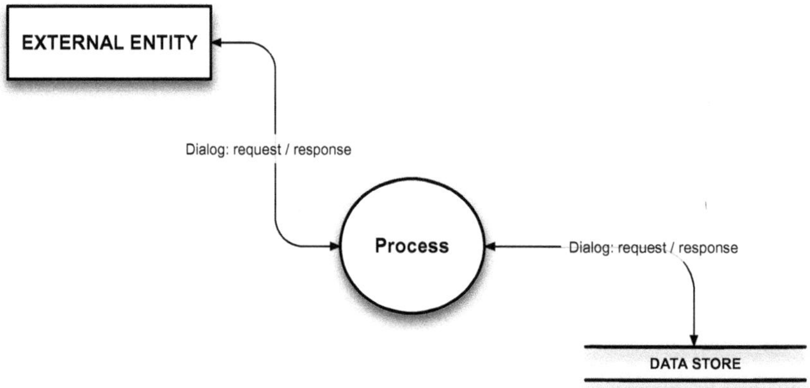

Figure 1.3 DFD dialog format.

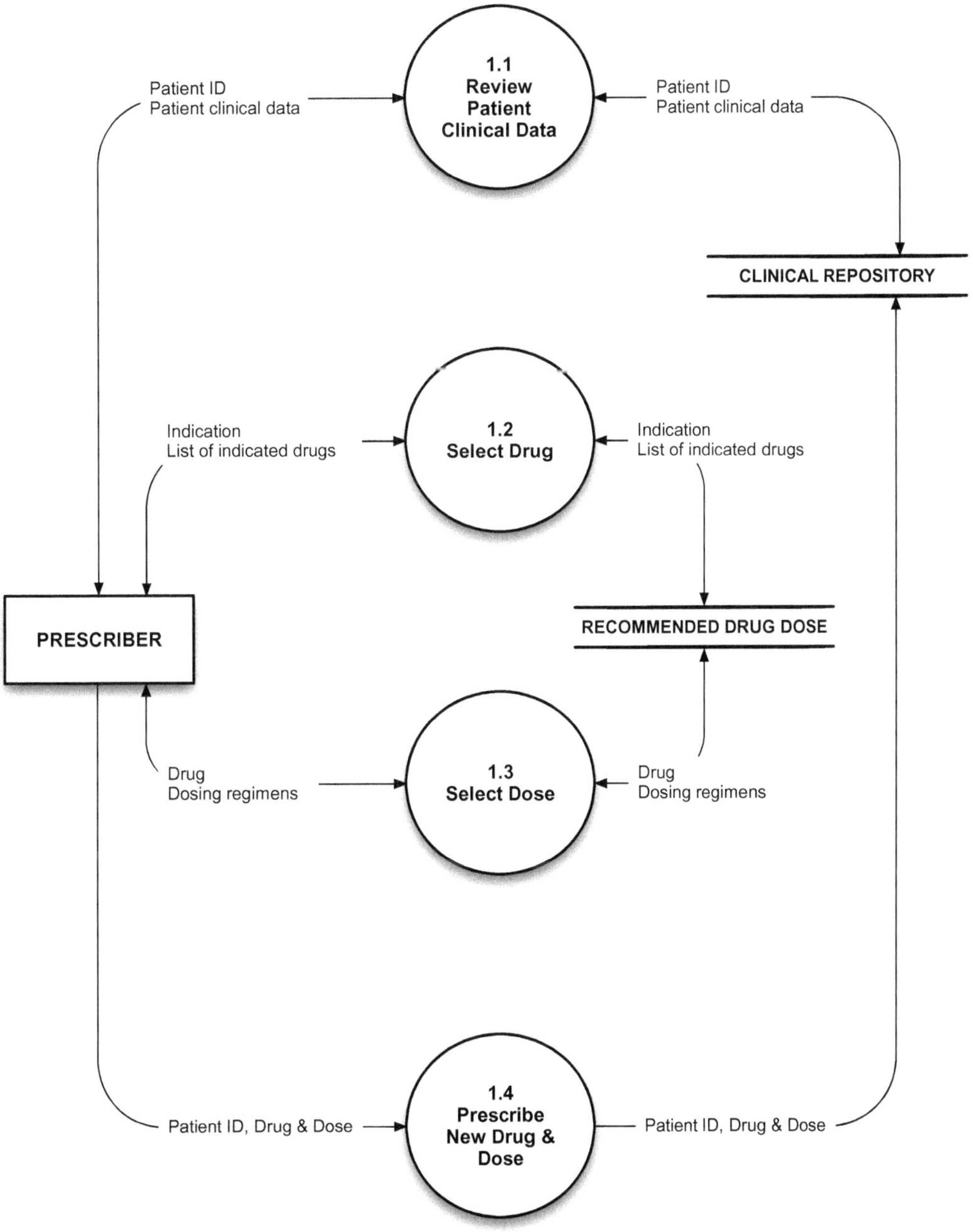

Figure 1.4 DFD example: New Rx (prescription).

Once *PRESCRIBER* selects a drug, it is sent to process 1.3 *Select Dose.* Process *1.3 Select Dose* queries *RECOMMENDED DRUG DOSE* data store and retrieves the dosing regimens for the specific drug (usual dose, maximal dose, age and weight considerations, etc.) and displays the results to *PRESCRIBER.* Finally, *PRESCRIBER* sends the *Patient ID, Drug,* and *Dose* to *CLINICAL REPOSITORY,* where it is stored.

This DFD example can convey to both user and developer a complex process using a simple diagram. In addition, the DFD is easier to maintain than the text that explains it.

Let us discuss some aspects to consider when building a set of DFDs:

The DFD design process starts with a *context diagram*: In this simple, highest-level DFD, there are no data stores, and the system is represented as one bubble/megaprocess interacting with the external entities.

The next step in DFD decomposition is to draft the *DFD 0*: the black-box "system" represented by one bubble in the context diagram, also known as *process number 0,* is decomposed into its main workflows. We construct the context diagram and DFD 0 in the next chapter when we begin the top-down analysis of the medications domain.

The following are some recommendations to consider when drafting a DFD (6,7):

Avoid an overly complex DFD. A DFD should fit nicely into one page. As a rule of thumb, if there are more than a half dozen processes on a DFD, it usually means either the partitioning of the parent, higher-level DFD is incorrect or the current DFD needs a different decomposition approach.

The process of decomposing, partitioning, or exploding the system using DFDs is also called *leveling*. Leveling uses a series of increasingly (top-down) or decreasingly (bottom-up) detailed DFDs to represent a system from a functional perspective.

When is the work of decomposing DFDs supposed to end? The process of decomposing DFDs ends when we reach a stage at which a process, activity, or task is small enough to be detailed in one page. This detail level is called a *functional primitive,* and it is documented using a *flowchart, structured English* (*pseudocode*), or a *decision tree* in a manner that programmers can actually write software for it.

The system is responsible for all the interactions between external actors. Any other interaction between external actors that is outside system boundaries is not to be detailed in a DFD. The corollary is that there should be no direct dataflows between external entities.

Data stores cannot directly communicate with each other. Data stores are representations of data at rest; thus, data stores cannot initiate or complete a data transfer or data processing on their own volition. The corollary is that there should be no dataflow without a process between two data stores.

An external entity cannot communicate directly with a data store. The *PRESCRIBER* cannot retrieve the list of drugs directly from the *RECOMMENDED DRUG DOSE* data store, and the *PRESCRIBER* cannot directly retrieve information about the patient from the *CLINICAL REPOSITORY*. There must be a process, software code, that extracts the information from the data store. The corollary is that there should be no direct dataflows between external entities and data stores without a process in between.

Avoid *black hole* data stores: data stores that have only inputs and no outputs.
Avoid *spontaneous generation* data stores: data stores that have only outputs and no inputs.
Avoid *gray hole* data stores: data stores that have at least one input and one output, but the data are obviously insufficient or irrelevant to produce the output

A DFD represents a set of *asynchronous* processes. The processes numbers do not reflect their sequence. Thus, in Figure 1.4 the process 1.3 *Select Dose* may or may not be processed after process 1.2 *Select Drug*. Since processes are asynchronous, there is no guarantee the correct or expected sequence will be followed by the system. According to the DFD example in Figure 1.4, a dose may be selected before a drug was chosen, a situation that does not make much sense.

There are no components on a DFD to describe *conditional logic*. A DFD has no condition-action rules, such as "If condition X is fulfilled then act according to Y. Otherwise do Z." The time-dependent, sequential nature of a process and the conditional logic a process has to obey are defined with flowcharts, pre- and postconditions, and sometimes *control processes*: bubbles that control the behavior of other bubbles.

Then, why bother numbering processes? The numbering system of the processes is used to *level and balance* the DFDs (3,6). Since a system may have hundreds and thousands of processes and since we cannot clearly depict more than half a dozen "bubbles" on one diagram, we use the numbering system to level the DFDs.

Each and every diagram in the top-down analysis process is named according to the higher-level process it details. For example, the parent diagram named DFD 1 (Figure 1.4) has four processes: 1.1, 1.2, 1.3, and 1.4. The child diagram named DFD 1.2 *Select Drug* will represent the decomposition of the process (bubble) numbered 1.2 from the parent diagram. Assume DFD 1.2 *Select Drug* has its own three processes (not shown in Figure 1.4): 1.2.1, 1.2.2, and 1.2.3. A lower-level child diagram of DFD 1.2 can continue the partitioning, top-down analysis, for example, detailing the 1.2.3 bubble. That diagram will be named DFD 1.2.3 and its own bubbles will be numbered 1.2.3.1, 1.2.3.2, 1.2.3.3, and so on. Leveling keeps the processes (bubbles) in sync with the diagrams.

During the leveling activity, one has to *balance* the processes with the diagrams: The inputs and outputs shown on a high-level, parent bubble have to correspond to the same inputs and outputs detailed on the child DFD. Ignore the internal dataflows and consider only the inputs and outputs to and from external entities: If a process on a DFD has two inputs and three outputs to and from external entities, then the DFD detailing the process must have the same two inputs and three outputs to and from external entities for the process and diagram to be considered balanced.

We are done with data-in-motion analysis and its tools: the DFD. Let us introduce the ERD, an analysis tool for data at rest/databases.

Entity Relationship Diagram

The ERD (also known as an E-R diagram) is a tool used to model data structures used by the system and their relationships (4). Since the data structures and relationships may be quite complex, *SSA* recommends analyzing them in parallel to and separately from the system workflows.

The components of the ERD are as follows (Figure 1.5):

1. *Entity*: a rectangle (not to be confused with the external actor in DFD, which is also a rectangle) represents "something" that the system has to collect, store, and usually retrieve information about, something the system has to "remember." Usually, an entity will evolve during the design phase into a database table. A row in such a table is called an entity *instance* or *occurrence*. An important rule to remember is that each and every row or instance in an entity must be *uniquely identified*.
2. *Attribute*: data elements that describe the entity. The attributes are the characteristics of an entity. In a database table implementing the concept of an entity, the attributes are the table columns. Attributes can be optional or mandatory: A required mandatory attribute must have a value and cannot be left empty, while an optional attribute can be left empty with no value stored.
3. *Relationship*: lines between the rectangles describe associations between entities. The relationships can be defined according to *cardinality*: *one to many* (*1:M*), *many to many* (*M:N*), or *one to one* (*1:1*). Relationships can also be defined by *optionality*: whether the relationship is optional or mandatory.

Figure 1.5 should be read as follows: An instance (occurrence) of *Entity A* may have one or more relationship *name A to B* with *Entity B*. The small circle near *Entity B* means that the relationship is optional. The "crow's foot" near *Entity B* means a cardinality of many.

Now, we traverse the diagram in the opposite direction, from *Entity B* toward *Entity A*. This should be read as follows: An instance (occurrence) of *Entity B* must have only one relationship *name B to A* with *Entity A*. The lack of a crow's foot near *Entity A* means a cardinality of one. The small vertical line near *Entity A* means the relationship is mandatory in that direction.

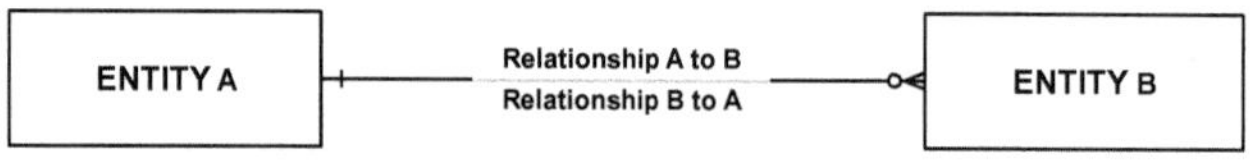

Figure 1.5 ERD symbols.

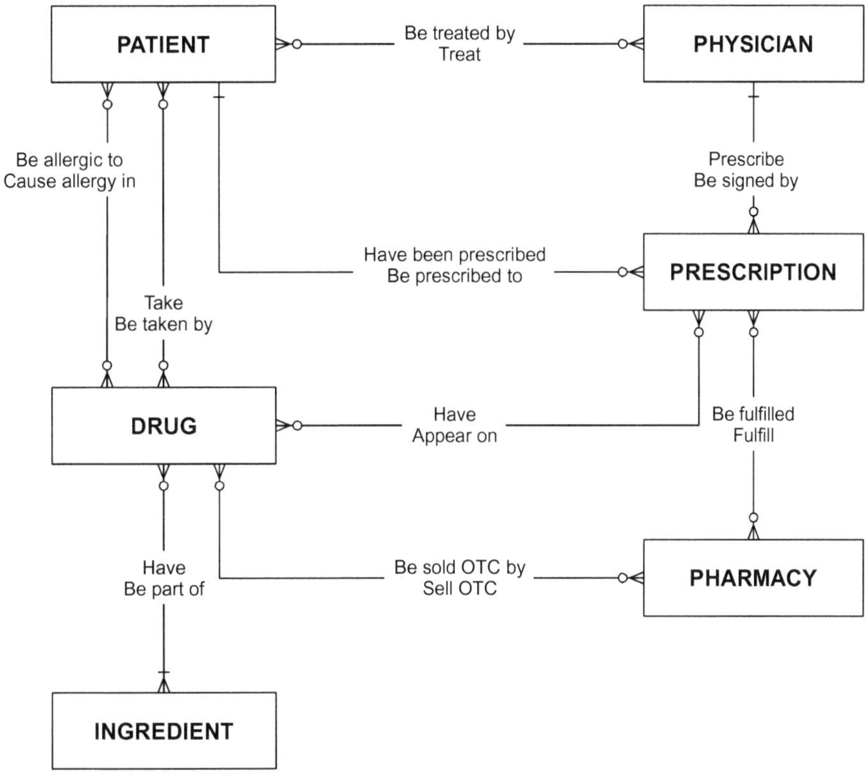

Figure 1.6 Conceptual ERD example.

Let us examine Figure 1.6, an example of a conceptual model. The example is a partial *conceptual* ERD model since it does not have all entities involved. Remember that a medium-size system may have hundreds of entities. It is a conceptual model and not a fully developed logical model ERD since the entities do not have at this stage any attributes defined.

What can we learn from this conceptual ERD?

A *PATIENT* may be treated by one or more *PHYSICIANs*. A *PHYSICIAN* may treat one or more *PATIENTs*.

A *PHYSICIAN* may prescribe one or more *PRESCRIPTIONs*. A *PRESCRIPTION* must be signed by only one *PHYSICIAN*.

A *PATIENT* may have been prescribed one or more *PRESCRIPTIONs*. A *PRESCRIPTION* must be prescribed to only one *PATIENT*.

A *PRESCRIPTION* may have one or more *DRUGs*. A *DRUG* may appear on one or more *PRESCRIPTIONs*.

A *PATIENT* may take one or more *DRUGs*. A *DRUG* may be taken by one or more *PATIENTs*. Note that a *DRUG* does not need a *PRESCRIPTION* since it can be purchased as over the counter (OTC).

A *PATIENT* may be allergic to one or more *DRUGs*. A *DRUG* may cause an allergic reaction in one or more *PATIENTs*.

A *DRUG* must have one or more *INGREDIENTs*. An *INGREDIENT* may be part of one or more *DRUGs*.
A *PHARMACY* may fulfill one or more *PRESCRIPTIONs*. A *PRESCRIPTION* may be fulfilled by one or more PHARMACY.
A *PHARMACY* may sell one or more OTC *DRUGs*. A *DRUG* may be sold OTC by one or more *PHARMACY*.

The following are some aspects to consider when building the ERD:

The conceptual model is the first, rough draft, and simplest modeling tool representing the system's data structures.
Entities are mentioned with the singular name, not the plural, hence: *PHYSICIAN* and not *PHYSICIANS*, *PATIENT* and not *PATIENTS*. It makes the task of constructing statements like, "An instance of *DRUG* must have one or more instances of *INGREDIENT*," much easier if naming entities is kept at the singular level.
Usually, the relationships shown on the conceptual diagram are the representation of business rules applying to the system. Relationships should always be named.

The conceptual model is translated into a *logical* data model, which specifies what are the system data structures and their relationships. There are more details on a logical model than on a conceptual model.

Attributes are added to the entities in the logical model. The attributes (table columns) can belong to several types or categories (also known as data types): Boolean (yes or no/true or false), integer or floating point number, currency, fixed- or variable-length string, or binary object type. The conceptual and logical models are usually technology independent.

The logical model is further developed into the *physical* model, which shows how the model actually is implemented in a database system. The physical model is more detailed than the logical model, and it considers the attributes' types, physical storage, performance, indexing, referential integrity, access mechanisms, and load distribution. The physical model is usually technology dependent.

Normalization

In the DFD example (Figure 1.4) there is a *CLINICAL REPOSITORY* data store. This is a representation of a cluster of several databases—a "mega" data store. For the sake of brevity and clarity, the DFD at this high-level view does not illustrate the various tables the *CLINICAL REPOSITORY* may have: patient demographics, patient diagnosis, surgical procedures, physical exams, immunizations, medications, allergies, lab values, consultant notes, and so on. As the DFDs are decomposed into more detailed views, it will reveal more granularity and uncover the actual tables that may comprise *CLINICAL REPOSITORY*. While it is okay to refer to a cluster of databases in the high-level DFDs as well as in the conceptual model, these repositories must be decomposed into more granular data structures—tables in the logical and physical models.

The developers are employing a formal and quite rigorous technique of *normalization* during the modeling process. The normalization rules, introduced by Edgar Codd (who conceived the relational database model in the 1970s), govern the manner in which data are allocated to the various tables. The purpose of the normalization is to eliminate *redundancy*, *inconsistency*, and *incompleteness* from the data model and prevent *anomalies* during database updates (3–8).

The normalization process of a data model usually starts with one large entity (such as *CLINICAL REPOSITORY*), and it divides this megastore into smaller tables according to a set of rules. These entities/tables in turn will have *primary keys* (*PKs*) defined. Attributes are then allocated to the tables and finally the tables are connected using *foreign keys* (*FKs*) based on the relationships—the dependencies discovered in the analysis of the business rules governing the system.

One of the normalization rules calls for the resolution of all the *many-to-many (M:N)* relationships. The *M:N* relationships that may still exist at the conceptual level model are not acceptable in a fully normalized model. In a normalized ERD, these *M:N* relationships are usually evolving into an *intersection entity* (also known as an associative, junction, bridge, or composite entity). The intersection entity is resolving the *M:N* relationship in a way that each original entity will have a *1:M* relationship with the newly created intersection entity.

Let us examine a more detailed ERD—a sample of a logical data model—evolved from the conceptual model (Figure 1.7). The main points to note in this example are as follows:

Each and every instance/occurrence of an entity must be uniquely identified. This is achieved using *PK*. The *PK* needs to be stable and not change during the system lifetime. Any other *non-PK* attributes may be updated/changed without disturbing the *referential integrity* of the database. For example, the *Patient last name* does not have to be repeated in the entity/table detailing patient's allergies. Nor does the *Drug name* need to be repeated in the table *PATIENT ALLERGY*. For the connectivity purpose, we use the *Patient ID* and the *Drug ID* as *FKs* in the *PATIENT ALLERGY* intersection entity. Thus, the *PK* from the table *DRUG* and the *PK* from the table *PATIENT* are used to connect the *PATIENT* and *DRUG* entities in the intersection entity named *PATIENT ALLERGY*, where they appear as *FK*.

Attributes have been added in the logical model: *Patient gender* in the *PATIENT* table, *Drug name* in the *DRUG* table, *Encounter actual start time* in the *ENCOUNTER* table, and so on. The ellipsis (…) symbol in the depicted entities is there to remind us that the list of attributes shown is partial, tentative, and by no means exhaustive. Reiterating through the

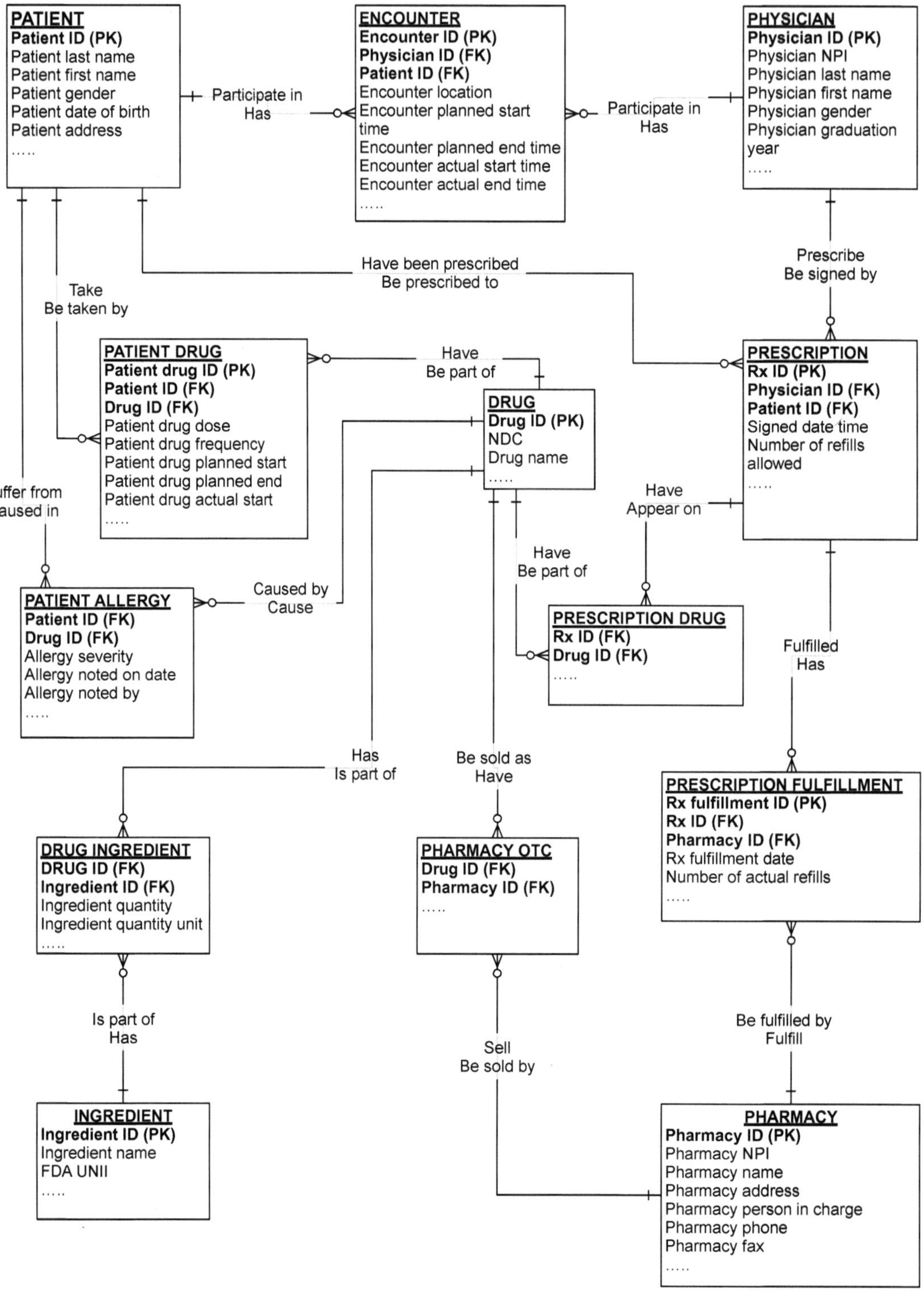

Figure 1.7 Logical model: ERD example.

analysis and refining the ERD, attributes may be added, removed, or relocated from one entity to another.

There is no need to repeat attributes between entities. Any redundant replication of data is a potential source for inconsistency. If the address of a patient changes, for example, we need update only the *Patient address* in the *PATIENT* table. The unchanged *Patient ID* (the *PK* and unique identifier for the patient) will remain the same in all the related, referenced tables (where it appears as an *FK*), and we need not worry about updating the *Patient address* in any other tables.

Sometimes, using the combination of two or more foreign keys forms a unique identifier. If the combination of the foreign keys *Patient ID* and *Drug ID* in the table *PATIENT ALLERGY* uniquely identifies an instance of this entity, an occurrence of a specific allergy in a specific patient, then it is perfectly okay to use the combination as the *compound PK* for this table.

The *M:N* relationships have been resolved using intersection entities. For example, instead of the *M:N* relationship between the *PHYSICIAN* entity and the *PATIENT* entity that was present in the conceptual diagram, the logical model shows an intersection entity named *ENCOUNTER*. Sometimes, there is no clear naming solution for an intersection entity; in these cases, we use the concatenation of the names of the two entities we are intersecting. For example, instead of the *M:N* relationship between the *PATIENT* and *DRUG* entities we had initially on the conceptual model (one *DRUG* may be taken by one or more *PATIENTs* and one *PATIENT* may take one or more *DRUGs*), the logical model will have an intersection entity named *PATIENT DRUG*. The intersection entities created usually will have a different optionality than the *M:N* relationship it has resolved. The initial many-many relationship between *PATIENT* and *PHYSICIAN* was optional in both directions ("may" have the relationship). Once resolved, it became a mandatory relationship in one direction: a *PATIENT* may participate in an *ENCOUNTER*, but an *ENCOUNTER* must have a *PATIENT* defined. In the same manner, a *PHYSICIAN* may participate in an *ENCOUNTER* with a *PATIENT*, but an *ENCOUNTER* must have a *PHYSICIAN* defined. In other words, for an *ENCOUNTER* occurrence to be properly defined according to the business rules, it must have both a *PATIENT* and a *PHYSICIAN*.

Avoid mandatory relationships in both directions. Usually these types of relationships should be regarded with suspicion since the business rules may have been represented incorrectly (4,8).

Usually, but not always, *one-to-one* relationships can be reduced, with one of the entities incorporated into the other. In most cases, one entity participating in this 1:1 relationship is just an attribute of the other.

The ERD examples we have used so far are simplifications of the medications domain. The data model in Figures 1.6 and 1.7 is incomplete: There is no representation of the relationship between the concepts of *Brand name* (*Tylenol*,

Aceta, Actamin) and *Generic name* (*Acetaminophen*); there is no *Drug Form* entity (*tablet, capsule, syrup, solution*), *Strength* information (*10 mg/ml, 50 mg/ tablet*), *Indications, Contraindications, Precautions;* and there is no information related to *Maximal dose or Usual dose.* We build more accurate and complete models of the data structures in the following chapters.

Data Dictionary

Although it is lacking the graphical appeal of the other modeling tools mentioned and it is also quite tedious to build one, the data dictionary (DD) is nevertheless an important part of the system specifications. Without a DD, the DFD and ERD are just sketches; beautiful as they may be, their usefulness remains limited.

The DD is a central repository of information about the data and processes of the system under consideration. It is to be used by both users and developers as a communication tool to clarify terms during the system life cycle: analysis, design, code build, testing, implementation, training, and maintenance.

To keep in sync the various diagrams with the DD, a *computer-aided software engineering (CASE)* tool is highly recommended. A CASE tool can help automate most of the documentation process; for example, when drafting an ERD, a CASE tool can automatically create an entry in the DD for a newly defined entity. When a data store is mentioned on a DFD, CASE can automatically create an entry in the DD. A CASE tool can also minimize the inconsistency that may be introduced in the system specifications while detailing the models and components in multiple diagrams and the DD. A CASE-based DD can be searched for a term like any other electronic document, and it can generate reports, such as all the workflows or all the entities of a system. What should a DD keep information about?

Fields are the discrete data elements (also known as attributes or stored data items). A field is the smallest piece of information that has any meaning to the user. It is an elementary or *atomic* item, so from a user perspective, it is meaningless to try decomposing it any further. A Social Security number (SSN) may be considered an indivisible or atomic data element. Even if one could break the SSN further into three groups of numbers, it would probably be meaningless to the user. Gender is another example of a discrete, atomic data element since it would be pointless to try breaking down the *M* or *F* symbol any further. In a database table, the fields are the columns. The documentation of fields in the DD should include

Field name: *Patient heart rate*
Field definition: *Patient heart rate per minute*
Field aliases, synonyms, or alternate naming convention: *HR, H.R., Patient HR, Patient H.R.*
Field type and length: *Numeric/Integer*

Whether a null value is allowed: Can the field contain nothing? Is it a mandatory or an optional field?
Allowed values: For example, the system may allow in the *Gender* field only *Male, Female*, or *Undefined*. A *Patient heart rate* field may allow values between 0 and 300.
Default value: If no input is provided, what value should the system assign by default to the field? For the *Date Rx signed*, the default value may be the current date: *Today*. For *Time Rx signed*, the default value may be the current time: *Now*.
Source: What are the possible sources for the field? The sources for the *Patient heart rate* field could be *manual* (nurse input) or *from monitor* (through a medical device interface)
Security: which system, individual, or role has *Read/Write* access to the field.
A couple of examples and any other notes that may be helpful clarifying the concept.

Records are the structured combination of data items. In a database table, the records are the rows. For example, the data items *Patient ID, Patient last name, Patient middle name, Patient first name, Patient gender, Patient date of birth, Patient SSN* could be considered as a meaningful combination of related data elements in a *Patient demographics* record (row).

Data stores are the records storage structures (also known as tables or entities). In this category, we could include the repositories, aggregated clusters of databases such as the *CLINICAL REPOSITORY* mentioned. A DD will document the following for each data store:

Data store name: the name of the data store/entity as it appears on the DFD and ERD: *PHYSICIAN*
Data store definition and its purpose: *ENCOUNTER is the intersection entity between PATIENT and PHYSICIAN and contains information about the planned and actual location, planned and actual date and time frames for these encounters.*
Aliases and synonyms: *PATIENT ALLERGY* may be a.k.a. *Allergies, Drug allergies, Patient drug allergies.*
Table name: the name of the table as it appears on the physical model. The table name is not always the same as the entity name or data store in the conceptual model. The *CLINICAL REPOSITORY* data store in the DFD example in Figure 1.4 is actually a high-level representation of a cluster of data stores: *Demographics, Present and past diagnosis, Active problems list, Allergies, Lab values, Medications* etc. — to be implemented in the physical model in a number of tables within a number of databases.
Primary Key (PK)
Foreign Key(s) (FK)
Dataflows that may enter or leave the data store.

Volume: is it a high-volume data store (such as *Patient vital signs* captured every 30 seconds from monitors) or low-volume data store expected to rarely change — *Patient date of birth* or a slow changing dimension such as the *Patient name* or *address.*
Frequency: whether data store is accessed frequently or infrequently.
Security: which system, individual or role has *Read/Write* access to the data store.

Dataflows should be documented in the DD with:

Dataflow name as it appears on the DFD: *Rx details*
Dataflow description and purpose: *prescription details*
Dataflow aliases or synonyms: *Prescription*
Dataflow point of origin: *Process 1.2.1*
Dataflow destination: *Pharmacy*
Dataflow volume: *Low*
Dataflow frequency: *High*

As mentioned in the DFD section of this chapter, the system *processes* are decomposed systematically in a top-down fashion from a high-level view of workflows into more detailed processes, activities, and tasks to the point of low-level, high-detail *functional primitives* (see next section). A DD should document the following for each process:

Process number and name as it appears on the DFD: *1.2.4. Consider Formulary*
Process description and purpose: *Check with Pharmacy Benefit Management (PBM) whether drug is covered by patient's pharmacy benefits plan and the corresponding drug formulary tier*
Process aliases or synonyms: *Query PBM on quantity limitations, Check drug tier, Check PBM coverage of drug*
Process input: *Patient ID, Drug ID*
Process output: *Drug coverage by PBM, Drug tier, Patient's co-pay, Limitations on quantity*

Functional Primitives Specification

In partitioning the workflows into processes, activities, and tasks, the *functional primitive* is the bottom-most bubble, which is reached when further decomposition would yield the function's detailed logic rather than additional child bubbles. The *functional primitive* is one step before the actual software code. It is used to detail tasks that should not be further decomposed into other activities or tasks. A functional primitive can be documented using flowcharts or *structured English* (also known as pseudocode). It is good practice to state for each functional primitive the *preconditions* to be satisfied before the actual processing as well as the

Precondition:
· A valid Rx exists

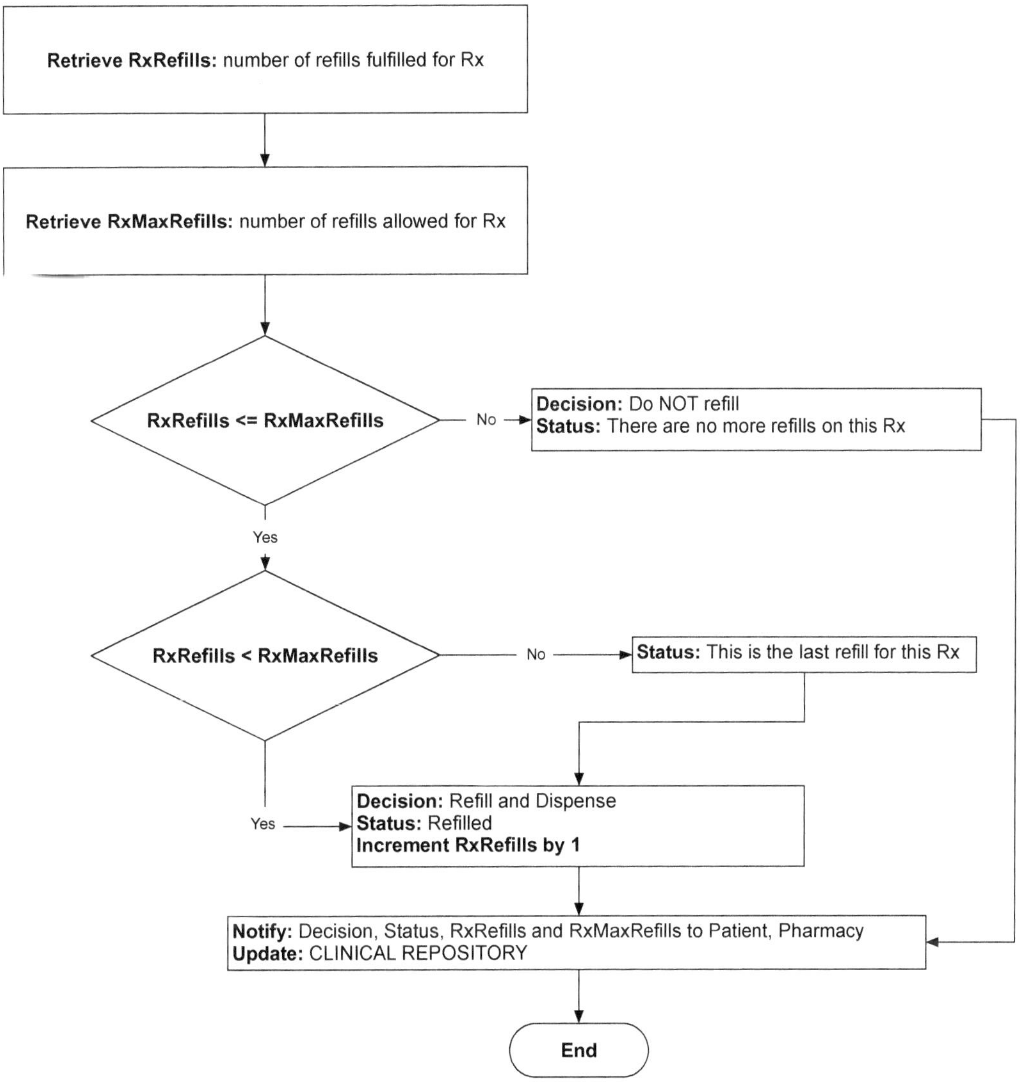

Post conditions:
· If Rx was refilled, then RxRefill was incremented by 1.
· If Rx was not refilled, RxRefill has not changed.
· The decision, status, RxRefill and RxMaxRefills have been notified to Patient and Pharmacy.
· CLINICAL REPOSITORY has been updated.

Figure 1.8 Flowchart example: Rx refill.

postconditions the system is expected to fulfill and accomplish once the processing is done.

Flowcharts can be used to document a functional primitive using three logical control structures: *sequence, selection,* and *iteration.*

What can be learned from the flowchart in Figure 1.8?

Before proceeding with the *Rx refill*, a precondition must be satisfied: There has to be a valid Rx.

System retrieves (*sequence*) from data stores two parameters for the uniquely identified Rx: number of refills allowed and number actually fulfilled/refilled.

System decides (*selection*) whether there are still refills on the Rx.

If there were more than one prescription to refill for the same patient, the system would also loop (*iteration*) through the process (not shown in Figure 1.8).

System updates *CLINICAL REPOSITORY* data stores and notifies *PATIENT* and *PHARMACY* on the refill status.

At the end of the flowchart, there is a set of postconditions that must be fulfilled by the system. If the postconditions have been fulfilled, then the processing of the *functional primitive* is to be considered correct and complete. Think about the postconditions as a safeguard checklist of facts the goals system needs to achieve once the flow through the chart is done. Pre- and postconditions are also a good starting point for designing test scenarios.

Structured English is a functional primitive so near the actual software code that it is also known as pseudocode:

> *Structured English is the use of the English language with the syntax of structured programming. Structured English aims at getting the benefits of both the programming logic and natural language. Program logic helps to attain precision while natural language helps in getting the convenience of spoken languages. (9)*

There are a couple of points to remember regarding structured English:

Use only the three control structures mentioned: *sequence, selection,* and *iteration.*

Use a (very) limited vocabulary of verbs and nouns and have all terms defined in the DD.

A functional primitives specification in pseudocode is usually less than one page long.

The highly detailed specification can be used to develop actual software code.

The following is an example of pseudocode:

```
For a given RxID
Preconditiion: A valid Rx exists
Retrieve RxRefills and RxMaxRefills
If RxRefills <= RxMaxRefills then
        If RxRefills < RxMaxRefills then
```

Dispense:
Decision: Refill and Dispense
Status: Refilled
Increment RxRefills by 1
Proceed: Notify and Update
Else
Status: This is the last refill on this Rx
Proceed: Dispense
Else
Decision: Do NOT refill
Status: There are no more refills on this Rx
Proceed: Notify and Update
Notify and Update:
NoteRx: Date and time (now), Decision, Status, RxRefills, Rx MaxRefills
Send NoteRx to Patient
Data Stores: Clinical repository
Store NoteRx in Data Stores
End

Since the book does not use this level of granularity, *structured English* is not discussed further.

Balancing the Models

When modeling a system from different perspectives (data, functions, sequence, conditional logic) as well as maintaining a dictionary and a set of functional primitives specs, there is always a concern with developing *inconsistent* interpretations of the reality the system is supposed to model. A fact mentioned in one model may be missing from the others. Terms and concepts appearing in a DFD may be missing from the DD or vice versa. Naming conventions may not be strictly enforced, and the same concept may have slightly different names (singular vs. plural is a frequent "suspect" issue, such as *PATIENT* vs. *PATIENTS*).

Errors introduced at an early phase such as during systems analysis are likely to be propagated and magnified during the system life cycle. These errors become notoriously more expensive to correct further down the road. So, it makes sense to try eliminating at an early stage as many errors as possible from the models we create. Balancing the models is one way to minimize errors and inconsistency in the analysis deliverables. There are several straightforward techniques recommended for balancing the different models (3): DFD versus DD, ERD versus DD, DFD versus functional primitives' specifications, and ERD versus DFD and functional primitives' specifications.

DFD versus DD

Every dataflow, process, and data store appearing in a DFD must be defined in the DD.

Every dataflow, process, and data store defined in the DD must appear in a DFD.

ERD versus DD

Every data store appearing in the ERD must be defined in the DD.

Every data store defined in the DD must appear in the ERD.

DFD versus Functional Primitives' Specifications

Every bubble in the DFD must be associated with a lower-level DFD or a functional primitive specification, but not both.

Every functional primitive specification must have a bottom-level bubble in a DFD.

Inputs and outputs related to a bubble must match the inputs and outputs (READ and WRITE statements) in the functional primitive spec.

ERD versus DFD and Functional Primitives' Specifications

Every data store in a DFD must have a representation in the ERD.

An entity in the ERD must have a data store representation in a DFD.

Every data store in a functional primitive specification must have a representation as an entity in the ERD.

An entity in the ERD must have a representation on a functional primitive spec.

The techniques mentioned for balancing the models require little intelligence or creativity. Still, balancing the models is a necessity for the modeling deliverables to be considered complete, correct, and consistent. It is a tedious, time-consuming activity and thus well suited for a machine: a CASE tool.

Summary

This chapter introduced a methodology and a set of tools. The methodology—structured system analysis or SSA—recommends focusing separately and in parallel on the data and functions aspects of a system. SSA provides the tools to model a system from these two perspectives: ERD and DFD. In addition, the technical specifications of a system (but not this book) should define a DD, a set of control processes, state transition diagrams as well as flowcharts, and structured English/pseudocode.

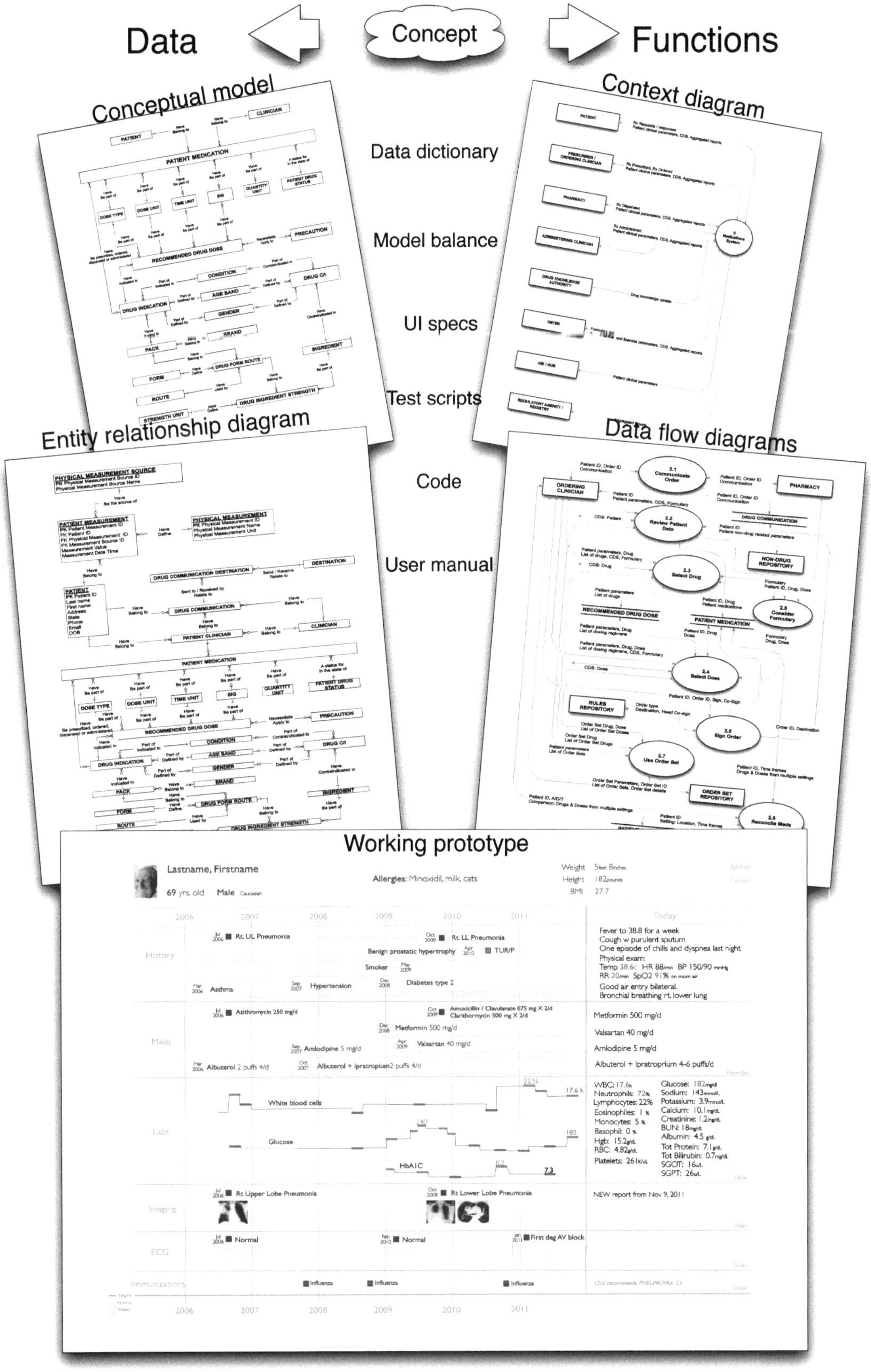

Figure 1.9 Summary of SSA.

There is an increasing level of details while evolving from the conceptual to logical to the physical data model and from the context diagram to DFDs and to functional primitive specifications. Thus, the audiences for these models may differ: The conceptual model, context diagram, and first level of a DFD may be considered appropriate high-level communication tools for business stakeholders, directors, and product/project managers. The logical, physical models, low-level DFD, and pseudocode delve into more details and should be used for in-depth discussions between users, programmers, application engineers, and database designers.

Modeling a system is an iterative, recursive activity. Models are rarely perfect the first time drafted, and usually they need to be refined several times before they can be considered correct, complete, and consistent. Once done, models are not static artifacts, and they need to be updated as often as the system requirements may change during the life cycle.

In the next chapters, we use the principles and tools introduced here while systematically analyzing the medications domain. We show that this ever-daunting analysis job is much more fun with diagrams than text and spreadsheets.

References

1. Merriam-Webster online dictionary. http://www.merriam-webster.com/dictionary/system (accessed December 2, 2010)
2. Wiegers, K.E. 1989. *Software requirements.* Microsoft Press.
3. Yourdon, E. 1989. *Modern structured analysis.* Upper Saddle River, NJ: Prentice-Hall.
4. Simsion, G.C., and G.C. Witt. 2005. *Data modeling essentials.* New York: Elsevier.
5. Fowler, M., and K. Scott. 2000. *UML distilled.* Boston: Addison-Wesley.
6. Shelly, G.B., T.J. Cashman, and H.J. Rosenblatt. 2008. *Systems analysis and design.* Boston: Thomson Course Technology.
7. Rob, P., and C. Coronel. 2009. *Database systems. Design, implementation and management.* Boston: Thomson Course Technology.
8. Silverston, L., and P. Agnew. 2009. *The data model resource book—Universal patterns for data modeling.* New York: Wiley.
9. Wikipedia. Structured English. http://en.wikipedia.org/wiki/Structured_English (accessed December 2, 2010).

Review Questions

1. List five reasons why narrative text is not the best tool for systems analysis.
2. List six characteristics of structured systems analysis.

Find the errors in the following diagrams:

3.

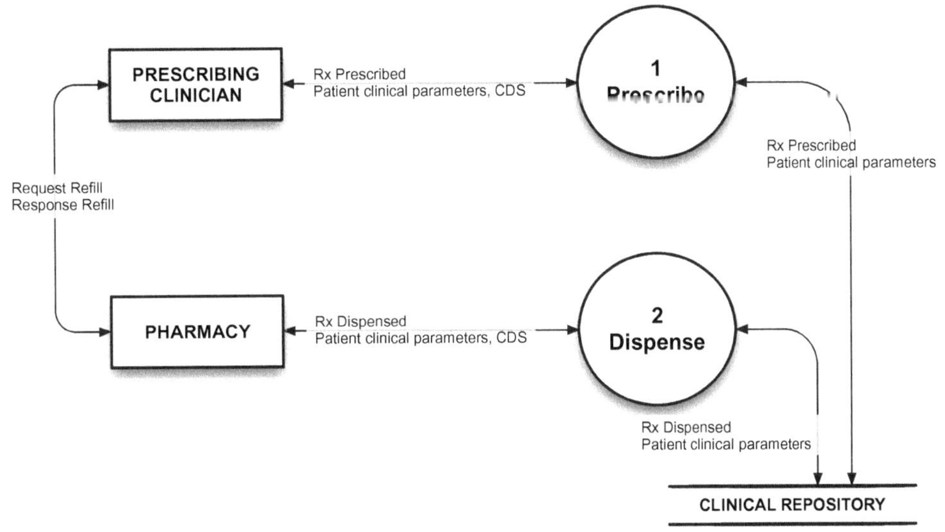

4.

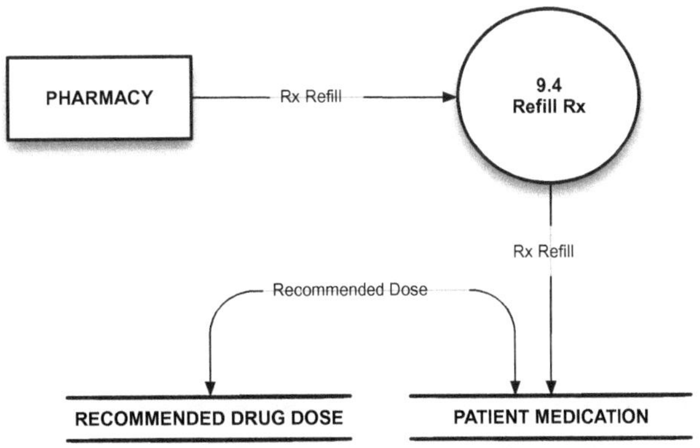

5.

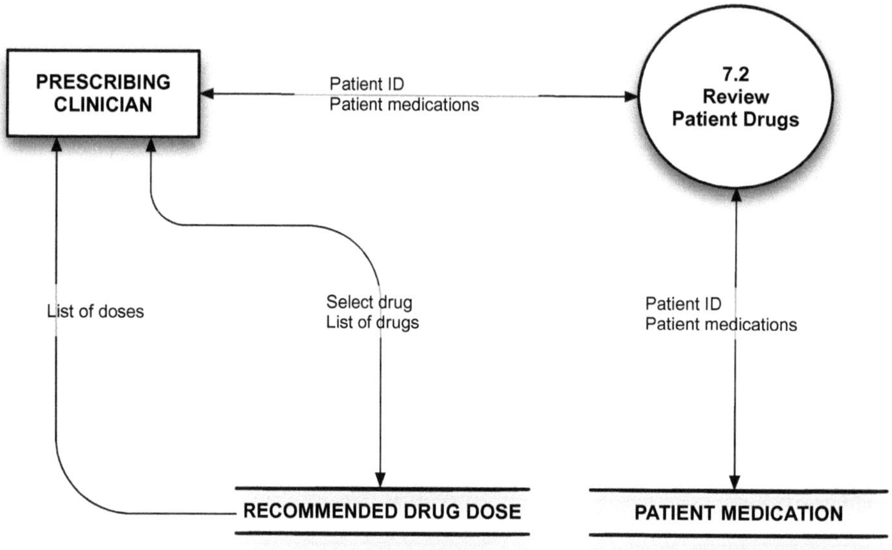
PRESCRIBING CLINICIAN
Patient ID
Patient medications
7.2
Review
Patient Drugs
List of doses
Select drug
List of drugs
Patient ID
Patient medications
RECOMMENDED DRUG DOSE
PATIENT MEDICATION

6.

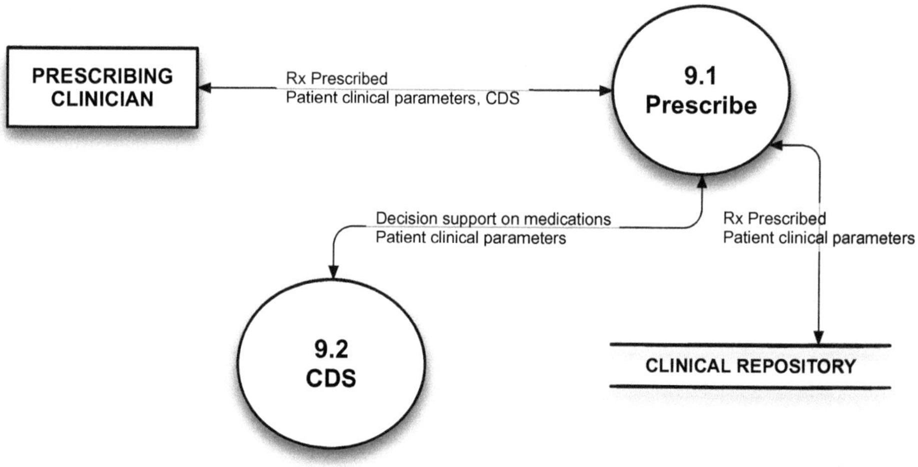
PRESCRIBING CLINICIAN
Rx Prescribed
Patient clinical parameters, CDS
9.1
Prescribe
Decision support on medications
Patient clinical parameters
Rx Prescribed
Patient clinical parameters
9.2
CDS
CLINICAL REPOSITORY

7.

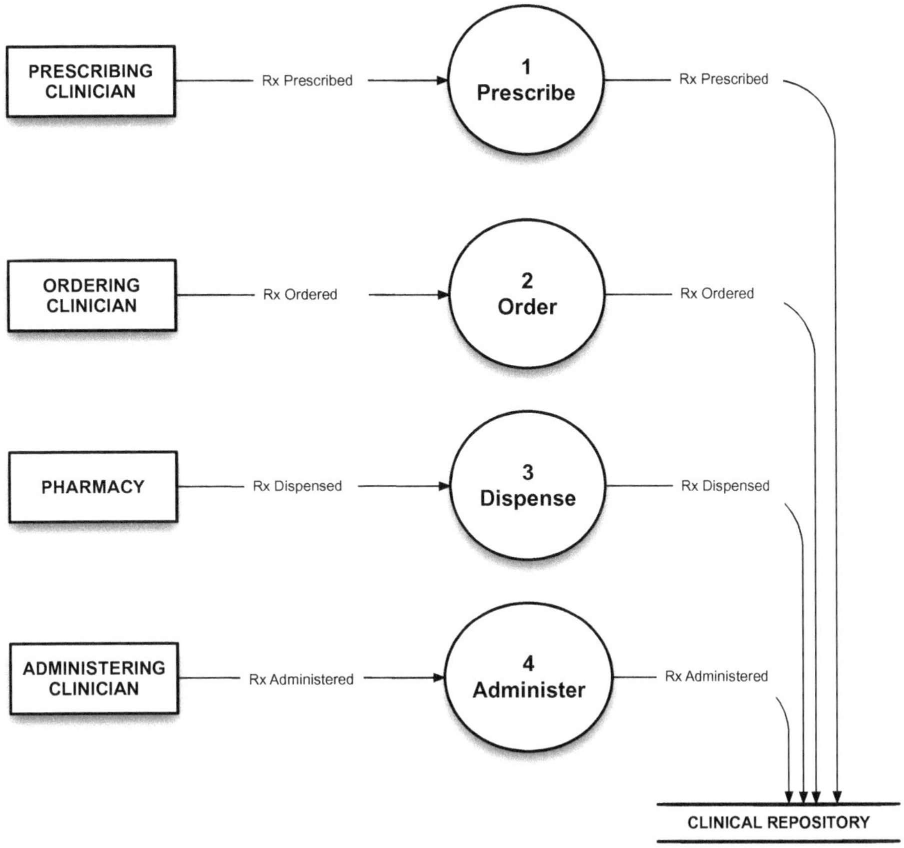

8. List the processes in the order defined in Figure 1.4.
9. Find all the errors in:

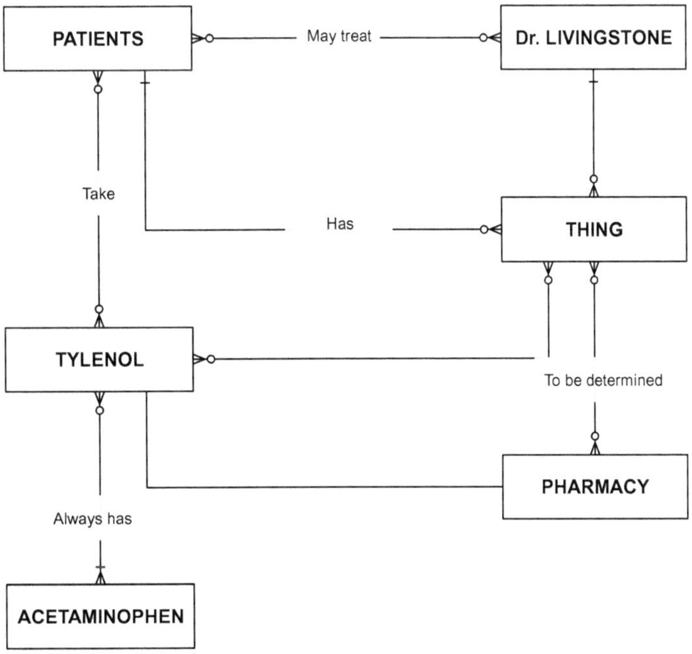

10. Draft the following relationship at the conceptual level and then resolve it for a logical model:
    - An ingredient may be part of many drugs.
    - A drug may have many ingredients.
    - How is the PK of the new entity called?

Chapter 2

The Medications Domain Workflows and Data Structures

Methodology is the product of common sense applied to circumstances.

Samuel E. Morison

In this chapter, we initiate the top-down analysis of the medications domain—"the system"—from two main perspectives: workflows and data structures. The tools we use— dataflow diagram (DFD) and entity-relationship diagram (ERD)—were introduced in Chapter 1.

We identify the external entities interacting with the medications system in the context diagram, and this in turn will help us define the system boundaries.

We then expose the *Medications* system main workflows: prescribe, order, dispense, and administer in DFD 0. Each of these workflows is detailed in a dedicated chapter further in the book.

Finally, we draft the main data structures related to the medications system while building a conceptual data model. The data model is the precursor of a database to be used by the system's workflows to store and retrieve data elements.

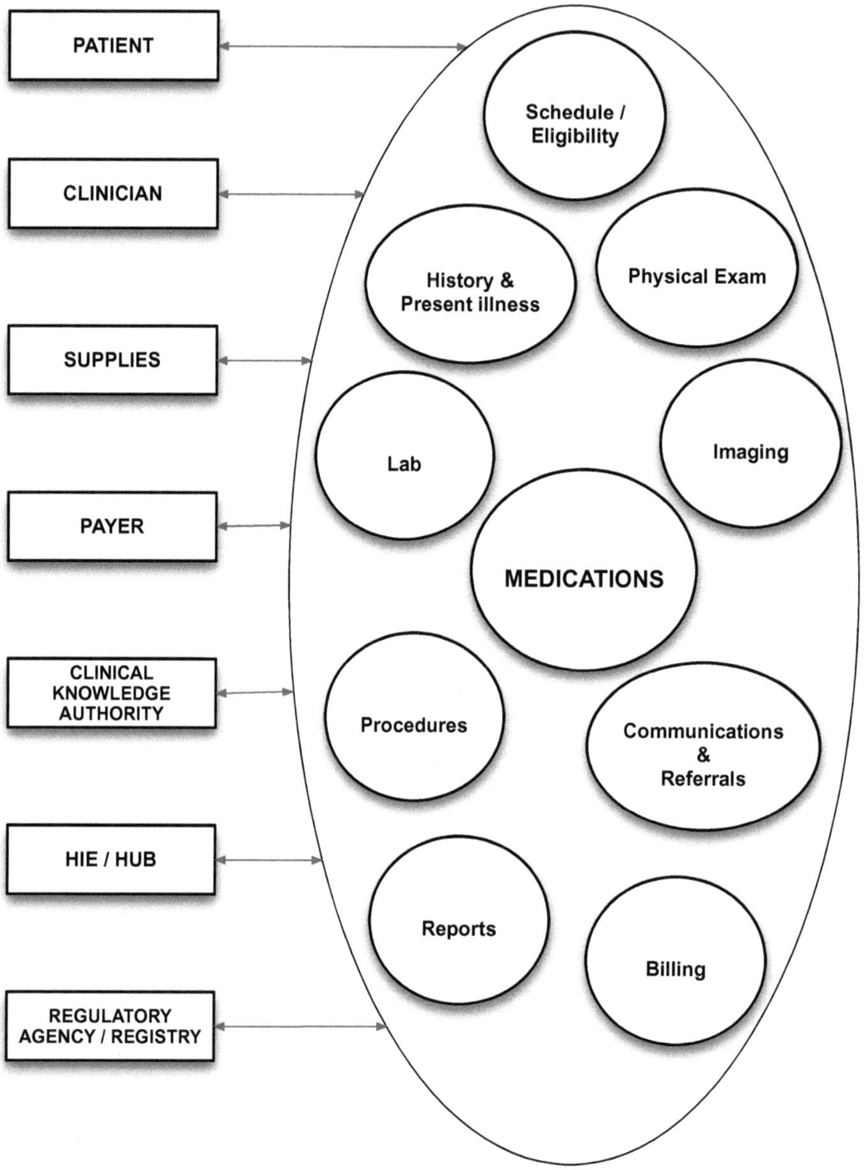

Figure 2.1 Medications system in perspective.

Let us proceed with a top-down analysis, starting with the *Medications* system in perspective of the much larger healthcare/clinical domain (Figure 2.1). The large, elliptic, megabubble in the diagram of Figure 2.1 represents the healthcare domain. The *Medications* domain is one of many systems:

- Scheduling encounters and eligibility check (also known as patient access management)
- History & Present illness (maintenance of the problems list, diagnosis, main complaint, symptoms, physiological systems review, allergies, etc.)
- Physical exam
- Laboratory tests

- Imaging (x-rays, computed tomography [CT], magnetic resonance imaging [MRI], ultrasound, etc., and the resulting imaging reports)
- *Medications* (prescribe, order, dispense, and administer drugs)
- Procedures (diagnostic, treatment, surgical, endoscopies, etc., and the resulting procedure reports)
- Communications (with patient and other providers) and referrals
- Reports (quality indicators, pay for performance, registries)
- Billing (coding, charge capture, etc.)

The high-level view of the healthcare domain in Figure 2.1 provides a general idea about the potential complexity of this megaworkflow. It also illustrates what should be in scope and what should be out of scope from the *Medications* system perspective. One should be careful when deciding about the system scope since at this early stage it may change by orders of magnitude.

Information moving into and out of the *Medications* system does not necessarily imply the workflows or processes generating this information are within the system boundaries. For example, even though the *Medications* system may consume information that *History & Present illness* workflow produces, capturing and maintaining the patient history, is not in scope; it is outside the *Medications* system boundaries. In the same spirit, the *Billing* system is not in scope, even though information the *Medications* system produces is consumed by the *Billing* system.

Let us take a closer look at the *Medications* system functions.

Context Diagram

The context diagram is used for setting the system boundaries and identifying the external entities interacting with the system. As mentioned in Chapter 1, interactions between these external entities are by definition within the system boundaries and are to be analyzed with structured functional decomposition. Thus, while defining the external entities interacting with the system, we are actually setting the system boundaries.

Who are these external entities? Usually, they are users, preferably good-natured humans and other computer systems. What is common to all the external entities is the fact that they are consumers or producers of information relevant to the medications system, which is represented by the circle on the right side of Figure 2.2.

- Patient may have, for example, a request on a prescription *Rx Request* or may receive an update *Rx Response*.
- Prescriber/Ordering clinician is a physician, a dentist, or a nurse licensed to prescribe. Prescribing is usually done in outpatient, ambulatory settings using an electronic prescription tool (*eRx*). Ordering is done in-hospital

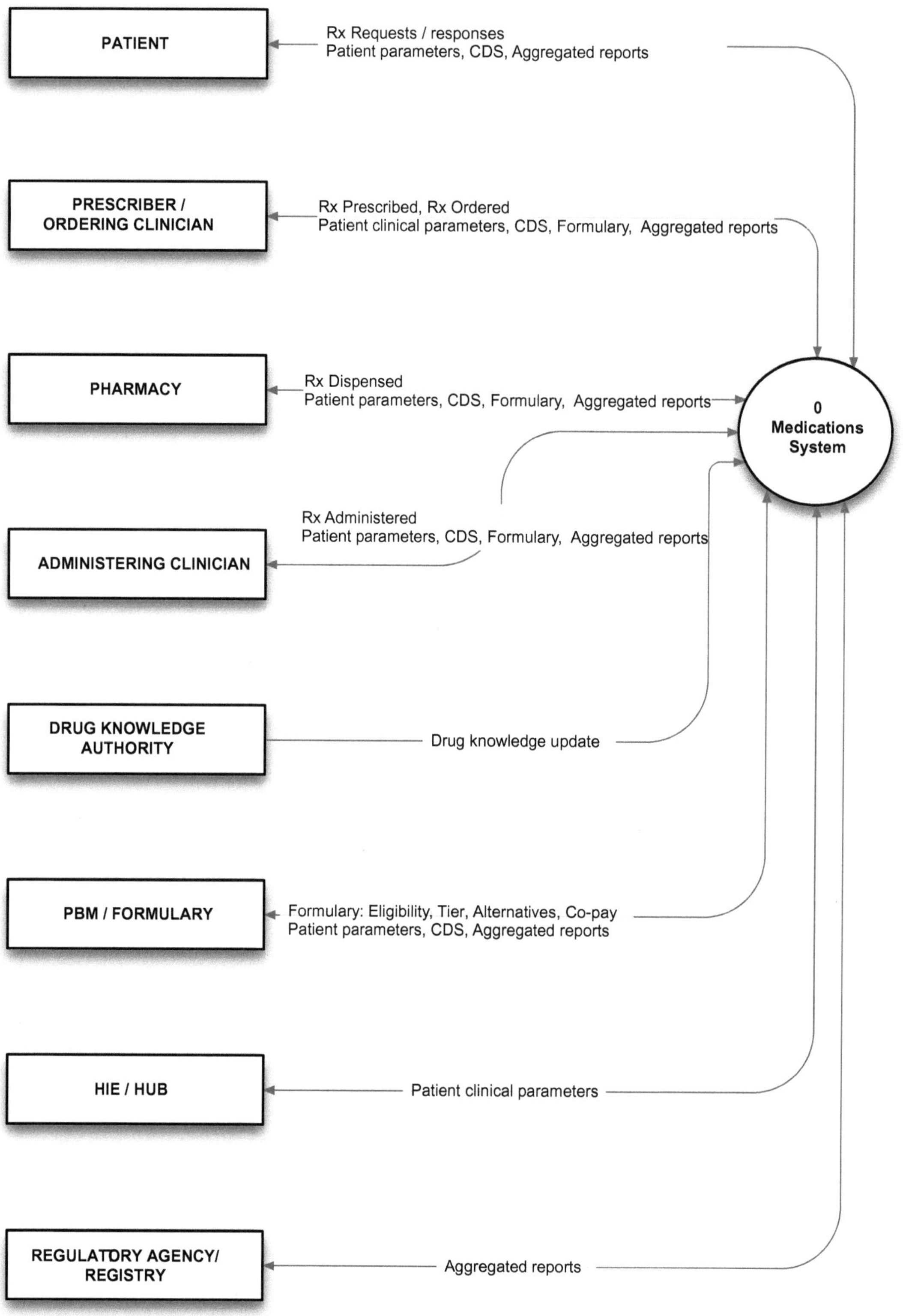

Figure 2.2 Medications context diagram.

using a computerized physician order entry (*CPOE*) system. Whether in a role of prescribing or of ordering clinician, this entity consumes information on patient clinical parameters, recommendations from the clinical decision support (*CDS*) system and produces the details of the prescription or order: *Rx Prescribed* or *Rx Ordered.*

- Pharmacy is a community-based or an in-hospital entity dispensing medications. Based on patient clinical parameters and *CDS* recommendations received, it will respond with the details of the *Rx Dispensed.*
- Administering clinician is usually a nurse in charge of the administration of a drug. An external entity that consumes information about the patient and *CDS* recommendations and produces information on *Rx Administered.*
- Drug knowledge authority provides periodically updated information on medications and evidence-based medical knowledge related to drugs.
- *PBM/FORMULARY* is either a pharmacy benefit management (*PBM*) system—the patient's health insurance plan governing the medications benefits for the outpatient—or the hospital formulary. Given the patient and drug identifiers, this entity provides information related to formulary, eligibility, drug tier, alternatives, and patient copay.
- Health information exchanges (*HIEs/HUBs*), regional health information organizations (*RHIOs*), and other national entities collect and share with other computer systems medical information on consenting patients. The patient clinical data coming from disparate sources and systems can be consolidated with the help of *RHIO/HIE/HUB* entities into "one patient—one record" to be used "anytime and anywhere."
- Regulatory agencies/Registries are regional, national, and international entities consuming aggregated reports on segments of the population, "anonymized" data grouped under certain criteria. This is done for quality control purposes, public health monitoring, quality indicators reports, pay for performance, immunization registries, and so on.

In the context diagram, the medications system is defined as workflow number 0. As such, it is the highest-ranking bubble, the top workflow. At this early analysis stage, workflow number 0 is depicted as a black box: There are no other workflows or processes.

There are also no repositories or data stores on a context diagram. Interactions of the system with the external entities are depicted in a high-level/low-detail fashion, and thus it may seem vaguely defined at the moment. The details are there; we just cannot see them clearly from this high level. So, let us analyze the medications workflow # 0 a bit further.

DFD 0

Dataflow diagram 0 (DFD 0) shows the decomposition of workflow number 0—the medications system itself. The focus moves now from *System* to *Workflows* on the scale presented in the first chapter:

System → *Workflow* → *Process* → *Activity* → *Task*

First, we draft the external entities as rectangles and the main workflows as bubbles, and then we add the dataflows (Figure 2.3). The differences between the workflow of a *PRESCRIBING CLINICIAN* and an *ORDERING CLINICIAN* might justify having two external entities. Other than this difference, these are the same external actors mentioned in the context diagram. The *Medications* system's internal workflows are now visible as bubbles:

1. *Prescribe* medications (outpatient, eRx)
2. *Order* medications (inpatient, CPOE)
3. *Dispense* medications (pharmacy)
4. *Administer* medications (including self-administration by the patient)
5. Provide *CDS* (clinical decision support)
6. *Report* on grouped segments of the patients' population
7. *Update/Sync* the clinical repository with respect to other systems: HIE, RHIO, prescriptions hub, and so on

At this resolution, the *CLINICAL REPOSITORY* represents all the clinical data stores of the system—medications and nonmedication related. It represents a cluster of databases, other data repositories, or collection of services. Some are internal data stores to which the *Medications* system has direct read/write access. These internal data structures are discussed in detail in this chapter. Other data stores, being nonmedication related, are external to the medications system, and we represent them as external actors during further analysis. Some patterns one should look for when checking or drafting a DFD:

Pattern of read/write:

The dataflows follow a pattern of input/output, between entities and workflows as well as between workflows and data stores. For example, an *ORDERING CLINICIAN* may read *Patient clinical parameters* and may write a new *Order*.

All interactions between actors are through the system:

Interactions between actors relevant to the medications must be part of the system. For example, all the interactions a prescribing clinician may have with the pharmacy relevant to the medication system are by definition

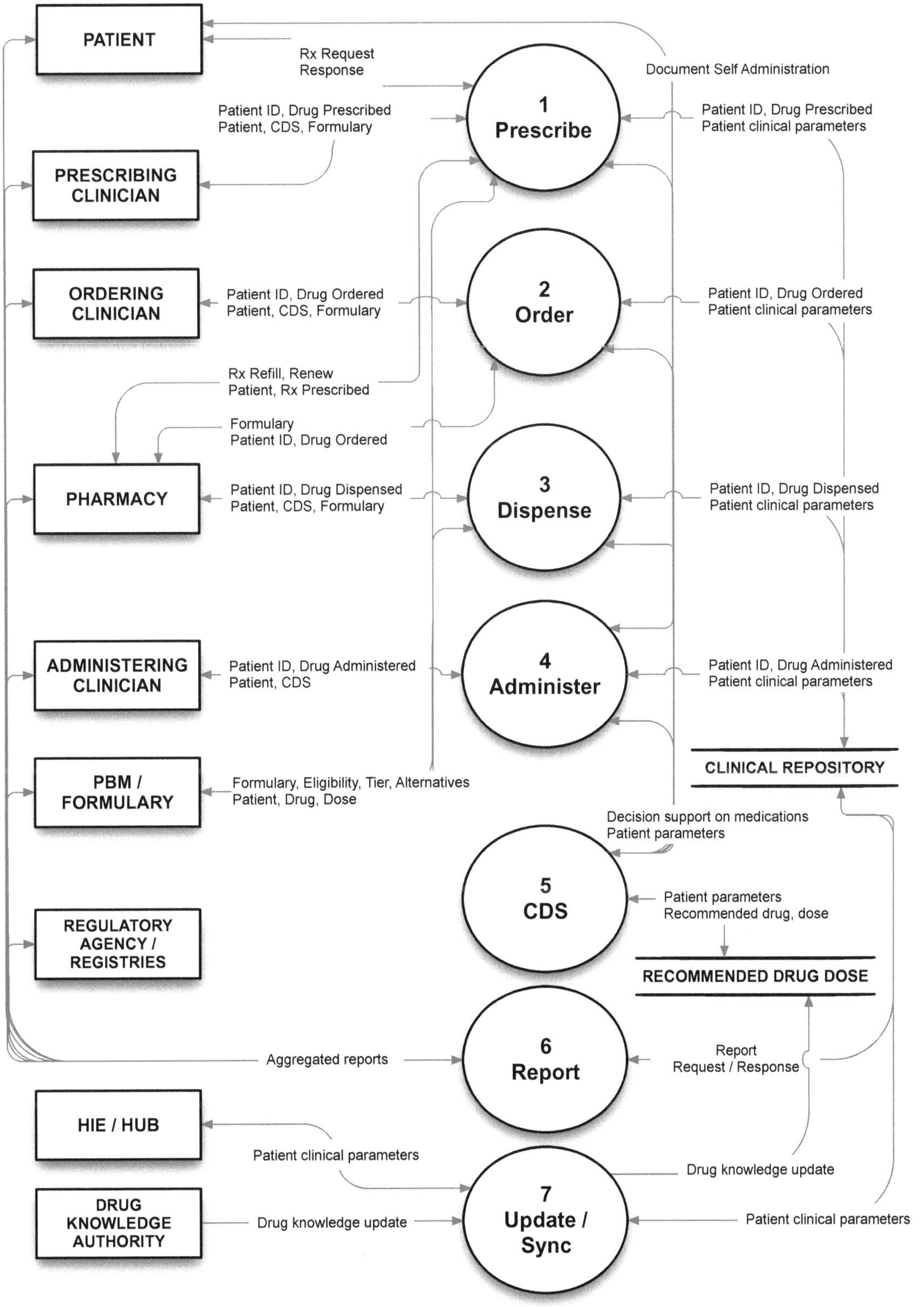

Figure 2.3 DFD 0: Medications system.

within the system's boundaries; thus, there should be no direct dataflows between *PRESCRIBING CLINICIAN* and *PHARMACY.*

Entities interact with the system workflows and not directly with data stores:
The entities may provide input into or may consume the output of the system, but only via system workflows or processes. For example, there is no direct access from *PRESCRIBING CLINICIAN* to the *CLINICAL REPOSITORY.* In between, there must be a workflow or process: in this case it is workflow number 1, *Prescribe* workflow. *Prescribe* is responsible for reading and displaying the *Patient clinical parameters* and correctly storing the *Rx Prescribed* details in the *CLINICAL REPOSITORY.*

One external entity may interact with many workflows:
An entity may provide information services to several workflows. For example, *PBM/FORMULARY* provides *Formulary* data to *Prescribe, Order,* and *Dispense* workflows.

One workflow may interact with many external entities:
A workflow (e.g., *Report*) may serve several entities with aggregated reports (*PATIENT, PRESCRIBING CLINICIAN,* etc).

A workflow may share information with other workflows:
For example, *CDS* provides *Decision support on medications* to *Prescribe, Order, Dispense,* and *Administer* workflows.

Workflow Responsibility

Many entities and workflows use *Patient clinical parameters* dataflow—no doubt an important component of the medications system. Still, it is not the responsibility of the medications system to capture *Patient clinical parameters* completely. The system is responsible only for maintaining the data relevant to the medications domain. Other systems are responsible for capturing the rest of the *Patient clinical parameters.*

We continue the top-down analysis of the system's functions in the following chapters. During the analysis we may discover new actors, workflows, and data stores, and most probably we will need to iteratively change the dataflow diagrams. Each workflow mentioned has its own chapter in the book:

- Workflow 1 *Prescribe* is analyzed in the next chapter starting with DFD 1.
- Workflow 2 *Order* is decomposed in another chapter using DFD 2.
- Workflow 3 *Dispense* is discussed using DFD 3.
- Workflow 4 *Administer* is discussed using DFD 4, and the remainder follow the same process.

We have discussed so far the system functions and main workflows—data in *motion*. Let us switch now to modeling the data at *rest*—the *Medications* system data model, the precursor of the database.

Data Model

The workflows defined—*Prescribe, Order, Dispense, Administer, CDS,* and so on—all need discrete data elements, which must be stored in, and retrieved from, a data structure: a database. The conceptual data model is the first raw draft of a database (1,2). What are the goals when building a conceptual model and the subsequent logical model (2,3)?

- Identify the *concepts* involved from the perspective of the workflows
- Identify the *business rules* that accompany the concepts
- Identify the *relationships* between concepts
- Define the *discrete data elements* that cover these concepts
- Decide whether a data element is an *entity* or an *attribute*
- Identify possible combinations of parameters for each business rule
- Partially resolve the *many-to-many* relationships into *intersection* entities and eliminate redundancy and data anomalies from the model

When drafting a conceptual model, one should check whether it satisfies the business rules of the system (2,5). "*A Business rule is a statement that defines or constrains some aspect of the business*" (6). We draft the conceptual data model using the following method:

Real-life examples will help us derive the initial concepts and business rules. For each business rule, we define the discrete data elements and possible combinations of parameters the model needs to cover as far as data structures are concerned. In some cases, we discuss considerations that have influenced the specific modeling path chosen.

The *CLINICAL REPOSITORY* has a number of data stores related to patients (Figure 2.4): diagnosis, lab results, imaging, procedures, medications, and so on. For simplicity sake, let us illustrate only the patient side. The participation of clinicians in diagnosis, orders, procedures, and so on is not shown in Figure 2.4. At this high level, the relationships between concepts are many to many (*M:N*):

- A *PATIENT* may suffer from or be diagnosed with one or more *PROBLEM DIAGNOSIS*. A *PROBLEM DIAGNOSIS* may be attributed to one or more *PATIENT*.
- A *PATIENT* may undergo one or more *PROCEDURE*, and a *PROCEDURE* may be performed on one or more *PATIENT*.

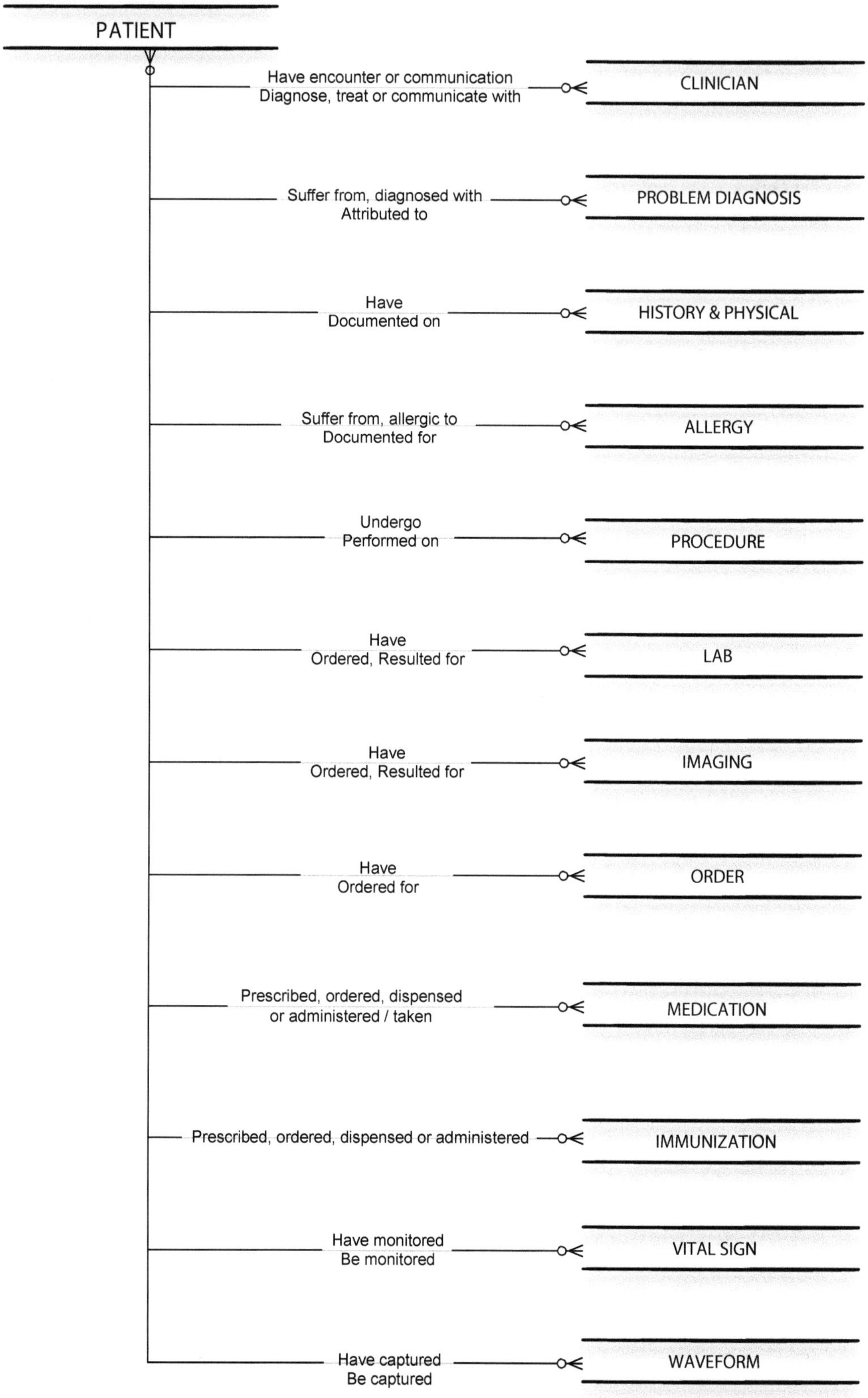

Figure 2.4 Clinical repository data structures.

- A *PATIENT* may take or be administered, prescribed, ordered, dispensed one or more *MEDICATION*. A *MEDICATION* may be prescribed, ordered, dispensed, or administered to one or more *PATIENT*.
- This continues in the same manner.

Clarification: Do not confuse external actors in the DFD with entities/data stores in a conceptual model/ERD. In DFD (used in functional, workflow analysis), external actors are defined as entities consuming or producing information relevant to the system. In conceptual models (used in data structures analysis), entities are concepts about which the system needs to keep/store information.

For example, *PHARMACY* is an external entity in the DFD mentioned, but there is no *PHARMACY* entity in the *CLINICAL REPOSITORY* model. *PHARMACY* data structure exists and even thrives in a different repository somewhere. In this model *PHARMACY* is not a data store. It is an actor.

Consider the opposite situation: *MEDICATION* appears as an entity in the *CLINICAL REPOSITORY* model, but there is no such external entity in the DFD because *MEDICATION* is not an actor interacting with the system. It is a data store.

From Figure 2.4, let us isolate the concepts that are related to the medications domain and add the *CLINICIAN* concept (Figure 2.5). We focus on four concepts: *PATIENT*, *CLINICIAN*, *MEDICATION*, and an intersection entity between them: *PATIENT MEDICATION*. In this diagram, the *M:N* relationships have been resolved. The statements from Figure 2.5 should be

- A *PATIENT* may have one or more *MEDICATION* prescribed, ordered, dispensed, or administered as one or more instances in the *PATIENT MEDICATION* entity. A unique instance in the *PATIENT MEDICATION* entity must belong to only one *PATIENT*.

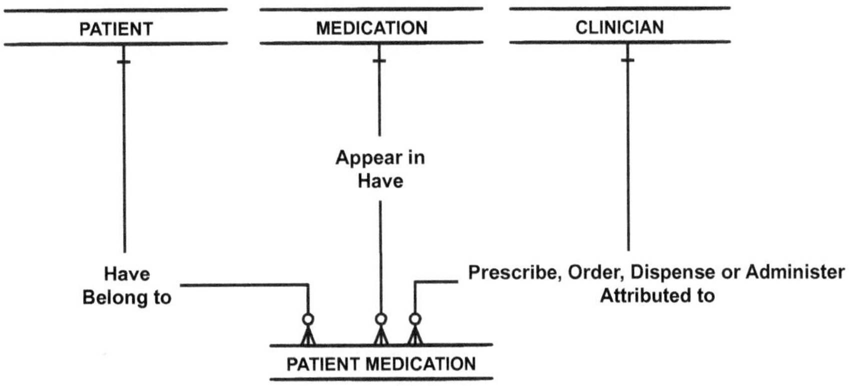

Figure 2.5 Medications-related concepts.

- A *MEDICATION* may appear in many instances of the *PATIENT MEDICATION* entity. A unique instance of the *PATIENT MEDICATION* entity must have only one *MEDICATION*.
- A *CLINICIAN* may prescribe, order, dispense, or administer one or more *MEDICATION*. A unique instance in the *PATIENT MEDICATION* entity must be attributed to only one *CLINICIAN*.

Obviously, we should expect a large and diverse medications domain, so we are going to build the conceptual model in three steps. At each step, we will add several entities and their relationships to the model. We use examples from real life to "test" model compliance with the business rules: brands with many forms and routes, multiple ingredients, ingredient with different indications, contraindications, and so on. For most of the entities we mention a minimum number of attributes in order to be able to develop the examples.

Conceptual Model Step 1

A medication may be clinically defined, at least initially, with several concepts:

- Brand (proprietary) name: *Tylenol, Advil, Centrum*
- Pack or package: *Leena 28 days, Centrum Silver, Triphasil*
- Form of the medication: *tablet, syrup, solution, capsule*
- Route medication should be taken or administered: *PO (by mouth), IV (intravenous)*
- Ingredient: *Acetaminophen, Diphenhydramine, Vitamin K*
- Strength and Strength unit to measure the active ingredient strength or concentration: 50 *mg/tab, 1.5 mg/ml*
- Ingredient class/category: *alpha-blocker, beta-blocker, local anesthetic*

We can now analyze the *MEDICATION* entity from Figure 2.5 using this list of concepts. Most concepts will become entities in the data model, and these entities later evolve into tables in the database, during logical model development: *Concepts → Entities → Tables → Fields.*

Since concepts are such good candidates for entities, we proceed by drafting them and their relationships (Figures 2.6 and 2.7). Each entity has multiple instances or occurrences (rows). It is usually a precursor to a table in a relational database, and it comes with some rules attached. For example, each row/instance/occurrence must be uniquely identified. The concepts/tables have relationships, and these relationships are identified by the thin lines connecting the entities, as explained in the first chapter.

What are the business rules the data model has to satisfy?

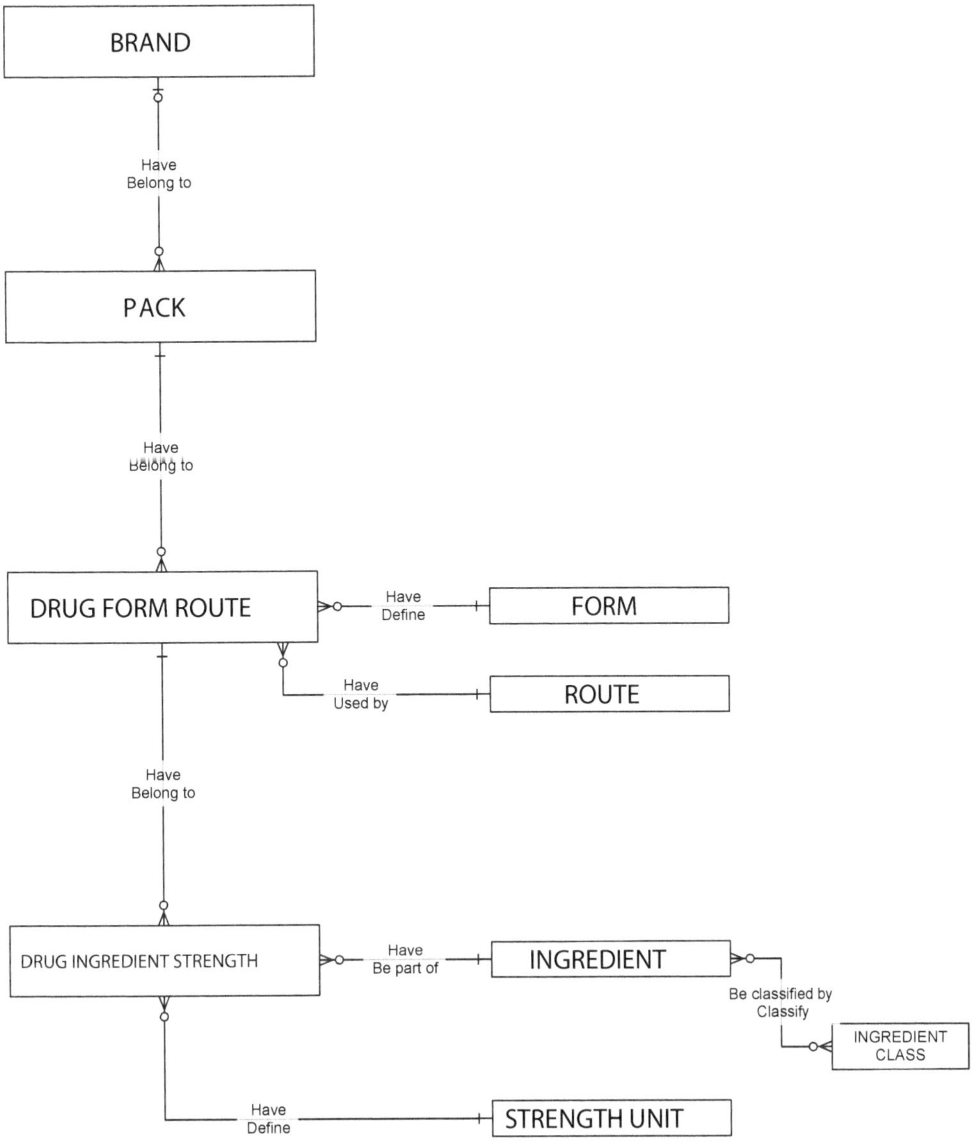

Figure 2.6 Step 1: Conceptual model.

One Brand—Many Packs

Consider the following brand name: *Tylenol*. What does the brand name *Tylenol* mean?

Tylenol Extra Strength, Tylenol PM, Infants' Tylenol, Children's Tylenol Plus Cold: There are many *Tylenol* packs. The concept of *Brand* has a one-to-many relationship with the concept of *Pack*. Pursuant to this business rule, the relationship between the *BRAND* and *PACK* entities is thus defined as one-to-many on the diagram. One *BRAND* may have many instances related in the *PACK* entity, and one *PACK* instance may belong to only one *BRAND*. It is optional in both directions, allowing a *PACK* to exist without a *BRAND*, and a *BRAND* to have no *PACK*.

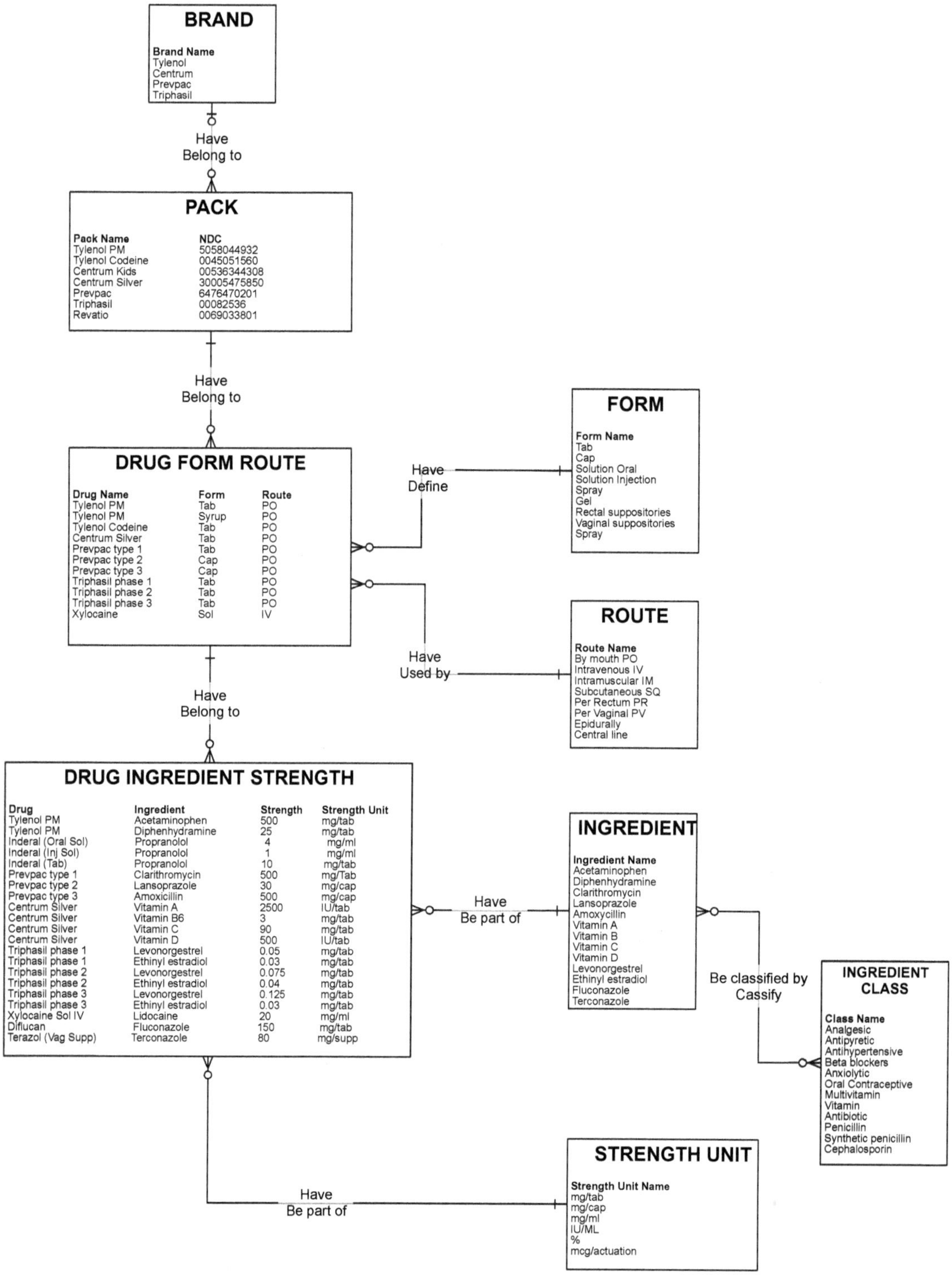

Figure 2.7 Step 1: Conceptual model example.

One Pack—Many Items

Prevpac is a package of a brand name with the following items:

Prevpac type 1 Clarithromycin Tablets
Prevpac type 2 Lansoprazole Capsules
Prevpac type 3 Amoxicillin Capsules

Thus, a pack may contain many types of drug items, and each may have a different form (*tablet, capsule*). How does the conceptual model accommodate this specific business rule? We need an entity to connect the concepts of pack and drug item in a way that one pack may have many items, and each item may be a different drug form. And, while we discuss the concept of drug form, is the concept of drug *route* closely related? Shall we allocate these concepts—form and route—to the same entity/table?

One Drug—Many Forms and Many Routes

Voltaren tablet to be taken *by mouth*
Voltaren gel to be applied *topically (on the skin)*
Voltaren drops for *ophthalmic (eyes)* use
Voltaren suppositories to be used *per rectum (PR)*

The conceptual model must satisfy this requirement for different forms and routes for the same brand name. Consider the following statements:

Tablet is usually PO (per os, by mouth).
Capsule is usually PO.
Rectal suppository is always PR.
Vaginal suppository is always PV (per vagina).

One may be induced to err, thinking the drug form actually implies or determines the drug route. But, this is not the case: A drug form does not imply the drug route. For example, *Lidocaine Solution* (one form) may be used via the following routes:

Intravenous
Subcutaneous
Intra-articular
Epidural

Thus, one form may have many routes. Let us summarize the business rules related to item, form, and route:

- A brand may have many packs.
- A pack may have many items.
- Each item may have one form and one route.
- Form does not imply route.

The concepts of *Brand* and *Pack* are related by now, so we need to create a relationship between the *Form, Route,* and *Pack* concepts. As explained in the systems analysis primer, we are going to use an intersection entity. What is the best name for it? We already know *Pack* may have many items, so maybe we should name this entity "*ITEM*"? *ITEM* is too generalized a concept and may mean too many things to too many people. It is better to name this entity by the other concepts it connects: *DRUG FORM ROUTE*. One instance in this new table is a unique combination of *Pack, Drug item, Form,* and *Route*. Let us check a real-life situation for which these issues of form and route may be important, like the following prescription for a nasty vaginal infection:

Tab. Diflucan 150 mg single dose PO
Vaginal Suppository Monistat 1200 mg PV

A misunderstanding about the form or route in this case could prove highly embarrassing and quite unhealthy for the patient. And, we do not want the data model to be blamed for such a snafu. The conceptual model should be able to cover correctly and unambiguously a prescription with different drug items, forms, and routes.

Diflucan as a brand name can be tracked according to the model, through a unique pack to a unique drug item: *Diflucan Tab PO*. It is the same with *Monistat*: a brand name, a unique package, and within it a unique item as a row in a table: *Monistat Vaginal suppository PV*. The model correctly identifies the drug item, form, and route in its entities.

One Drug Item—Many Ingredients and Strengths

Tylenol PM Tab PO has two ingredients: *acetaminophen* and *diphenhydramine*. *Centrum Silver Tab PO* has 31 active ingredients. The safe prescription, order, dispensation, or administration of drugs necessitates the consideration of allergy to ingredients, ingredient-ingredient interactions, ingredient-contraindications, and so on. The part of the medications system responsible for these safety considerations is workflow number 5: *CDS* (Figure 2.3). To do a great job, *CDS* needs to know all ingredients. The model should accommodate for many, actually an unlimited number of, ingredients in one drug item. An ingredient has a strength and a strength unit attached: *50 mg/tab, 1 IU/ml, 250 mg/capsule,* and so on. Ingredients within the same one drug item may have different strength units; for example: one tablet of *Centrum Silver* has *Vitamin A* defined with the strength unit of *IU/tab* and another ingredient, *Vitamin B6* defined with the unit of *mg/tab*.

We need *DRUG INGREDIENT STRENGTH* as an intersection entity between the concepts of drug item (*DRUG FORM ROUTE*), *INGREDIENT*, and *STRENGTH UNIT*.

The model uniquely identifies many ingredients in one item prescribed, ordered, dispensed, or administered. The model also accommodates different strength units for ingredients within the same item (*mg/Tab* and *IU/Tab*).

Ingredients could be inert or inactive and still may have detrimental pharmacological effects like allergic reactions caused by the (inert) color of the pill. Can we accommodate both active and inactive ingredients with this model? We can define another attribute (column) *Inactive Ingredient* in the table *INGREDIENT* to define each and every instance whether active or not. Each time this attribute is set to *Yes*, the system knows it is an *Inactive Ingredient*. When the attribute is set to *No*, the system knows it is an *Active Ingredient*. Alternatively, we could decide on having all the inactive ingredients allocated in their own table: *INACTIVE INGREDIENT*. These modeling decisions can be left for a later analysis phase.

Another issue to consider is that an ingredient may have a precise name: the salt name of the ingredient. For example, the precise name of *Fluoxetine* is *Fluoxetine hydrochloride*. An attribute can hold the ingredient name, and another column may store the salt name. To summarize the drug name issue, a drug may have the following (7):

- A generic name (*Acetaminophen*)
- Brand name(s) (*Tylenol, Acamol*)
- Chemical name (*N-acetyl-p-aminophenol*).
- Molecular formula (*C8H9NO2*)
- A structural formula—usually a picture of the drug structure and atomic arrangements, bonds, rings, and so on
- An image of the drug item
- An image of the package
- The imprint on the tablet or capsule

The decision on how to have this information stored—whether as an attribute or another entity/table—is postponed at this time. Currently, only the active ingredients appear in the model with their generic names.

National Drug Code

Since we are at the level of a conceptual model, we can delay setting the attributes of the entities at the moment; this activity is expected during the logical model analysis. One attribute related to *PACK* that needs to be mentioned is the *NDC*, the *National Drug Code* (8). The *NDC* is a 7- to 10-digit unique code assigned to medications at the level of the product package. It has three parts:

The Food and Drug Administration (FDA) assigns the first part to the manufacturer; the second part identifies the drug, form, and strength; and the last one is for the package size. The manufacturer assigns the second and third parts of the *NDC*. It could be argued that the *NDC*, uniquely identifying a row, can be the primary key (*PK*) of the *PACK* entity; *NDC* is thus considered a *candidate PK*.

Another potential candidate *PK* is the *UPC*, *universal product code*, widely used in the United States for tracking items in stores (9). The *PACK* unique identifier—*PK* (*NDC*, *UPC*, or some number generated by the system)—helps us identify a specific, unique row in the *PACK* table. This in turn helps us tie the *PACK* table to the *BRAND* table on one side and to the *DRUG FORM ROUTE* table on the other side. The *PACK PK* can also potentially connect *PACK* to *MANUFACTURER, FORMULARY, PACK PRICE,* and other entities that are not shown on this diagram. Whether to use *NDC*, *UPC*, or some other code as the *PK* for *PACK* is out of the discussion scope at the moment. These issues should be part of logical model development.

Just having many items, forms, routes, ingredients, and strengths modeled is not enough. We have to check whether the system can correctly navigate or traverse the data model from one concept to another, especially since prescriptions can be written with the brand name (not the ingredients) in mind.

Consider the following example, a drug item in a prescription: *Tab Triphasil PO*. To provide *CDS*, the system needs to know all the ingredients and their strengths that *Tab Triphasil* contains. And, it is hard to believe, but the doctor is not in a mood to pick ingredients when prescribing, so the system should be able to do it all by itself—identify all ingredients and their strengths with only the brand and form (or route) given. Using the model as is, can the system correctly link the brand name *Triphasil* to its ingredients and strengths? *Triphasil* is uniquely identified at the *BRAND* level as *Triphasil* and at the *PACK* level as *Triphasil* (*NDC 0008-2535*). Using the *PACK PK*, the system can uniquely identify three different rows in the *DRUG FORM ROUTE:*

Triphasil phase 1 Tab PO
Triphasil phase 2 Tab PO
Triphasil phase 3 Tab PO

Since there are three rows in the table with the same *PACK* identifier, the system assumes they all belong to the same one package.

Triphasil phase 1 Tab PO has two ingredients and thus two occurrences (rows) in the *DRUG INGREDIENT STRENGTH:*

Triphasil phase 1 Levonorgestrel 0.05 mg/Tab
Triphasil phase 1 Ethinyl estradiol 0.03 mg/Tab

Since both rows have the same drug item name *Triphasil phase 1 Tab PO* and same unique identifier, the system assumes the two ingredients belong to same one drug item: a tablet to be taken PO. (The system actually uses the *PK/FK* mechanism mentioned to make the connections between tables and not the brand name or pack name for this relationship and connectivity purpose.)

Triphasil phase 2 has the same two ingredients as *Triphasil phase 1* at different strengths, so it has two occurrences/rows of its own in the *DRUG INGREDIENT STRENGTH* table:

Triphasil phase 2 Levonorgestrel 0.075 mg/Tab
Triphasil phase 2 Ethinyl estradiol 0.04 mg/Tab

It continues in the same way with *Triphasil phase 3*.

Centrum Silver (Tab PO) with 31 ingredients is thus expected to have 31 rows in the *DRUG INGREDIENT STRENGTH* table, and each row identifies a different ingredient, strength and strength unit.

If a unique package is given, the system can navigate the conceptual model and correctly identify all ingredients and their strengths and vice versa: With a given ingredient, the system can identify all packages and brands containing the ingredient.

One Ingredient—Many Brands

So far, we have seen one brand and many ingredients. Yet, there are situations for which one ingredient has many brands: *Sildenafil* is the active ingredient in *Viagra* and *Revatio*.

Can the conceptual model deal with the issue of one ingredient—many brands? Can it correctly identify the ingredient if it participates in two brand names? There is no limit to the number of brands for one ingredient. The concept of ingredient in the model does not define a brand and is not related in a one-to-one relationship with the brand concept. On the contrary, an ingredient is actually defined by the brand concept: A brand defines the pack, the pack defines an item, and an item eventually defines the ingredients.

One Ingredient—Many Classes

Acetaminophen is an antipyretic. Acetaminophen is an analgesic.
Propranolol is an antihypertensive. Propranolol is an anxiolytic (anti anxiety).
Lidocaine is a local anesthetic. Lidocaine is an antiarrhythmic.

One Class—Many Ingredients

Beta-blockers: Propranolol, Atenolol, Esmolol, and so on
NSAID (nonsteroidal anti-inflammatory drugs): Ibuprofen, Diclofenac, Naproxen, and so on

Since the Ingredient-Class have a *M:N* relationship, it must be resolved with an intersection entity (*INGREDIENT CLASS*) in the next modeling iteration.

One Class—Many Parents and Many Children

The entity *INGREDIENT CLASS* is a bit more complicated than the solution to the *M:N* relationship mentioned. The concept of *classification* is usually a complex one. For example, a *Beta-blocker* may be a parent to other children, which are also members of the *INGREDIENT CLASS*: *Non-selective beta-blocker, Selective beta 1 blocker, Beta-blocker with intrinsic sympathomimetic activity, Beta and Alpha-1 blocker.* Each of these children may be a parent to other children and so on. In addition, *Beta-blocker* may be itself a child to more than two parents: *Anti hypertensive, Anti anxiety, Anti migraine* classes.

Now, imagine all these parents and children, uniquely identified instances, with all their relationships, as one happy family: the *INGREDIENT CLASS*. This is a typical hierarchy pattern in modeling, and it is usually solved with a *self-referencing* entity. At the moment, though, it is beyond the scope of this discussion, and in the following diagrams the *INGREDIENT CLASS* is not illustrated.

Tall Man Letters

A number of drugs may have look-alike (or sound-alike) names that may cause confusion and mixups, usually with catastrophic results. For example, *chlorpropamide* may be confused with *chlorpromazine*, *cyclosporine* may be confused with *cycloserine*, *daunorubicine* with *doxorubicine*, and so on.

The *Institute for Safe Medication Practices* (ISMP) has identified a list of drug names that have been involved in accidental mishaps or have the potential to cause harm because of the look-alike names (10). The FDA has recommended the practice of using capital letters within the drug names—*Tall Man letters*—to help differentiate between similarly named medications: *cyclocycloSERINE-cycloSPORINE, chlorproPAMIDE-chlorproMAZINE, DAUNOrubicine-DOXOrubicine*, and so on. The letters in the drug name that are different are capitalized, while those similar are left as lowercase. This issue may apply to ingredient, generic name, or brand name (*CeleBREX-CeleXA*). The tables of *BRAND, PACK*, and *INGREDIENT* may need to have an attribute (column) for the Tall Man letters. In keeping the diagrams as simple as possible and since we are not at the logical model level yet, the Tall Man letter naming convention is not shown in the conceptual model.

Conceptual Model Step 2

Let us add to the model another layer of concepts defining a recommended drug dose (Figures 2.8 and 2.9):

- Drug—a combination of parameters: pack, form, route, potentially many items, ingredients, and strengths. Each parameter is by now, after step 1, uniquely identified in the model.

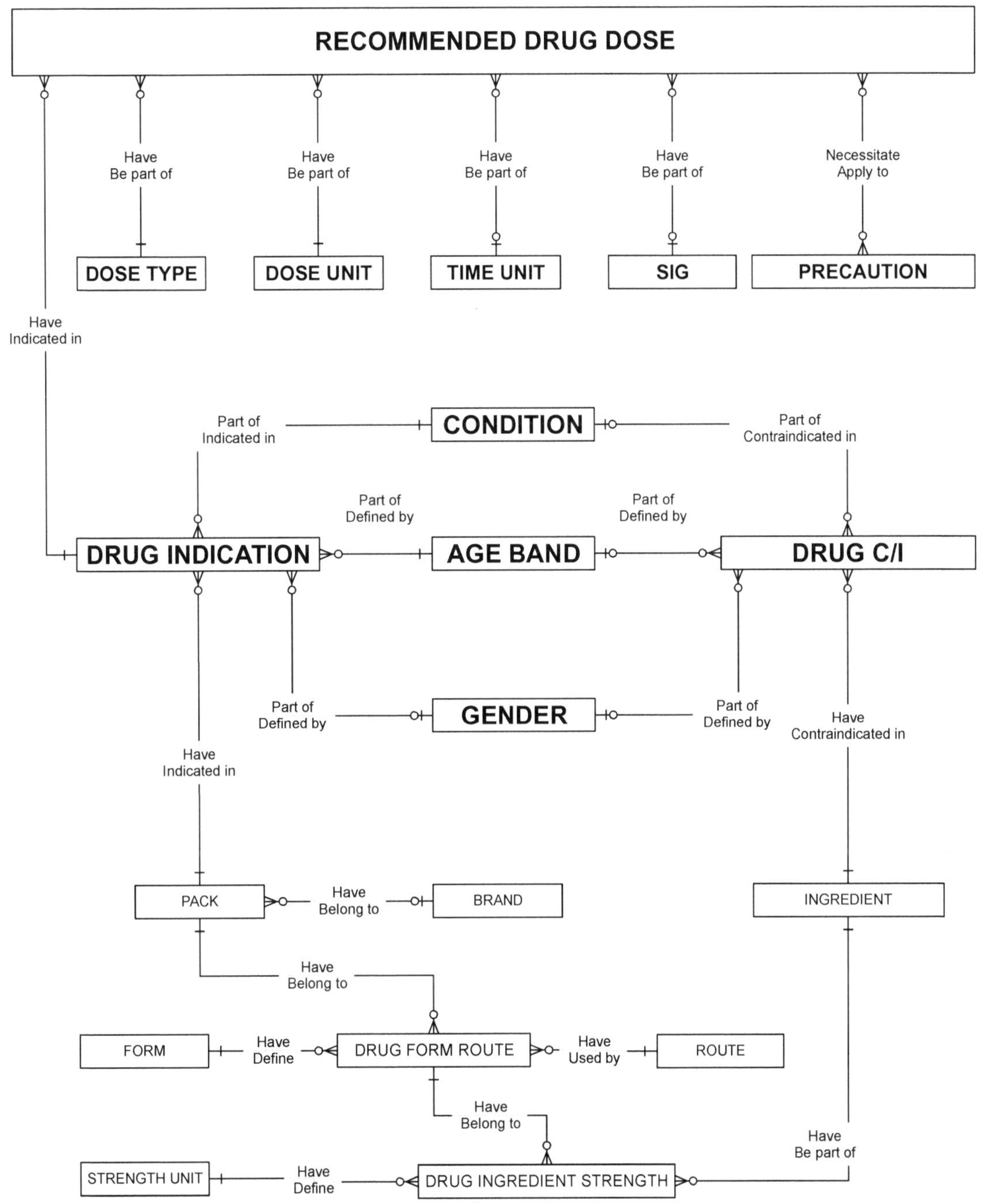

Figure 2.8 Step 2: Conceptual model.

- Condition—for which the drug is indicated or contraindicated: *Diabetes, Hypertension, Pregnancy.*
- Age band—for which the drug is approved: *Neonate, Child, Adult.*
- Gender— for which the drug is indicated or contraindicated.
- Dose type—the dosing regimen intended for the combination of drug-condition-age: *Initial, Maintenance, Maximal.*
- Dose and Dose Unit—a number with a unit: *1 Tab/dose, 2 Cap/dose, 0.5 mg/kg, 0.1 mcg/kg/min, 10 drops/min.*

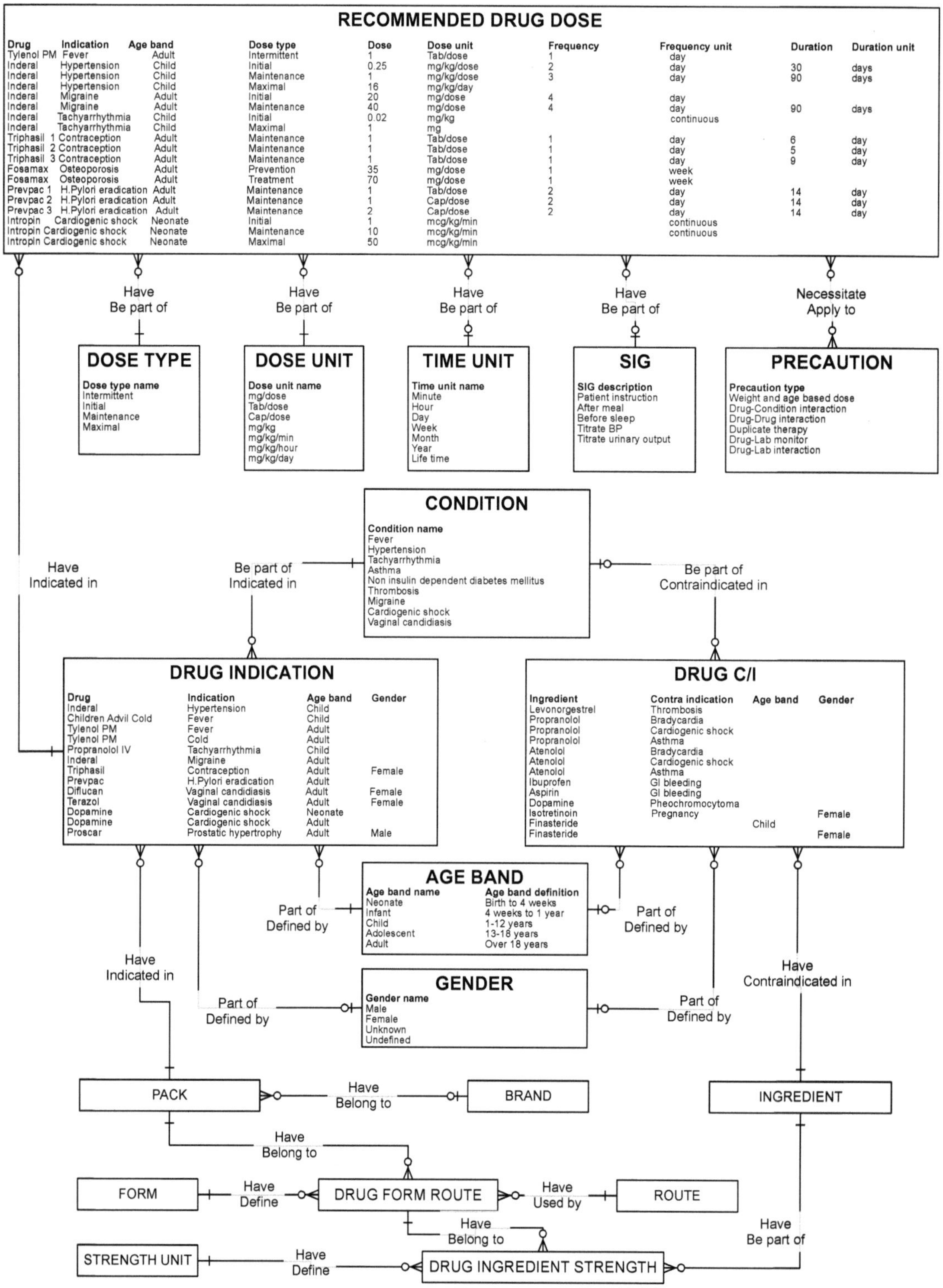

Figure 2.9 Step 2: Conceptual model example.

- Frequency and Frequency unit—doses per time frame: *3 Doses/day, 1 Dose/week, Continuous.*
- Duration and Duration unit—a number with a time unit: *12 hours, 14 days, 1 week, 6 months.*
- Time unit—used in defining both the frequency unit and the duration unit: *minute, hour, day, week, month, year, lifetime, and so on.*
- SIG—(from signature) semistructure instructions: *Take after meals, Take before sleep, Per patient instruction sheet, Titrate to urinary output of 1 ml/kg/hour.*

One Pack—Many Indications

Drugs may be indicated (approved for use) for many conditions or diseases:

Inderal Tab PO is indicated for treating hypertension in children.
Inderal Tab PO is indicated for treating migraine in adults.

The model has an entity named *CONDITION*, which is a list of symptoms, diseases, and other medically relevant conditions: *Pain, Fever, Hypertension, Asthma, Diabetes, Pregnancy, Lactation.* For each indication, there is also an age band for which the drug is approved for use.

AGE BAND is basically a list of age categories and their definitions:

Neonate: Birth to 4 weeks
Infant: 4 weeks to 1 year
Child: 1–12 years
Adolescent: 13–18 years
Adult: over 18 years
Senior: over 65 years

Sometimes, a medication may be approved for a specific gender:

Proscar: Prostatic hypertrophy Adult Male
Triphasil: Contraception Adult Female

The important concept to grasp from this is that drugs are not approved for use as individual ingredients. A drug is approved/indicated for specific combinations of:

- Drug (as package)
- Condition (as indication)
- Age band
- Gender (optional)

The conceptual model adapts to this business rule with the *DRUG INDICATION* entity (table), which stores the unique combination of a pack, an indication (a condition), an age band, and a gender. This allows one drug to be indicated for many conditions and many age bands and for one condition-age band to be treated by many drugs. The model also accommodates gender-specific indications.

One Ingredient—Many Contraindications

An ingredient may have many conditions for which its use is contraindicated. A condition may contraindicate the use of many ingredients. Thus, a many-to-many relationship needs to be solved.

Propranolol is contraindicated in Asthma.
Atenolol is contraindicated in Asthma.
Propranolol is contraindicated in Bradycardia.
Aspirin is contraindicated in GI [gastrointestinal] bleeding.

An ingredient may be contraindicated in a certain age band: *Finasteride is contraindicated in children.* Note that there is no condition specified in this last contraindication example: *Finasteride* is contraindicated in children for all and any condition.

A contraindication may also be related to gender: *Finasteride is contraindicated in females.* Thus, the concept of contraindication is related to

- Ingredient
- Condition
- Age band
- Gender

DRUG C/I is the intersection entity between *INGREDIENT, CONDITION* (as a contraindication), *AGE BAND,* and *GENDER.* It permits one ingredient to be contraindicated in many conditions and one condition to have many contraindicated ingredients. The model allows for drugs to be contraindicated in a specific age band or gender and accommodates contraindications based on age or gender, without any condition defined.

Why is the concept of *Contraindication* tied to the concept of *Ingredient* and not to the concept of *Pack* (like the previously discussed concept of *Indication*)? Once a *Pack* is approved, it means all the ingredients in the pack are approved for a combination of condition-age-gender. When it comes to contraindications, it is an ingredient that would cause the whole package to be contraindicated. Once an *Ingredient* is C/I for a combination of condition-age-gender, it means all packs containing the ingredient are contraindicated. Thus, drugs are contraindicated at the level of ingredient, not at the level of package. For example, all packages containing the ingredient *Propranolol* are contraindicated in *Asthma*. It does not necessarily mean that we need an entity/table to enumerate all the packages, with different forms and routes—and for each one—the fact that they are contraindicated in *Asthma*. *Propranolol* comes in many forms and routes, so storing data at the package level can prove highly redundant. A more efficient way of storing/retrieving the information about C/I would be if we use the relationships of *INGREDIENT*. Once the ingredient *Propranolol* is contraindicated in *Asthma*, all *Pack*(s) containing it become contraindicated in *Asthma*. Knowing that *Propranolol* is C/I in *Asthma* is a fact represented as a single row in the *DRUG C/I* table. From the *INGREDIENT* table via *DRUG INGREDIENT STRENGTH* and *DRUG FORM ROUTE* tables and the respective relationships, the system can identify all instances of *PACK* that contain *Propranolol*. Vice versa, starting from *PACK*, via *DRUG FORM ROUTE, DRUG INGREDIENT STRENGTH, INGREDIENT,* and *DRUG C/I, the* system can traverse the table relationships and identify packs that are contraindicated in *Asthma*.

Dosing Types

Inderal Hypertension Child Initial dose: 0.25 mg/kg/dose 2 doses/day
Inderal Hypertension Child Maintenance dose: 0.5 mg/kg/dose 2 doses/day
Inderal Hypertension Child Maximal dose: 2 mg/kg/day

Doses may have different *Dose types*. We create an intersection entity—*RECOMMENDED DRUG DOSE*—between *DRUG INDICATION* (which has the drug-condition-age-gender information) and multiple other entities: *DOSE TYPE, DOSE UNIT, TIME UNIT, SIG,* and *PRECAUTION*. One of the attributes in *RECOMMENDED DRUG DOSE* is *Dose type*. The *Dose type* defines whether the recommended dose is *Initial dose, Maintenance dose, Maximal dose,* and so on. The model accommodates for one drug-indication-age-gender to have many recommended dosing regimens with many *Dose types*.

Intermittent versus Continuous

The model supports both intermittent and continuous dosing regimens.

Intermittent: *Cap Amoxicillin PO 500 mg/dose 3 doses/day*
Continuous: *Solution Dopamine IV 5 microgram/kg/min*

The model also accommodates more than one instance of the same dose type. For example, there may be two *Maximal doses* for the same drug-condition-age combination:

Inderal Tachyarrhythmia Child Maximal dose 16 mg/kg/day
Inderal Tachyarrhythmia Child Maximal dose 60 mg/day

The *CDS* workflow will need to decide which *Maximal dose (16 mg/kg/day* or *60 mg/day)* to use.

Dose Units

The model allows the flexibility of many dose units: *2 mg/kg/day, 2 Tab/dose, 1 Cap/dose.* The *Drug dose* concept has two components: a *Dose* and a *Dose unit.* Both are attributes in the *RECOMMENDED DRUG DOSE* table. This allows the flexibility of different dosing units for the same drug-indication-age-gender combination: *Tab/dose, mg/kg/dose, microgram/kg/min,* and so on.

Time Units

The model accepts many units of time: minute, hour, day, week, and so on.

Intropin Cardiogenic shock Neonate Maintenance 10 mcg/kg/min
Vancomycin Pre-Operative prophylaxis Adult Continuous 15 mg/kg during one hour
Inderal Hypertension Child Initial dose 0.25 mg/kg/dose 2 Doses/day for 2 weeks
Fosamax Osteoporosis Adult Maintenance dose 70 mg/dose 1 Dose/week for 6 months
Actonel Tab PO Osteoporosis Adult Maintenance dose 150mg/dose 1 Dose/month for 1 year

The model allows different time units to be used in the same dose recommendation: one time unit for *Frequency* and if necessary a different time unit for *Duration.*

Frequency

The frequency recommended for a drug has two components: *Frequency* (a number), and *Frequency unit* (text)—both attributes in the *RECOMMENDED DRUG DOSE*:

Inderal Tab PO 0.25 mg/kg/dose 2 Doses/day
Fosamax Tab PO 70 mg/dose 1 Dose/week
Actonel Tab PO 1 Dose/month

Duration

The model allows each recommended drug dose to have a different *Duration* and *Duration unit* for each instance in the *RECOMMENDED DRUG DOSE*:

Triphasil phase 1 1 Tab/Dose 1 Dose/day for 6 days
Triphasil phase 2 1 Tab/Dose 1 Dose/day for 5 days

Prevpac type 1 Cap/dose 2 Doses/day for 14 days
Actonel Tab PO 1 Dose/month for 6 months

SIG

The actual instructions on how to use a drug are sometimes called SIG (from signature). *SIG* may be as simple as:

Take 2 Aspirins and call me in the morning
Take after meals
Take before sleep

Or as complex as:

Titrate Insulin rate to blood sugar between 180 and 250 mg% during next 4 hours

The model permits for each recommended drug dose to have a different *SIG* attached and for one *SIG* to be used in many dosing regimens. Note that due to space issues, *SIG* is not shown in Figure 2.9 as an attribute/column within *RECOMMENDED DRUG DOSE.*

Precaution

The *Precaution* entity is actually "hiding" concepts and issues that need further elucidation: *Side-effects, Drug-Drug interactions, Duplicate therapy, Adverse reactions, Drug-Lab, Renal & Hepatic considerations, Drug-Vital, and so on.* A drug dose may have many *Precautions.* A *Precaution* may be related to more than one drug dose. The intersection entity to solve this *M:N* relationship is developed in the *CDS* chapter.

Not All Concepts Are Entities

Note that not all concepts are automatically translated into entities (tables). Some concepts are represented by attributes (columns) within an entity:

Strength: The number defining the strength of a medication is not an entity. The actual *Strength* is an attribute, a defining part of an intersection entity named *DRUG INGREDIENT STRENGTH.* The actual *Strength* concept appears also in PATIENT *MEDICATION* as an attribute.

It is the same issue with *Dose*: There is no entity named *Dose.* The actual information about the medication *Dose* is an attribute in the *RECOMMENDED DRUG DOSE* and *PATIENT MEDICATION* intersection entities.

And, it is the same with *Frequency, Duration,* and *Quantity*—these are attributes and not entities.

Conceptual Model Step 3

Following Figures 2.5–2.9, let us add to the concept of *MEDICATION* we have developed so far the concepts of *PATIENT* and *CLINICIAN* and the intersection entity between the three: *PATIENT MEDICATION*. The actual dose is a different concept than the recommended dose; thus, the model shows two separate entities: *RECOMMENDED DRUG DOSE* and *PATIENT MEDICATION*. *PATIENT MEDICATION* contains information about real patients and drugs; it is the main repository for all actual drugs prescribed, ordered, dispensed, or administered. In contrast, *RECOMMENDED DRUG DOSE* is a repository of recommendations (Figures 2.10 and 2.11).

Both *RECOMMENDED DRUG DOSE* and *PATIENT MEDICATION* entities have similar columns and are related to the same tables. Both have relationships with *DRUG INDICATION*. The *PATIENT MEDICATION* model has the same flexibility detailed in step 2 for *RECOMMENDED DRUG DOSE* as far as *Dose type, Dose unit, Time unit (Frequency, Duration), and SIG* are concerned. Some assumptions and business rules are discussed next.

The Patient Is Uniquely Identified

It is one of the model's basic assumptions that the patient is uniquely identified in the *PATIENT* entity. With the *Patient ID,* the system can identify all relevant information about the patient in the system's own tables and in other systems as well, such as information stored in the *Lab results* part of the *CLINICAL REPOSITORY*.

The Clinician Is Uniquely Identified

The *CLINICIAN* entity provides the *Clinician ID,* the *CLINICIAN* unique identifier (physician, nurse, pharmacist, etc.) who may interact with the patient. With the *Clinician ID,* the system can find any pertinent information, such as clinician's name, phone number, address, credentials, privileges, schedules, outcomes, and so on. Again, it is a model assumption that the clinician is uniquely identified at the level of the *CLINICIAN* entity.

Medication Life Cycle

A prescription may have the following life cycle:

Prescribed → Dispensed → Self-Administered → Completed

An order may have a slightly different life cycle:

Ordered → Dispensed → Administered → Completed

A medication life cycle has steps and each step has a *PATIENT DRUG STATUS* attached.

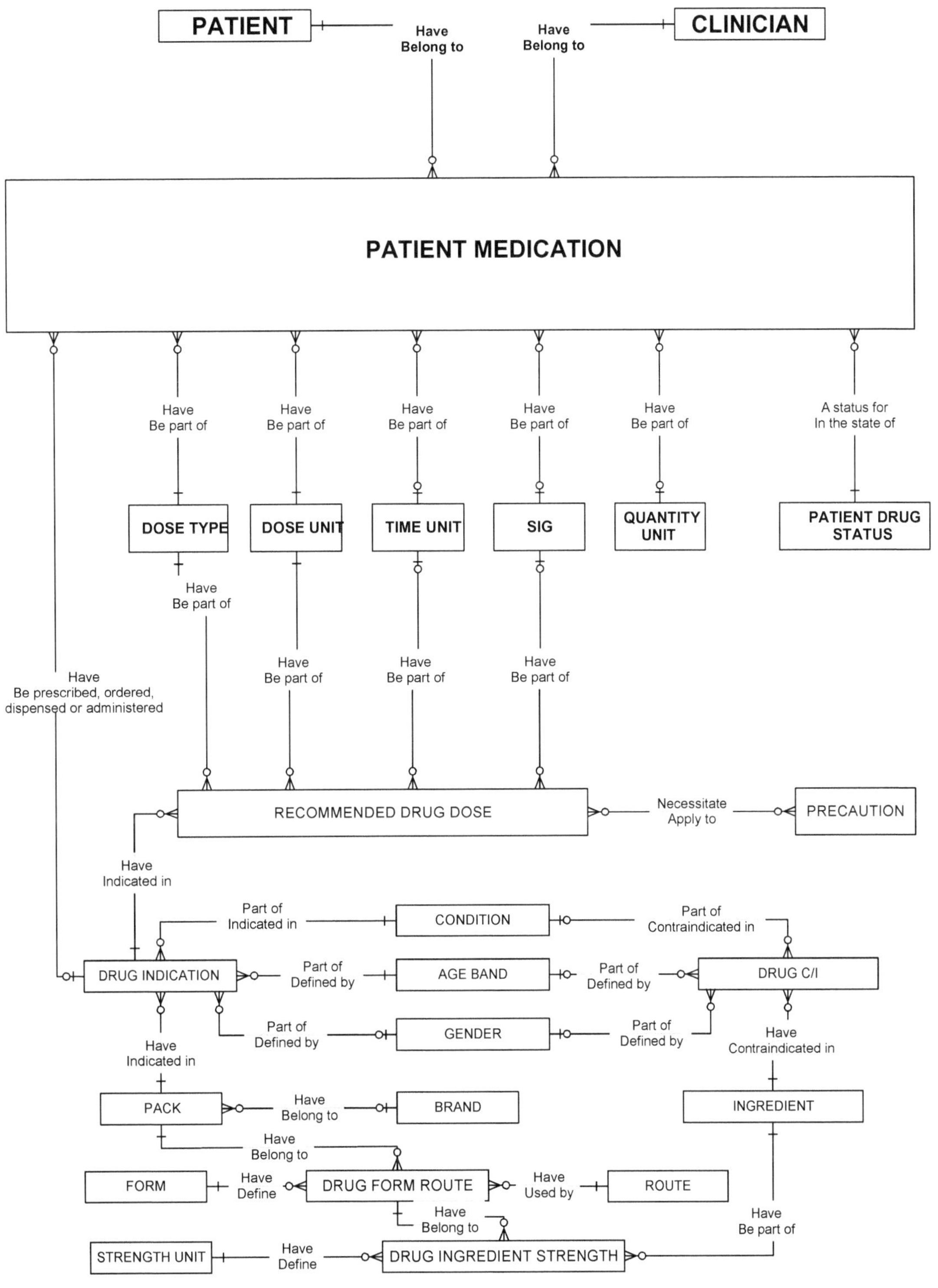

Figure 2.10 Step 3: Conceptual model.

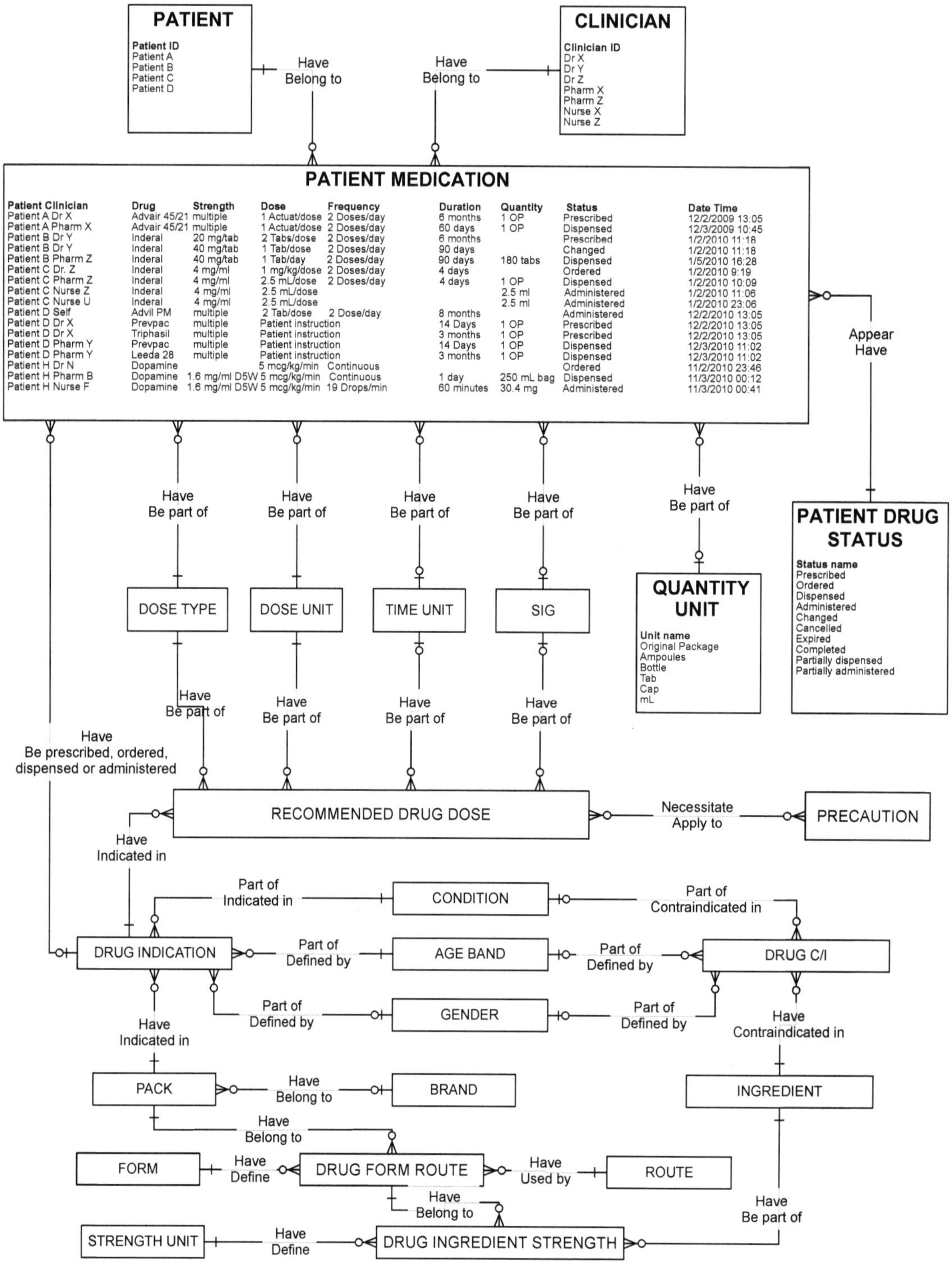

Figure 2.11 Step 3: Conceptual model example.

One Medication Life Cycle—Multiple Statuses

The *PATIENT DRUG STATUS* entity models the concept of different steps in the medication life cycle, and it is related to the *PATIENT MEDICATION* entity. Each item in a prescription (or order, dispensation, or administration) may have a different status during the medication life cycle; for example, some drugs may have been dispensed, while others on the same prescription, missing from stock, have not been dispensed. Consider statuses of a prescription: *Prescribed, Dispensed, Self-administered, Completed, Stopped/Cancelled, Expired, Partially dispensed.* Statuses of an order can be *Ordered, Dispensed, Administered, Completed, Stopped/Canceled, On hold.* The model allows one prescription (or order, dispensation, administration) to have some drugs in one status and others drugs in a different status. There is additional discussion about the *PATIENT DRUG STATUS* concept in the following chapters.

One Patient—Many Prescriptions, Orders, Dispensations, and Administrations

An obvious business rule for the *PATIENT MEDICATION* entity is to allow one patient to have many prescriptions, orders, dispensations, and administration of drugs. Each instance of these must be uniquely identified and belong to only one patient. The model satisfies the rule mentioned having *PATIENT* and *PATIENT MEDICATION* related and allowing multiple medication instances for same patient:

Patient X: Cap Amoxicillin 500 mg/cap, 1 Cap/dose, 2 Doses/day, 7 Days, Quantity 14 Caps, Prescribed by Dr. A on 11/2/2011
Patient X: Cap Amoxicillin 250 mg/cap, 2 Caps/dose, 3 Doses/day, 7 Days, Quantity 28 Caps, Dispensed by Pharm. B on 11/3/2011
Patient X: Dopamine 5 mcg/kg/min, Continuous, Ordered by Dr. C on 12/5/2011

One Clinician—Many Prescriptions, Orders, Dispensations, and Administrations

The model allows one clinician to relate to many prescriptions, orders, dispensations, or administrations. The model also permits many clinicians (doctors, pharmacists, nurses, etc.) to be involved with one patient during a medication life cycle.

One Prescription (Order, Dispensation, Administration)—Many Items

The model supports one prescription (or order, dispensation, administration) to have many items as instances in *PATIENT MEDICATIONS*:

Patient X:
Item 1: Cap Amoxicillin 500 mg/cap, 1 Cap/dose, 3 Doses/day, 7 Days, Quantity 14 Caps
Item 2: Tab Inderal 20 mg, 2 Tabs/day, 60 Tabs/month, for 6 months
Prescribed by Dr. A on 11/2/2010

Both *Inderal* and *Amoxicillin* can be uniquely identified in *PATIENT MEDICATION*. But, how can the system realize that both *Amoxicillin* and *Inderal* are part of one prescription; how can the model group them under one prescription or one order? We can assume that if several events happen at the same moment in time to the same entities, then these events most probably are related: If *Dr. X* treating *Patient Y* has prescribed several drugs at the same moment in time, the system assumes all drugs are part of one prescription. If *Pharm. H* has dispensed to *Patient Y* several drugs at the same moment in time, the system assumes all drugs are part of one dispensation. If *Nurse B* has administered to *Patient Y* several drugs at the same moment in time, the system assumes all drugs were part of one administration.

Not Indicated Is Not Equal to Contraindicated

When a drug is not indicated or is not approved for a certain condition and age band, then its use is considered *off label*. If there are no instances in the *DRUG INDICATION* for a unique combination of drug-indication-age-gender, it means the drug is not FDA approved for the given combination. Whoever prescribes, orders, dispenses, or administers a drug that is not approved should be aware its use is considered off label. The model allows storing off labels in *PATIENT MEDICATION*. Any combination of drug-condition-age-gender can be used, and the combination does not have to originate from existing instances of *DRUG INDICATION*.

Contraindicated means something different: *avoid* the combination. It is known that this ingredient and that condition, age, or gender do not mix well and may have deleterious effects on the patient.

The Actual Dose Has a Quantity

The concept of *Quantity* is needed to define the various aspects of medications as they are related to patients: *How many tablets have been prescribed or ordered? How many capsules have been dispensed? How many micrograms have been administered?* For example:

Patient Z: Inderal Tab PO Migraine Adult, Maintenance, Strength 20 mg/tab, Dose 1 Tab/dose, Frequency 2 Doses/day, Duration 6 months, Quantity 6 refills, Prescribed by Dr. Y 3/1/2010

Patient Z: Inderal Tab PO Migraine Adult, Maintenance, Strength 20 mg/tab, Dose 1 Tab/dose, Frequency 2 Doses/day, Duration 1 month, Quantity 60 tabs, Dispensed by Pharm. W 3/1/2010

Quantity as a concept may change its meaning, depending on the *Status* of the medication: *Prescribed quantity, Ordered quantity, Dispensed quantity, Administered quantity*. These discrete data elements reside in the *PATIENT MEDICATION* table, in the *Quantity* column with the *Quantity unit* attached.

The model allows many quantity units to cover the medication life cycle. For each status, there may be a different quantity and quantity unit.

Dosing Regimens

The same as with the *RECOMMENDED DRUG DOSE*, multiple dosing regimens (*Initial, Maintenance, Maximal, etc.*) are allowed by the model for *PATIENT MEDICATION*. Note that *DOSE TYPE, DOSE UNIT, TIME UNIT,* and *SIG* have the same relationships with *PATIENT MEDICATION* as they have with *RECOMMENDED DRUG DOSE*. Thus, the same functionality detailed regarding the recommended dose is replicated for actual dosing regimens. For example:

Patient Q: Propranolol Solution for Hypertension Child, Strength 4 mg/ml (strawberry-mint flavored)
Initial dose 0.25 mg/kg/dose, Frequency 2 doses/day, Duration 7 days
Maintenance dose 0.5 mg/kg/dose, Frequency 3 doses/day, Duration 3 months
Maximal dose 2 mg/kg/day
Prescribed by Dr. S 3/1/2010

Actual Dose May Be Different from the Recommended One

The model permits an actual dose to be different from the recommended one. These are instances in different entities in the model:

RECOMMENDED DRUG DOSE
Inderal Tab PO Migraine Adult, Strength 40 mg/Tab, Maintenance dose: 1 Tab/dose, 2 Doses/day
PATIENT MEDICATION (Actual patient drug dose)
Patient Z, Inderal Tab PO Migraine, Adult, Strength 20 mg/Tab, Maintenance dose: 1 Tab/dose, 2 Doses/day

Drug Name and Other Parameters May Change during Medication Life Cycle

The model allows for a drug to be prescribed or ordered with a specific name and be dispensed with a different name, unless a clinician insists on *DAW (Dispense As Written)*. For example:

Patient D: Inderal Tab PO, Strength 20 mg/tab, Prescribed by Dr. B on 10/2/2010
Patient D: Generic Propranolol Tab PO, Strength 20 mg/tab, Dispensed by Pharm. E on 10/2/2010

The model allows the parameters of strength, dose, frequency, duration, and quantity to change during the medication life cycle. A drug on a prescription (or order, dispensation, or administration) may have many instances in the model. Each instance may have:

- One dose and one of many dose units
- One frequency and one of many frequency units
- One duration and one of many duration units
- One quantity and one of many quantity units.

For example:

Patient X: Cap Amoxicillin, Strength 500 mg/cap, Dose 1 Cap/dose, Frequency 2 Doses/day, Duration 7 Days, Quantity 14 Caps, Prescribed by Dr. A on 11/2/2010
Patient X: Cap Amoxicillin, Strength 250 mg/cap, Dose 2 Caps/dose, Frequency 2 Doses/day, Duration 7 Days, Quantity 28 Caps, Dispensed by Pharm. B on 11/3/2010

Not having the prescribed *500 mg/cap* in stock, *Pharm. B* dispenses twice the number of *Amoxicillin* capsules at half the strength. This changes the following parameters:

Strength: from *500 mg/cap* to *250 mg/cap*
Dose: from *1 Cap/dose* to *2 Caps/dose*
Quantity: from *14 Caps* to *28 Caps*

Note: *Frequency* (*2 Doses/day*) and *Duration* (*7 days*) did not change.

Half Tablets

The model supports half tablets as well as any other half-dosing unit. For example:

Patient X: Tab Tenormin, Strength 25 mg/tab, Dose 0.5 Tab/dose, Frequency 1 Dose/day
Duration 6 months, Quantity <null>, Prescribed by Dr. A on 11/2/2010

This flexibility may actually present a problem since there may be dosing units that cannot be halved. The implementation of a business rule such as *"Capsules should not be cut in half"* should be handled during further analysis.

Daily Dose versus Maximal Dose

Using discrete data elements, the system can calculate parameters that may have never been defined in the model, such as *volume/dose, daily dose per kg, daily quantity*, and so on and compare the resulting numbers to recommended dosing regimens. From these comparisons, *CDS* concludes information about drug safety.

For example, consider a simple calculation related to a *20 kg hypertensive child*:

Patient G: Inderal Sol (strawberry flavored) PO Hypertension Child, Maintenance dose
Strength <null>, Dose 0.5 mg/kg/dose, Frequency 2 Doses/day, Duration 6 months, Prescribed by Dr. W 3/1/2010

CDS can calculate whether this prescription is safe from a maximal dose perspective. In the *RECOMMENDED DRUG DOSE* table, *CDS* finds

Inderal Sol PO Hypertension Child Maximal (recommended) dose 2 mg/kg/day

CDS calculates:

Dose × Frequency = Dose per time frame
Prescribed 0.5 mg/kg/dose × 2 Doses/day = 1 mg/kg/day

Since patient G has been prescribed less than the maximal dose, CDS concludes it is a safe dose. A different case would be the following:

Patient G: Inderal Sol (strawberry flavored) PO Hypertension Child, Maintenance dose
Strength <null>, Dose 1.5 mg/kg/dose, Frequency 2 Doses/day, Duration 6 months
Prescribed by Dr. W 3/1/2010

CDS calculates:

Dose × Frequency = Dose per time frame
Prescribed 1.5 mg/kg/dose × 2 Doses/day = 3 mg/kg/day

In this case, patient G has been prescribed 50% more than the maximal dose; CDS concludes it is not a safe dose. All calculations are based on discrete data elements retrieved from the *Medications* system data model.

Let us see the CDS calculation during the dispensing process:

Patient G: Inderal Sol (strawberry flavored) PO Hypertension Child, Maintenance dose
Strength 4 mg/ml, Dose 2.5 ml/dose, Frequency 2 Doses/day, Duration 3 months, Quantity 1 Bottle 500 ml, Dispensed by Pharm. W 3/1/2010

The pharmacist defines the strength as *4 mg/ml* and can translate the *Dose* unit from *mg/kg/dose* to the more familiar *ml/dose*, which is used as a measuring unit by worried parents:

Patient weight X Dose = Quantity/dose
20 kg X 0.5 mg/kg/dose = 10 mg/dose
Quantity/dose/Strength = Volume/dose
10 mg/dose/4 mg/ml = 2.5 ml/dose

Quantity dispensed becomes

Patient needs 2.5 ml/dose X 2 doses/day = 5 ml/day
One bottle = 500 ml
Quantity/Daily needs = Duration
500 ml/5 ml/day = 100 days (approximately 3 months)

How would *CDS* react to the following?

Patient G: Inderal Sol (strawberry flavored) PO Hypertension Child, Maintenance dose
Strength 8 mg/ml, Dose 3 ml/dose, Frequency 2 Doses/day, Duration 3 months, Quantity 1 Bottle 500 ml, Dispensed by Pharm. W 3/1/2010

Actual dose dispensed becomes

Strength X Dose X Frequency = Quantity/day
8 mg/ml X 3 ml/dose X 2 Doses/day = 48 mg/day

For a 20-kg child,

Quantity/day/Patient weight =Daily dose per kg
48 mg/day/20 kg = 2.4 mg/kg/day

Since

Inderal Sol PO Hypertension Child Maximal (recommended) dose 2 mg/kg/day

CDS would alert that the maximal recommended dose for *Inderal-Hypertension-Child* has been exceeded.

Strengths, doses, frequencies, durations, and quantities—even drug names—may change during the medication life cycle, and the model allows this flexibility to support various clinicians with different views and different units collaborating on the same medication life cycle.

Not All Drug Parameters Are Clinically Relevant

There are other parameters to consider when defining a medication: manufacturer, price, formulary status, and so on. These parameters are not clinically relevant and thus did not make it to the diagram. It does not mean this information is not important or the system does not need it. To the contrary: The decision whether to include or exclude certain entities in the diagram is an exercise in setting the system boundaries: What should be the system responsibility? What should the *Medications* system capture, update, store, and retrieve? In this case, it is an arbitrary setting of the system boundaries on my behalf.

We have just finished the first draft of the medications system conceptual model. Using two dozen concepts, we have zoomed in from the huge concept of *CLINICAL REPOSITORY* to the tiny microgram/kilogram/minute dosing unit.

One picture in this case (Figure 2.12) is actually worth more than the proverbial thousand words. The following statements (1,277 words to be precise) can be read from the conceptual model. If you feel comfortable with the diagram in Figure 2.12, you can safely skip this tedious part.

- One *BRAND* (*Tylenol*) may have one or more *PACK* (*Tylenol PM, Tylenol Codeine, Tylenol Cold*).
- One *PACK* (*Tylenol PM*) may belong to only one *BRAND* (*Tylenol*).
- One *PACK* (*Prevpac*) may have one or more instances in *DRUG FORM ROUTE* (*Prevpac type 1 Tab PO, Prevpac type 2 Cap PO, Prevpac type 3 Cap PO*).
- One instance of *DRUG FORM ROUTE* (*Prevpac type 2 Cap PO*) must belong to only one *PACK* (*Prevpac*).
- *FORM* is a list of the drug forms: *Tablet, Capsule, Solution, Nasal spray, Gel, and so on.*
- A *FORM* (*Tab*) may define one or more instances of *DRUG FORM ROUTE* (*Centrum Silver Tab PO, Tylenol PM Tab PO*).
- A unique instance in *DRUG FORM ROUTE* (*Tylenol PM Tab PO*) must have only one *FORM* (*Tab*).
- *ROUTE* is a list of drug routes: *Oral, Intravenous, Intramuscular, Per Rectum, and so on.*
- A *ROUTE* (*PO*) may define one or more instances of *DRUG FORM ROUTE* (*Centrum Silver Tab PO, Tylenol PM Tab PO*).
- An instance in *DRUG FORM ROUTE* (*Tylenol PM Tab PO*) must have only one *ROUTE* (*PO*).
- A *PACK* may have more than one item, and each item in a *PACK* may have a different *FORM* and a different *ROUTE*.
- *INGREDIENT* is a list of active drug ingredients: *Acetaminophen, Diphenhydramine, Clarithromycin, Amoxicillin, and so on.*
- A row in the *DRUG FORM ROUTE* table (*Tylenol PM Tab PO*) may have one or more related instances in the *DRUG INGREDIENT STRENGTH* table (*Tylenol PM Acetaminophen 500 mg/Tab, Tylenol PM Diphenhydramine 25 mg/Tab*).

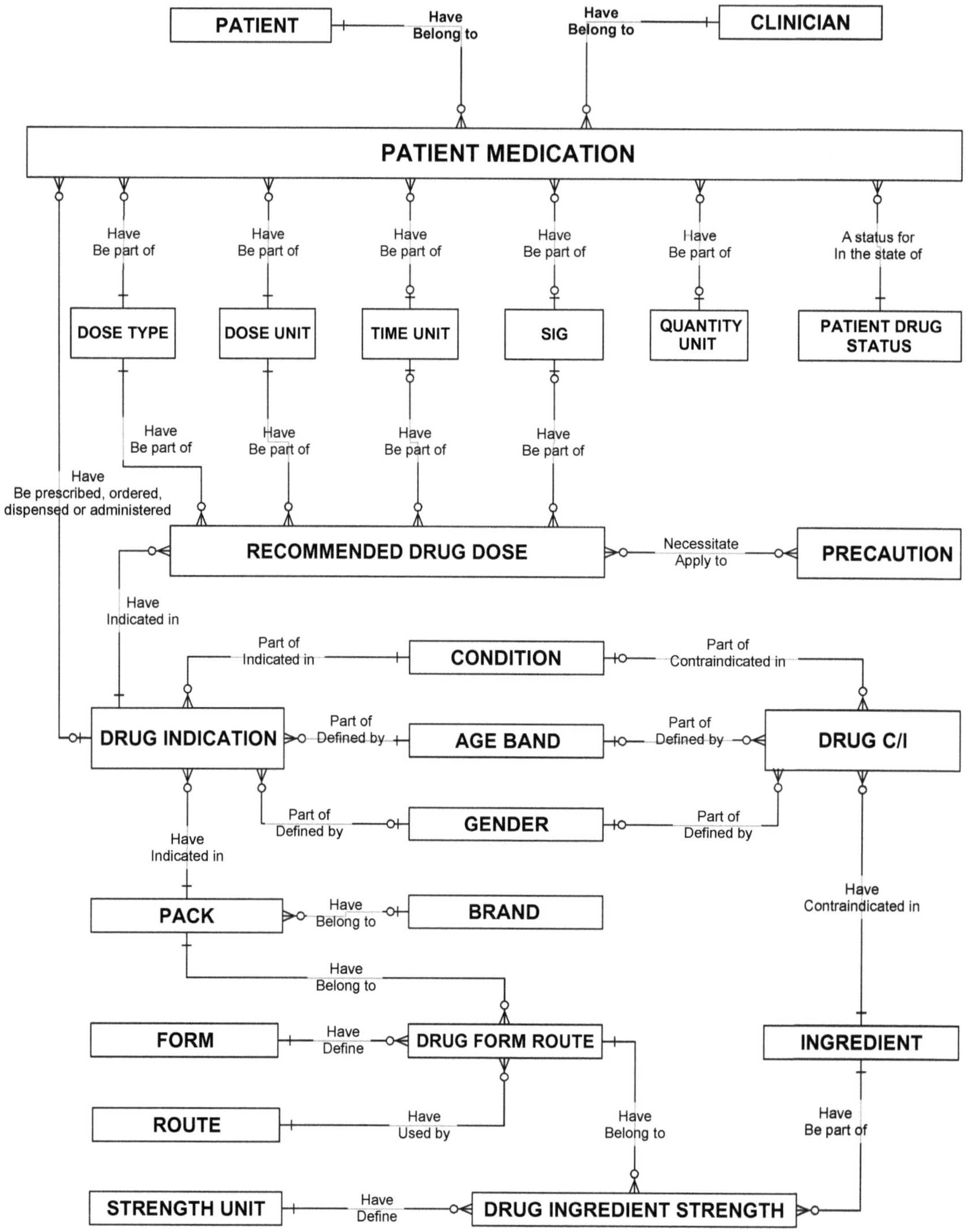

Figure 2.12 Medications system conceptual model draft.

- A row in the *DRUG INGREDIENT STRENGTH* table (*Tylenol PM Acetaminophen 500 mg/Tab*) must belong to only one instance of *DRUG FORM ROUTE* (*Tylenol PM Tab PO*).
- An *INGREDIENT* (*Acetaminophen*) may be part of one or more instances of *DRUG INGREDIENT STRENGTH* (*Tylenol PM Acetaminophen 500 mg/Tab, Tylenol-Codeine Acetaminophen 300 mg/Tab*).

- A row in the *DRUG INGREDIENT STRENGTH* table (*Tylenol PM Acetaminophen 500 mg/Tab*) must have only one instance of *INGREDIENT* (*Acetaminophen*)
- *STRENGTH UNIT* is a list of the drug strength units: *mg/Tab, mg/Cap, mg/ml, Unit/ml, IU/Tab, and so on.*
- A *STRENGTH UNIT* (*IU/Tab*) may define one or more instances of *DRUG INGREDIENT STRENGTH* (*Centrum Silver Vitamin A 2500 IU/Tab, Centrum Silver Vitamin D 500 IU/Tab*).
- A unique instance in *DRUG INGREDIENT STRENGTH* (*Centrum Silver Vitamin D 500 IU/Tab*) must have only one *STRENGTH UNIT* (*IU/Tab*).
- An *INGREDIENT* (*Acetaminophen*) may belong to one or more *INGREDIENT CLASS* (*Analgesic, Antipyretic*).
- An *INGREDIENT CLASS* (*Beta blocker*) may classify one or more *INGREDIENT* (*Propranolol, Atenolol, Esmolol*).
- *AGE BAND* is a list of age categories and their definitions: *Neonate Birth to 4 weeks, Infant 4 weeks to 1 year, Child 1–12 years, Adolescent 13–18 years, Adult over 18 years, and so on.*
- *CONDITION* is a list of diseases—*Fever, Hypertension, Asthma, Cardiogenic shock, etc.*—and other medically relevant conditions, such as *pregnancy, lactation, smoking, and so on.*
- One instance of *AGE BAND* (*Child*) may be an approved age band for many instances of *DRUG INDICATION* (*Inderal Hypertension Child, Inderal Tachyarrhythmia Child*).
- One instance of *DRUG INDICATION* (*Inderal Hypertension Child*) must have only one *AGE BAND* (*Child*).
- A *CONDITION* (*Hypertension*) may be part of many instances of *DRUG INDICATION* (*Inderal Hypertension Child, Atenolol Hypertension Adult*).
- One instance of *DRUG INDICATION* (*Inderal Hypertension Child*) must have only one *CONDITION* (*Hypertension*).
- One instance of *GENDER* (*Male*) may be approved for many instances of *DRUG INDICATION* (*Proscar Prostatic hypertrophy Male, Propecia Prostatic hypertrophy Male*).
- One instance of *DRUG INDICATION* (*Proscar Prostatic hypertrophy Male*) may have only one *GENDER* (*Male*).
- An *INGREDIENT* (*Atenolol*) may be contraindicated in many instances of *DRUG CONTRA-INDICATION* (*Atenolol Asthma, Atenolol Bradycardia*).
- An instance in *DRUG CONTRA-INDICATION* (*Atenolol Asthma*) may have only one *INGREDIENT* (*Atenolol*).
- A *CONDITION* (*Asthma*) may be contraindicated in many rows of *DRUG CONTRA-INDICATION* (*Atenolol Asthma, Propranolol Asthma*).
- One instance of *DRUG CONTRA-INDICATION* (*Propranolol Asthma*) may have only one contraindication as *CONDITION* (*Asthma*).
- A *GENDER* (*Male*) may be contraindicated in many instances of *DRUG CONTRA-INDICATION* (*Levonorgestrel Male, Ethynil estradiol Male*).

- An instance in *DRUG CONTRA-INDICATION* (*Levonorgestrel Male*) may have only one *GENDER* (*Male*).
- One instance in *DRUG INDICATION* (*Inderal Hypertension Child*) may appear many times in *RECOMMENDED DRUG DOSE* (*Inderal Hypertension Child Initial 0.025 mg/kg/dose 2 doses per day, Inderal Hypertension Child Maximal 2 mg/kg/day*).
- An instance in *RECOMMENDED DRUG DOSE* (*Inderal Hypertension Child Initial 0.025 mg/kg/dose 2 doses per day*) must have only one *DRUG INDICATION* (*Inderal Hypertension Child*).
- *DOSE TYPE* is a list of dosing regimens or types of doses: *Initial, Intermittent, Maintenance, Maximal, and so on.*
- One instance of *DOSE TYPE* (*Initial*) may appear in many instances of *RECOMMENDED DRUG DOSE* (*Inderal Migraine Adult Initial 20 mg/dose 4 doses per day, Intropin Cardiogenic shock Neonate Initial 1 mcg/kg/min*).
- One row in *RECOMMENDED DRUG DOSE* (*Inderal Migraine Adult Initial 20 mg/dose 4 doses per day*) must have only one *DOSE TYPE* (*Initial*).
- *DOSE UNIT* is a list of units for dosing: *mg/dose, Tab/dose, Cap/dose, mg/kg, mg/kg/min, and so on.*
- One instance of *DOSE UNIT* (*mg/dose*) may appear in many instances of *RECOMMENDED DRUG DOSE* (*Inderal Migraine Adult Initial 20 mg/dose 4 doses per day, Fosamax Osteoporosis Prevention 35 mg/dose 1 dose per week*).
- One row in *RECOMMENDED DRUG DOSE* (*Inderal Migraine Adult Initial 20 mg/dose 4 doses per day*) must have only one *DOSE UNIT* (*mg/dose*).
- *TIME UNIT* is a list of time units: *Minute, Hour, Day, and so on.* It is used in *RECOMMENDED DRUG DOSE*: defining the *Frequency unit* (*day*) in the *Frequency* concept (*2 doses per day*) and again in defining the *Duration unit* (*month*) in the concept of *Duration* (*6 months*) for a recommended drug.
- One row in *TIME UNIT* (*day*) may have many instances in *RECOMMENDED DRUG DOSE* (*Tylenol PM Fever Adult Intermittent 1 tab/dose 1 dose per day, Inderal Migraine Adult Initial 20 mg/dose 4 doses per day*).
- One instance in *RECOMMENDED DRUG DOSE* (*Inderal Migraine Adult Initial 20 mg/dose 4 doses per day*) may have up to two *TIME UNITs*: one as the field (column) defining the time frame in *Frequency unit* and/or one as the field defining the *Duration unit.*
- An instance of *RECOMMENDED DRUG DOSE* may have one *SIG* (*Take after meal, Take before sleep, and so on*).
- An instance of *SIG* may appear in many rows of *RECOMMENDED DRUG DOSE.*
- An instance of *RECOMMENDED DRUG DOSE* may have one or more *PRECAUTION.*
- One *PRECAUTION* may apply to many instances of *RECOMMENDED DRUG DOSE.*
- One *PATIENT* may have many instances of *PATIENT MEDICATION.*
- One unique row in the *PATIENT MEDICATION* must belong to only one *PATIENT.*

- One *CLINICIAN* may have many instances of *PATIENT MEDICATION*.
- One unique row in the *PATIENT MEDICATION* must belong to only one *CLINICIAN*.
- One *PATIENT DRUG STATUS (Prescribed)* may appear in many instances of *PATIENT MEDICATION (Patient X Dr. Y Atenolol … Prescribed 2/1/2009 13:32, Patient Y Dr. Z Lipitor … Prescribed 3/2/2010 14:20)*
- One instance of *PATIENT MEDICATION (Patient X Dr. Y Atenolol … Prescribed on 2/1/2009 13:32)* may have only one *PATIENT DRUG STATUS* defined *(Prescribed)*.
- *QUANTITY UNIT* is listing the units for prescribing, ordering, dispensing and administering drugs such as *Original Package (OP), Tablets, Ampoules, Bottle, and so on.*
- One instance of *DRUG INDICATION* may have many instances of *PATIENT MEDICATION*.
- One instance of *PATIENT MEDICATION* may have only one *DRUG INDICATION*.

In the following chapters, we will discover new entities, relationships, and attributes. Some concepts may need to be remodeled; some relationships and attributes may change as we continue the analysis process.

At this stage, the conceptual model is incomplete: There are still concepts, such as *PRECAUTION*, that are hiding other concepts within. There is more about side effects, drug-drug interactions, duplicate therapy, renal and hepatic considerations in the chapter on *CDS*.

Conditions, diseases, and their many vocabularies and classification systems are not represented at all. Chapter, *"Interoperability*," is devoted to classifications, standards, and vocabularies relevant to the medications domain.

While building the conceptual model in this chapter, we did not use some of the more refined modeling techniques, such as supertype/subtype, arcs, self-referencing entities, and the like. Not using these techniques may cost in model expressivity: how well the model is adapted to the nuances of the business rules. The conceptual model is a draft we will use in the following chapters as we iteratively refine the logical model.

Summary

In this chapter, the medications domain was analyzed from two perspectives: workflows and data structures. We have identified the main actors, system boundaries, and the main data stores. During the process, we have realized there are situations the *Medications* system needs and uses certain pieces of information that are not within the system's responsibilities to maintain. Other systems are expected to provide this information as a real-time service, off-line feed, regular update, or the like.

The deliverables of this initial analysis phase are:

- Workflows: context diagram and DFD 0
- Data structures: conceptual data model

Each one of the *Medications* system workflows mentioned has a chapter dedicated in the book in which we decompose the workflow further into processes, activities, and tasks.

These processes use data structures to store and retrieve discrete data elements. The conceptual model just defined is the basis for further development of the logical model, which in turn will evolve into a database. We have actually tasted a bit of logical modeling by setting a couple of attributes in our examples.

In the next chapter, we proceed with the analysis of workflow number 1: *Prescribe/eRx*. We use the same dual paths used in this chapter: DFD for the decomposition of the workflow into processes and the logical model for the analysis of the data structures to support these processes.

References

1. Yourdon, E. 1989. *Modern structured analysis*. Upper Saddle River, NJ: Prentice-Hall.
2. Simsion, G.C., and G.C. Witt. 2005. *Data modeling essentials*. New York: Elsevier.
3. Shelly, G.B., T.J. Cashman, and H.J. Rosenblatt. 2008. *Systems analysis and design*. Boston: Thomson Course Technology.
4. Rob, P., and C. Coronel. 2009. *Database systems. Design, implementation and management*. Boston: Thomson Course Technology.
5. Silverston, L., and P. Agnew. 2009. *The data model resource book—Universal patterns for data modeling*. New York: Wiley.
6. http://en.wikipedia.org/wiki/Business_rule (accessed December 5th, 2010)
7. A pharmacological overview of Tylenol® (acetaminophen). http://www.tylenolprofessional.com/pharmacology.html (accessed December 5, 2010).
8. Wikipedia. National Drug Code. http://en.wikipedia.org/wiki/National_Drug_Code (accessed December 5, 2010).
9. Wikipedia. Universal product code. http://en.wikipedia.org/wiki/Universal_Product_Code (accessed December 5, 2010).
10. Institute for Safe Medication Practices. FDA and ISMP lists of look-alike drug names with recommended Tall Man letters. http://www.ismp.org/tools/tallmanletters.pdf (accessed March 5, 2011).

Further Reading

Beaver, K. 2003. *Healthcare information systems*. Boca Raton, FL: Auerbach.

Donnelly, A.J, F.E. Cunningham, and V.L. Baughman. 2000. *Anesthesiology and critical care drug handbook*. Hudson, OH: Lexi-Comp.

Shortliffe, E.H., L.E. Perreault, G., Wiederhold, et al. (editors). 2001. *Medical informatics—Computer applications in health care and biomedicine*. New York: Springer.

Review Questions

Find the errors in the following five diagrams:

1.

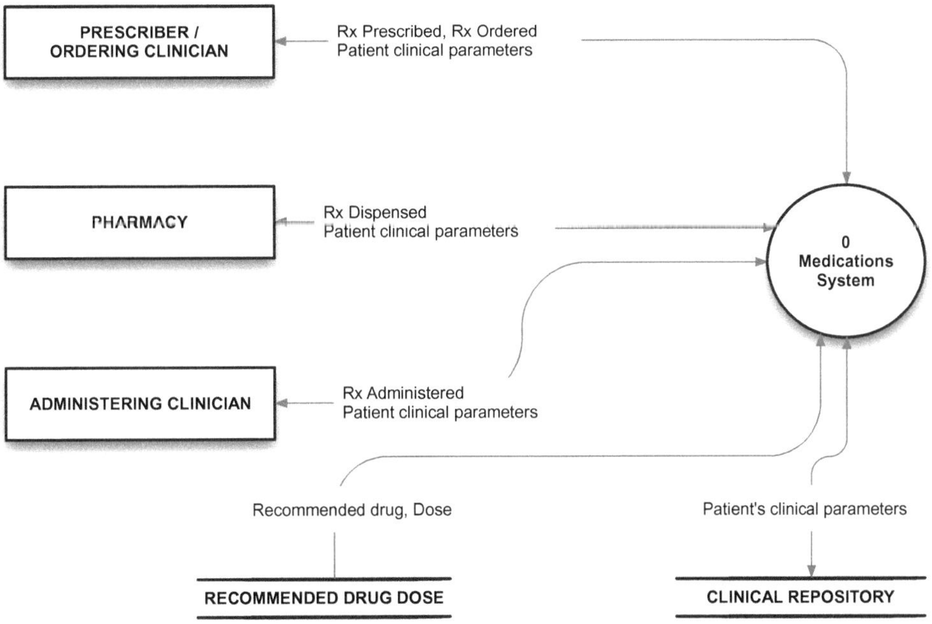

2.

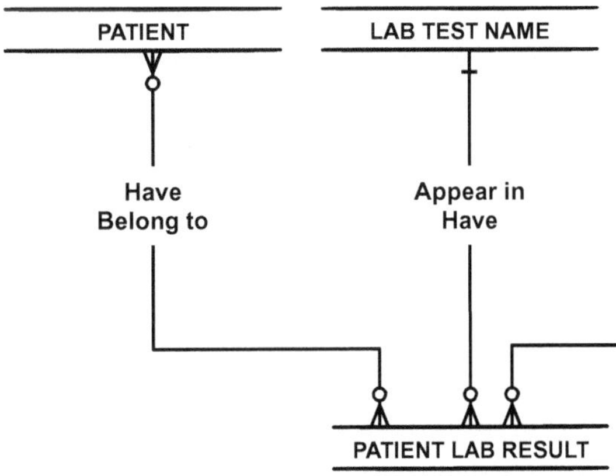

3.

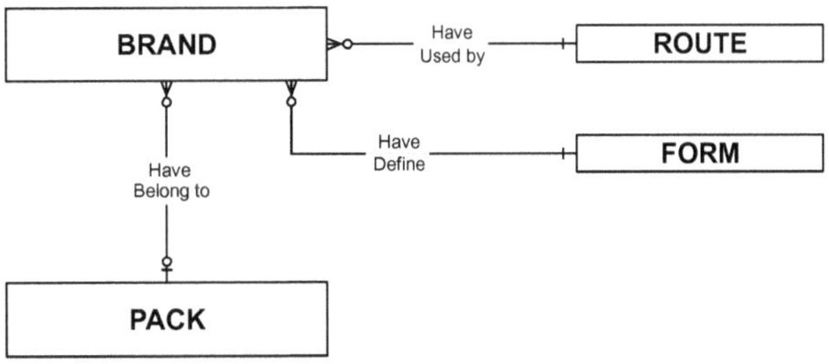

4.

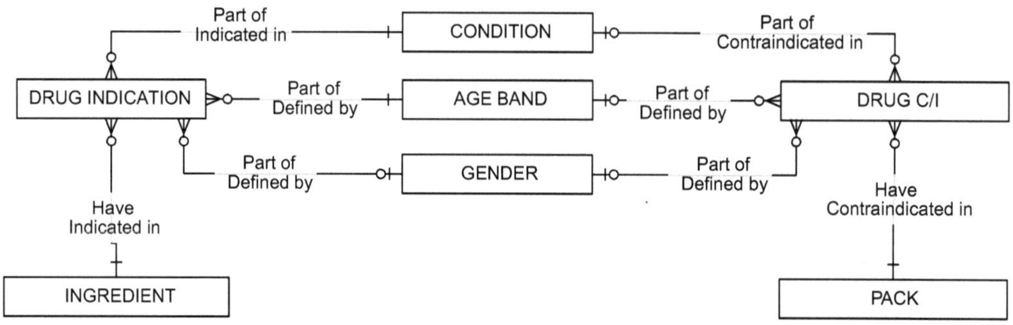

5.

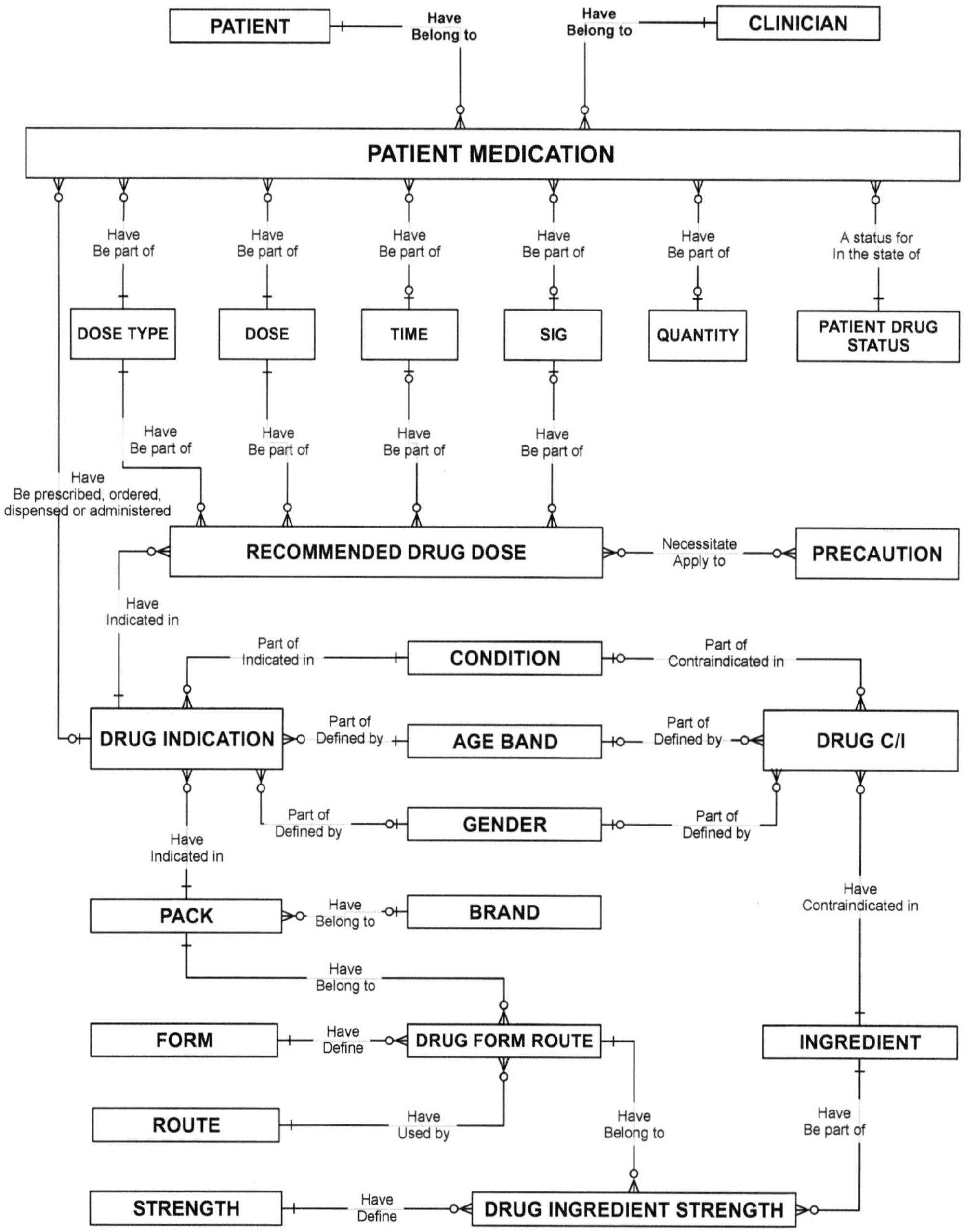

The next three questions refer to Figure 2.12:

6. Explain why the relationship (near the CONDITION side) between CONDITION and DRUG INDICATION is mandatory while that between CONDITION and DRUG C/I is optional.
7. Explain why the relationships between GENDER and DRUG INDICATION as well as between GENDER and DRUG C/I are optional near the GENDER entity.
8. Explain why the relationships of TIME UNIT, SIG, and QUANTITY UNIT are optional and not mandatory near these entities.
9. The recommended dose of acetaminophen for infants is up to 60 mg/kg/day (10–12 mg/kg/dose and up to 5 doses/day). The child has a fever and weighs 22 pounds. If the acetaminophen solution is 100 mg/ml for use by mouth and is dispensed as a bottle of 30 ml,
   - How many milligrams can be given in each dose?
   - How many milliliters for each dose?
   - What is the total number of milligrams per day that can be administered safely to this child?
   - What is the maximal number of milliliters per day that can be given?
   - How many doses are there in one dispensed bottle?
10. In example 9, indicate the following for the acetaminophen:
    - Form
    - Strength
    - Route
    - Recommended dose
    - Frequency
    - Maximal dose
    - Dispensed original package

Chapter 3

Prescribe/eRx

A prescription isn't dressed properly until it wears an indication.

Michael Van Ornum

This chapter analyzes workflow number 1—*Prescribe/eRx* from two perspectives: processes and data elements. The top-down analysis is straightforward: Using the dataflow diagram (DFD) as the main tool, we isolate the *Prescribe* workflow from the higher-level diagram DFD 0 (see previous chapter) and decompose it into lower-level processes—DFD 1. These processes in turn are then dissected into lower subprocesses and activities as DFD 1.x (1.1, 1.2, 1.3, etc.)

While defining the dataflows on DFDs, we will discover data elements required by the *Prescribe* workflow to perform its functions. External actors supply some of these data elements, while others originate from within the boundaries of the conceptual model presented in the previous chapter. We add attributes to the entities, define the entities' relationships with primary keys (*PKs*) and foreign keys (*FKs*), and make some readjustment to the conceptual model while drafting the logical data model.

Processes

DFD 1 Prescribe Workflow

The same way we have isolated in Chapter 1 the *Medications* system from the much larger *Healthcare* "megasystem," we are now isolating the DFD 1 *Prescribe* workflow from the *Medications* system:

Healthcare → *Medications* → *Prescribe*

From DFD 0: Medications system (Figure 2.3), we carve workflow number 1—*Prescribe*—with its external entities, data stores, and other processes attached (Figure 3.1).

On the left side of Figure 3.1, we can see the external actors participating in the *Prescribe* workflow: *PATIENT, PRESCRIBING CLINICIAN, PHARMACY,* and *FORMULARY.* There is only one other workflow illustrated at this level: workflow number 5 *CDS (*clinical decision support). The data store shown is the poorly defined "megastore" *CLINICAL REPOSITORY.* All the dataflows are at the same high-level view as in DFD 0, but this is going to change once we dig deeper and define more granular processes and the discrete data elements that accompany

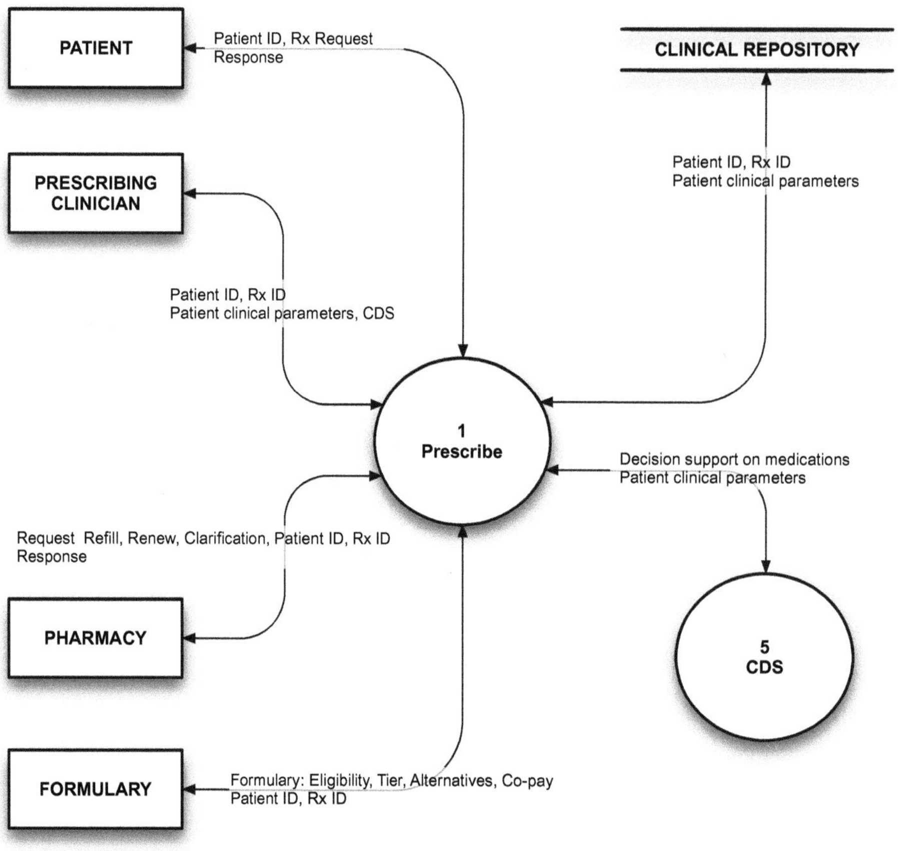

Figure 3.1 Prescribe workflow: Isolated from DFD 0.

them. The workflow a prescribing clinician usually follows can be decomposed into several processes (1–3):

- Review patient data
- Select a new drug or an existing drug (and then refill, modify, stop, or renew)
- Select a dose
- Sign and distribute prescription

During the *Prescribe* workflow, the clinician may need to communicate with other actors and consider *CDS* recommendations and formulary information.

Preconditions to DFD 1, which must be satisfied before the system can proceed with the *Prescribe* workflow, are as follows:

- Clinician is uniquely identified.
- Clinician is authorized to use system in lieu of prescribing.
- Patient is uniquely identified.
- All system's activities are logged.

With these points in mind, we can now draft DFD 1 (Figure 3.2).

The *Prescribe* workflow is decomposed into six main processes according to the user's needs and intentions as defined previously:

1.1 Communicate Prescription
1.2 Review Patient Data
1.3 Select Drug
1.4 Select Dose
1.5 Consider Formulary
1.6 Sign Rx (prescription)

Since DFD 1 shows only processes related to workflow number 1 (1.1 to 1.6), *CDS* as workflow number 5 is not visible anymore at this resolution. Higher-level processes are not represented on lower-level diagrams. Dataflows to and from *CDS* are still visible, as arrows from various *Prescribe*-related processes. *CDS* will continue to be invisible as a workflow on lower DFD 1.x.

Refill, Modify, Renew, and *Stop* are subprocesses of 1.3 *Select Drug*. Being lower-level processes, they also are not visible on DFD 1. They are analyzed as part of the decomposition of process 1.3 *Select Drug*.

Instead of the *CLINICAL REPOSITORY* entity on DFD 0, there is a new actor on DFD 1: the *NON-DRUG REPOSITORY*. It is the external entity to provide all clinical information that is not related to medications. Data stores that are related to medications are represented in the conceptual model by several tables. At this level, we can recognize two such tables from the previous chapter as the objects

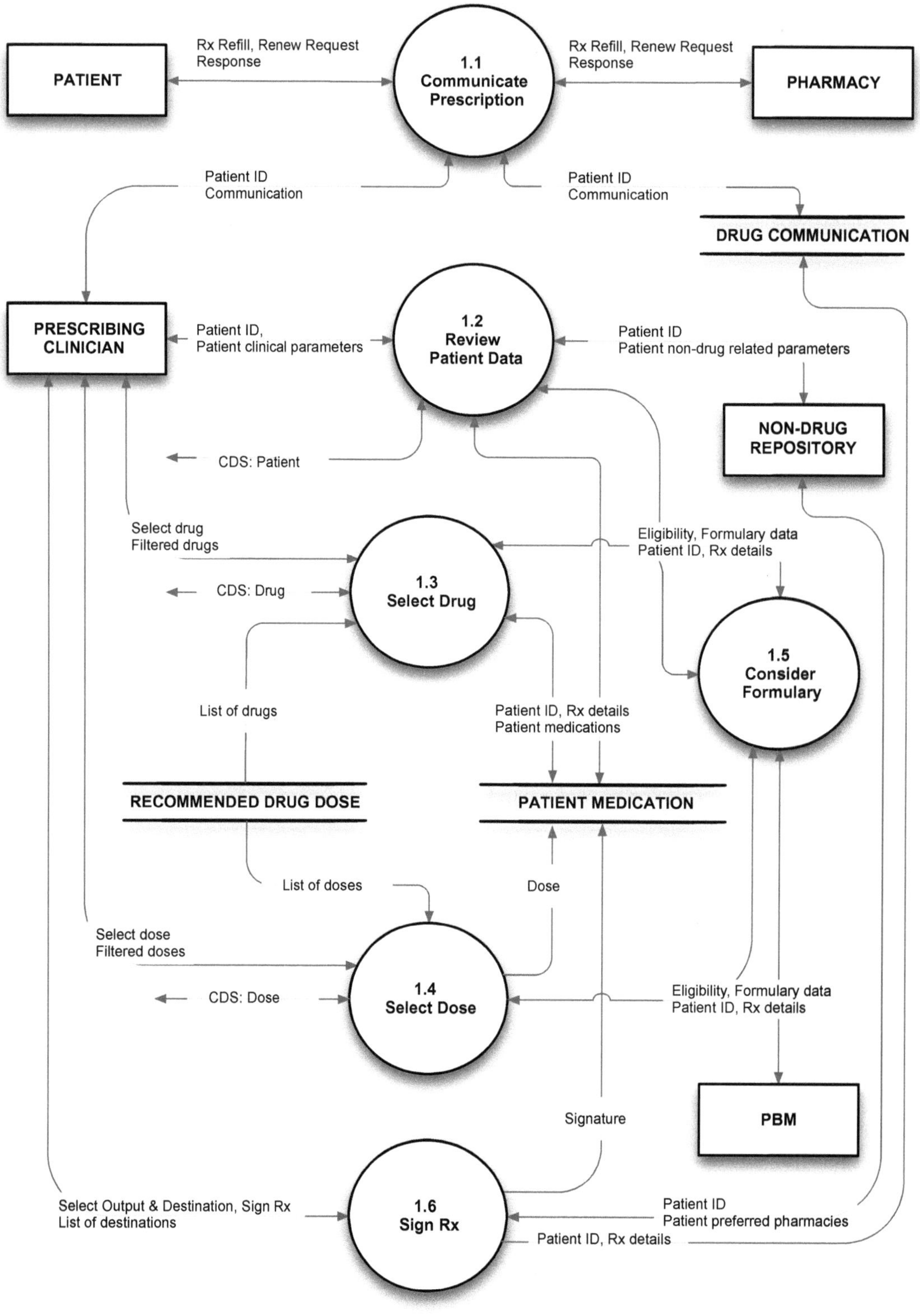

Figure 3.2 DFD 1: Prescribe workflow.

illustrated with the parallel lines: *RECOMMENDED DRUG DOSE* and *PATIENT MEDICATION*. The *CLINICAL REPOSITORY* is thus represented by a new actor (*NON-DRUG REPOSITORY*) and tables previously defined in the conceptual model.

Realizing the need for a data store for all communications related to prescriptions, there is a new table named: *DRUG COMMUNICATION*. It should not come as a surprise that this table is used by process 1.1 *Communicate Prescription*. Let us take a closer look at the 1.x processes.

DFD 1.1 Communicate Prescription

Out of DFD 1 *Prescribe* workflow, we isolate the first process: 1.1 *Communicate Prescription*. We keep attached the dataflows and any other processes and external entities related while we copy and paste (Figure 3.3). There are three actors interacting with process 1.1 *Communicate Prescription*: *PATIENT, PHARMACY*, and *PRESCRIBING CLINICIAN*. Most of the communications between these actors are expected to be in the form of request/response.

A patient or pharmacy may request a refill to be approved or a prescription to be modified, renewed, or clarified. A prescribing clinician may create a new prescription; approve (or not) a refill, modify, renew, stop, hold, or cancel a prescription; respond to a request for clarifications; or send a communication to another clinician.

The information related to prescriptions' communications is stored in the *DRUG COMMUNICATION* table, which may accept e-mails, phone calls, SMS (Short Message Service) messages, tweets, voice, video, and so on. At this stage of the analysis, there should be no restrictions on the communication channels or media to be used by system. According to the business rules, we can proceed drafting DFD 1.1 (Figure 3.4).

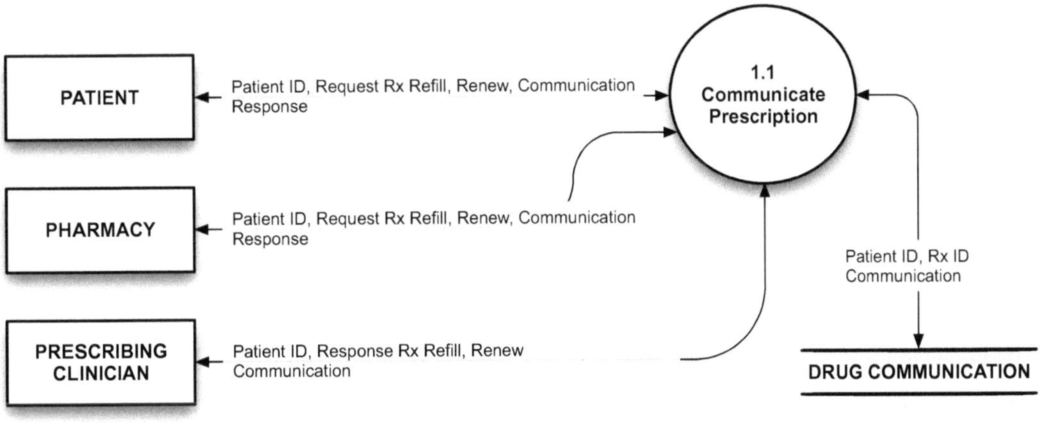

Figure 3.3 Communicate Prescription process: Isolated from DFD 1.

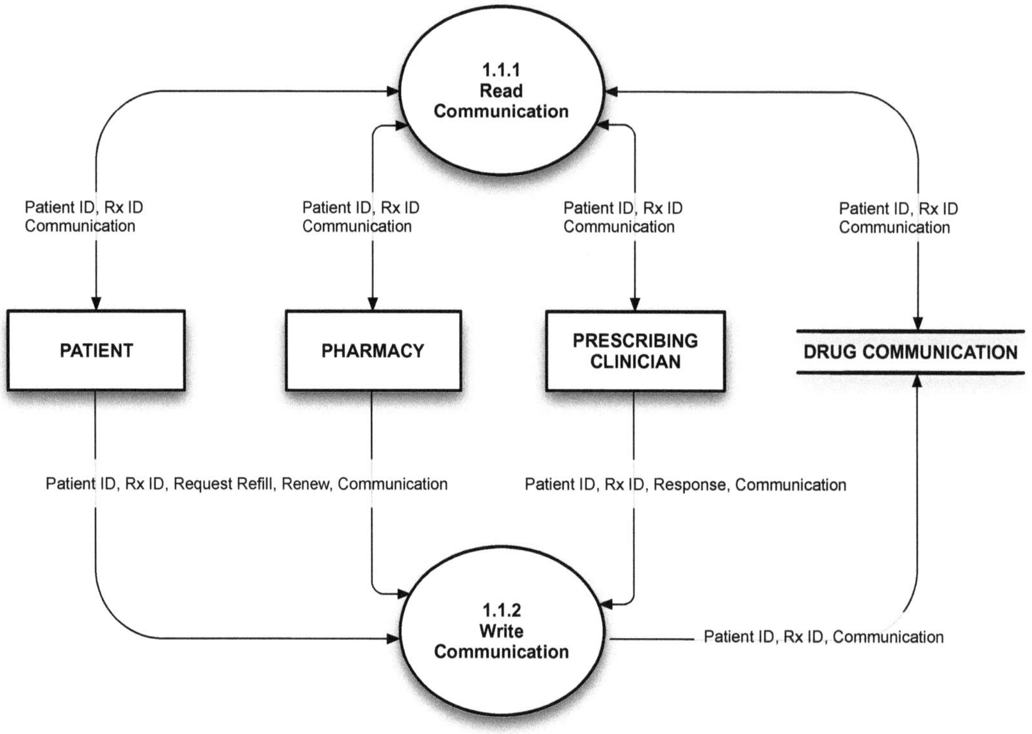

Figure 3.4 DFD 1.1: Communicate Prescription.

Due to the nature of the read/write the communications usually exhibit, the two main subprocesses of DFD 1.1 are accordingly

1.1.1 Read Communication:
Request is identified by *Patient ID* and *Rx ID.*
System reads and displays communications related to *Patient ID* and *Rx ID* from *DRUG COMMUNICATION.*

1.1.2 Write Communication:
A request or response for a prescription renew or refill, clarification, or the like is captured by the system.
Communication identified by *Patient ID* and *Rx ID* is stored in *DRUG COMMUNICATION.*

While we perform top-down analysis of system functions, from high-level workflows to processes, and from processes to activities and lower-level tasks, we may reach a point of a *functional primitive*: the stage when a process or activity can be defined in one page of *structured English* (also known as *pseudocode*), and there is no point in decomposing it further into lower DFDs (Chapter 1).

Processes 1.1.1 and 1.1.2 are at this *functional primitive* level: Each process has two or three activities. Each activity can be defined in less than one page of pseudocode. Together with the data elements identified as part of *DRUG*

COMMUNICATION detailed in the following material, these processes can be translated through design into actual code and queries in SQL (Structured Query Language, the lingua franca of relational databases). Thus, we can stop decomposing process 1.1 and move to process 1.2.

DFD 1.2 Review Patient Data

The clinician needs to review all pertinent information regarding a patient when prescribing: medications as well as non-medication-related. The clinician may also consider *CDS* recommendations and formulary information. Isolating the process 1.2 *Review Patient Data* from the original DFD 1, with its actors, other processes, and data stores attached is shown in Figure 3.5.

Process 1.2 receives information from the *NON-DRUG REPOSITORY* actor and the *PATIENT MEDICATION* table, shares the information with the *CDS* workflow and *PBM* actor, receives back *CDS* recommendations and formulary information, and displays it to the prescribing clinician. This is shown more succinctly in Figure 3.6.

There are two subprocesses (1.2.1 and 1.2.2) related to data retrieval and one subprocess (1.2.3) that integrates and consolidates data into a comprehensive display for the clinician:

1.2.1 Retrieve Non-drug Parameters
← Sends *Patient ID* to *NON-DRUG REPOSITORY*
→ Receives back non-drug-related parameters
← Sends non-drug parameters to 1.2.3

1.2.2 Retrieve Patient Drugs
← Sends *Patient ID* to *PATIENT MEDICATION* table
→ Receives patient's current and past medications
← Sends patient's medications to 1.2.3

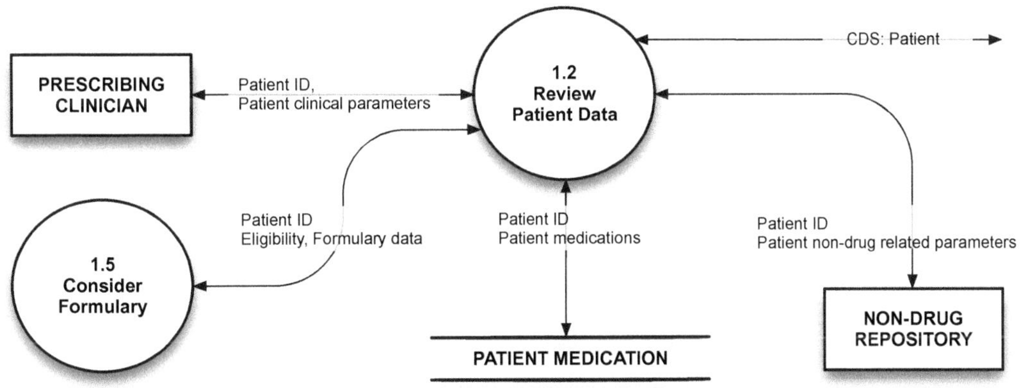

Figure 3.5 Review Patient Data process: Isolated from DFD 1.

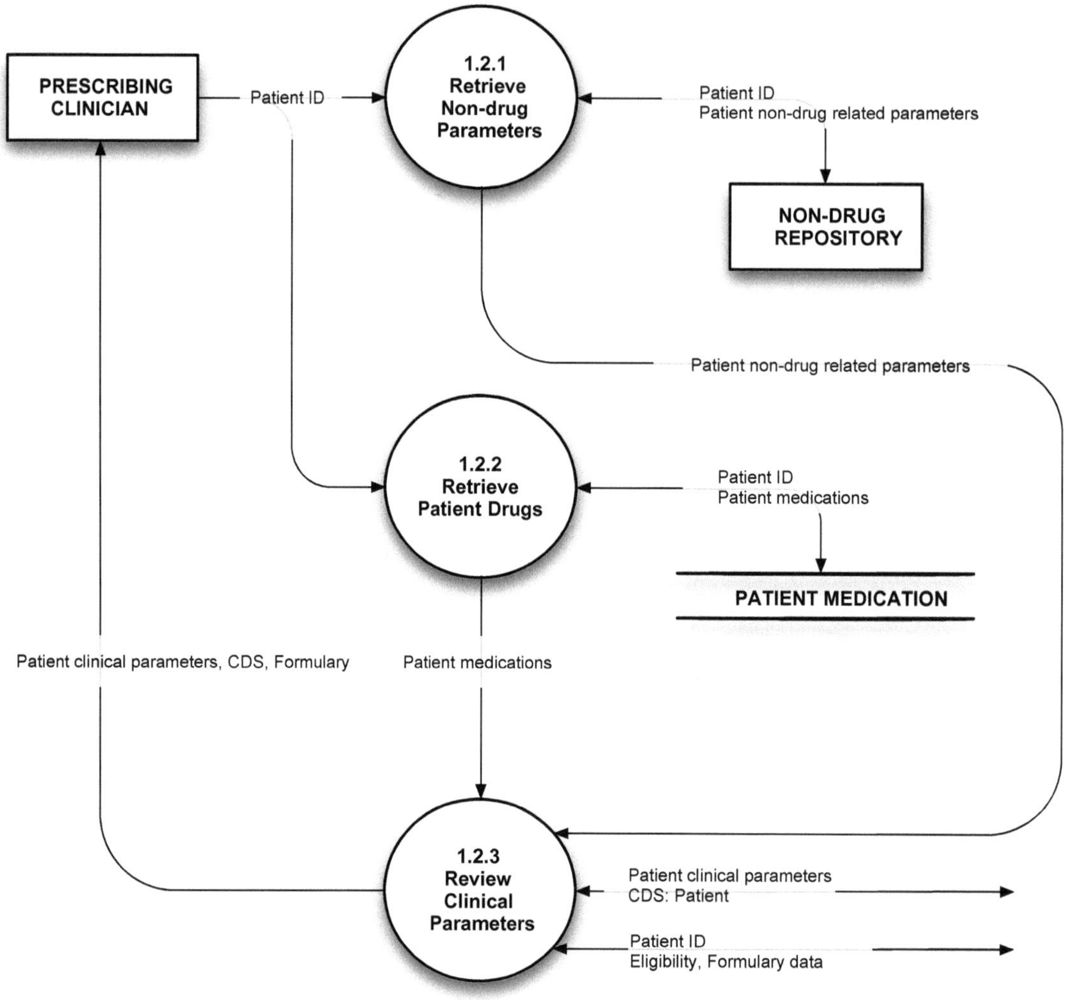

Figure 3.6 DFD 1.2: Review Patient Data.

1.2.3 Review Clinical Parameters
- → Receives non-drug-related parameters
- → Receives information on patient's medications
- ← Sends patient parameters to *CDS*
- → Receives recommendations from *CDS*
- ← Sends *Patient ID* and medications list to *PBM* (pharmacy benefit management)
- → Receives eligibility and formulary data from *PBM*
- ← Integrates and displays information to prescribing clinician

Since the lower activities expressed in 1.2.1, 1.2.2, and 1.2.3 can be further defined as tasks in pseudocode, we can focus on the next process.

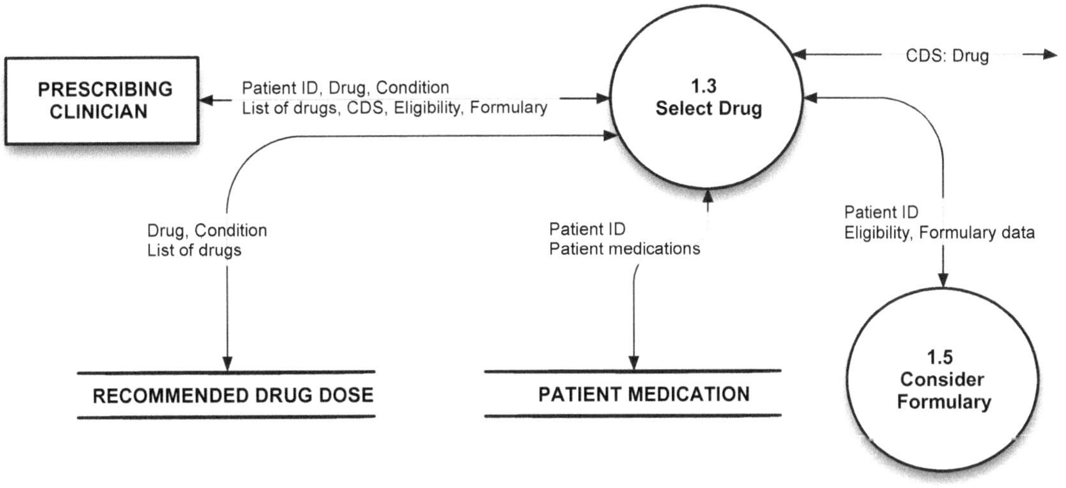

Figure 3.7 Select Drug process: Isolated from DFD 1.

DFD 1.3 Select Drug

From DFD 1, we isolate process 1.3 *Select Drug* (Figure 3.7).

Only one actor, the prescribing clinician, uses this process, which is supported by two data stores (*RECOMMENDED DRUG DOSE* and *PATIENT MEDICATION*) and one other process: 1.5 *Consider Formulary.* As a higher-level workflow (number 5), *CDS* is not shown at the level of DFD 1.x, but the double-headed arrow at the top right side represents the dataflow to and from *CDS*.

Process 1.3 *Select Drug* can be further decomposed into 1.3.1 *Select New Drug* and 1.3.2 *Select Existing Drug* (Figure 3.8).

At this level, process 1.5 *Consider Formulary* is not visible anymore, but the dataflows to and from 1.5 are depicted in the diagram. While 1.3.1 covers the aspects of selecting a new drug, 1.3.2 details the process of selecting an existing drug and then modifying, approving a refill or renewal, or stopping/canceling it. Let us examine these processes further (Figure 3.9).

A clinician may search for a new drug using an indication, condition, or disease or using a drug's generic or brand name, ingredient, class, or category. The clinician may consider *CDS* recommendations and information from the *PBM* when selecting a medication:

1.3.1.1 Search by Indication:
- ← Sends *Indication, Form, Route,* or *Strength* as query parameters (with patient's *age band* and *gender*) to *RECOMMENDED DRUG DOSE*
- → Receives back a list of drugs as query results
- ← Sends query results to 1.3.1.3

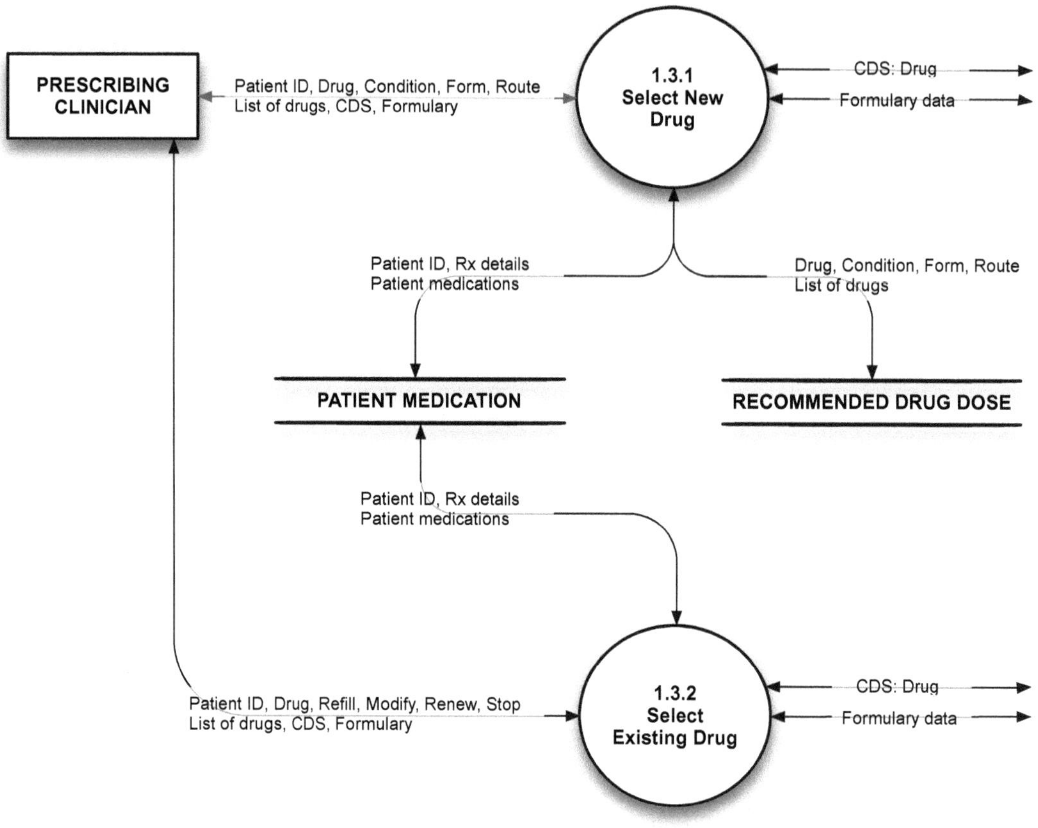

Figure 3.8 DFD 1.3: Select Drug.

1.3.1.2 Search by Drug:

← Sends one or more search parameters (*Generic name, Brand name, Ingredient,* or *Class)* to *RECOMMENDED DRUG DOSE*
→ Receives back a list of drugs satisfying search criteria
← Sends query results to 1.3.1.3

1.3.1.3 Review & Select New Drug

→ Receives list of drugs from processes 1.3.1.1 and 1.3.1.2
← Sends *Patient ID* to *PATIENT MEDICATION*
→ Receives patient medications
→ Captures a user-selected drug, form, and route
← Sends patient parameters and selected drug to *CDS*
→ Receives recommendations from *CDS* on selected drug
← Sends selected drug to *PBM*
→ Receives formulary data from *PBM* on selected drug
← Integrates and displays information to prescribing clinician

When a clinician selects an existing drug from the patient's medication list, one can assume the clinician may approve (or not) a refill request, modify the

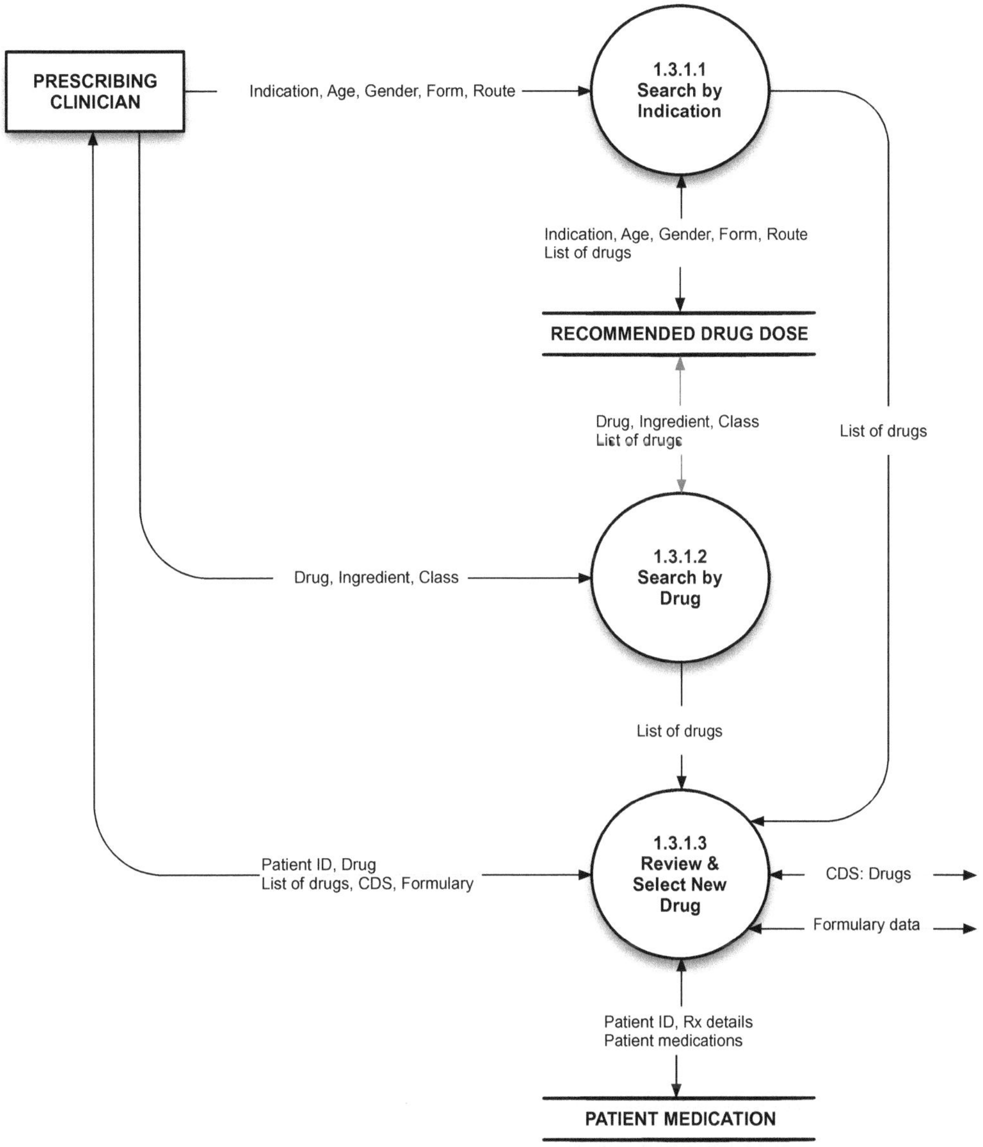

Figure 3.9 DFD 1.3.1: Select New Drug.

dose, stop/hold medication, or renew an existing prescription. The processes of 1.3.2 are accordingly shown in Figure 3.10.

1.3.2.1 Review & Select Existing Drug:
← Sends patient parameters and selected drug to *CDS*
→ Receives recommendations from *CDS*
← Sends *Patient ID* and selected drug to *PBM*
→ Receives formulary data from *PBM* on patient and selected drug
← Integrates and displays information to prescribing clinician

All processes within 1.3.2.x (refill, modify, stop, or renew) may trigger *CDS* and *PBM*.

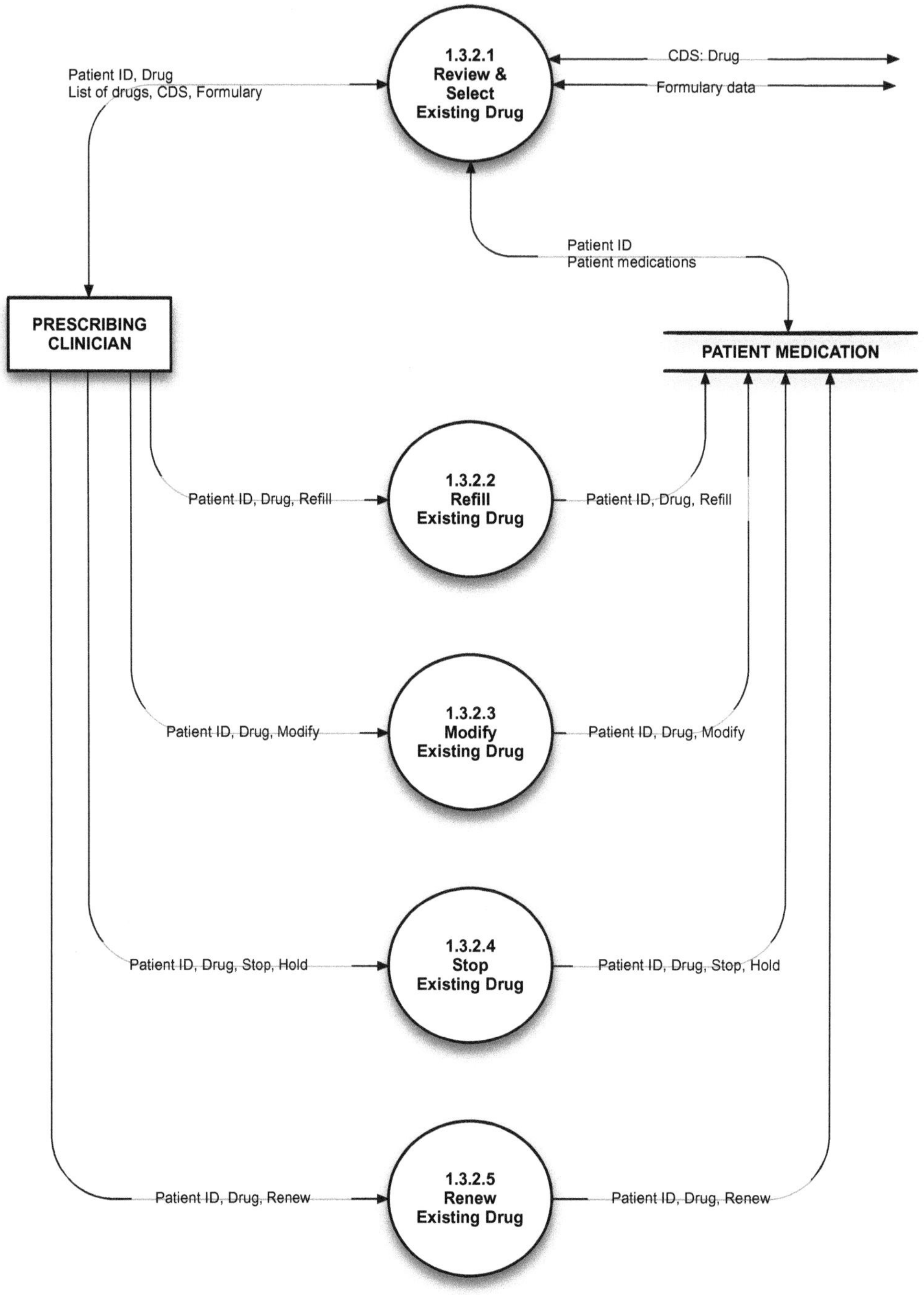

Figure 3.10 DFD 1.3.2: Select Existing Drug.

1.3.2.2 Refill Existing Drug:

→ Captures user decision whether to approve or not refill

→ If approved, captures *Number of Refills* in 1.4 *Select Dose* (changes of *Dose, Frequency* or *Quantity* are covered by process 1.4 *Select Dose*)

← Stores clinician decision in *PATIENT MEDICATION*

1.3.2.3 Modify Existing Drug:
→ Receives modifications to *Form, Route,* or *Strength*
← Stores modification details including notes in *PATIENT MEDICATION*

1.3.2.4 Stop Existing Drug:
→ Captures user decision to stop a medication and any notes on the reason; *On-Hold* is equivalent to *Stop* status but with a time frame attached, for example *Warfarin—On Hold—till Jan 22, 2011*
← Stores clinician decision in *PATIENT MEDICATION*

1.3.2.5 Renew Existing Drug:
→ Captures clinician decision to renew a medication
← Stores decision and details in *PATIENT MEDICATION*

It is important to distinguish between *Rx refill* and *Rx renewal.* Using the terms interchangeably may introduce confusion and errors (1). A prescription usually contains several refills; it can be refilled several times. When a prescription date is expired or a prescription has been refilled the maximum number of times, it needs to be renewed. A renewal is the generation of a new prescription based on a previous prescription with a new expiration date and a new number of refills.

Note the system retrieves information on all medications from all clinicians ever involved in the patient's treatment. There are restrictions that may apply, such as the permissions granted by the patient on who can see what information. There are situations (such as with psychiatric diseases or teenagers' healthcare) when a patient may decide to share only partial information with others. Depending on the specific law and state, the system needs to accommodate these privacy issues. A mechanism to allow a patient to grant permissions to others to see clinical information should be in place as well as a mechanism to bypass these privacy restrictions in emergency cases. These mechanisms are not shown in Figure 3.10.

The processes in Figure 3.10 are not final until signed. In addition to drug status of *Refilled, Modified, Stopped, or Renewed,* there is *Pending Signature* appended to the status *Refilled—Pending Signature, Modified—Pending Signature,* and so on. The process *1.6 Sign Rx* is responsible for changing the status of any *Pending Signature* to the final *Signed* as well as communicating a prescribing clinician's decision to *DRUG COMMUNICATION.*

DFD 1.4 Select Dose

This process has its own preconditions in addition to those mentioned that apply to all the *Prescribe* processes. The following parameters must be selected before setting the *Dose*: *Drug, Form, Strength, Route,* and *Indication* (optional). We isolate 1.4 *Select Dose* from the parent diagram DFD 1 (Figure 3.11).

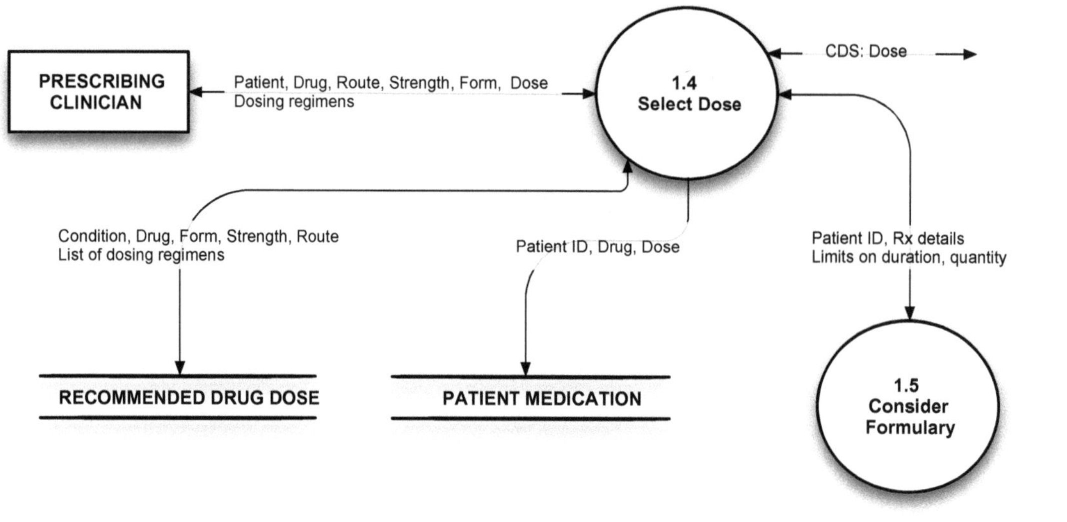

Figure 3.11 Select Dose process: Isolated from DFD 1.

The *PRESCRIBING CLINICIAN* is the only entity interacting with process 1.4 *Select Dose* (Figure 3.12). The clinician's decision on the actual dose, which may be different from what is recommended, is stored in *PATIENT MEDICATION*. *RECOMMENDED DRUG DOSE* and *CDS* support the process on dosing regimens, and *PBM* supplies information on quantity or duration limitations.

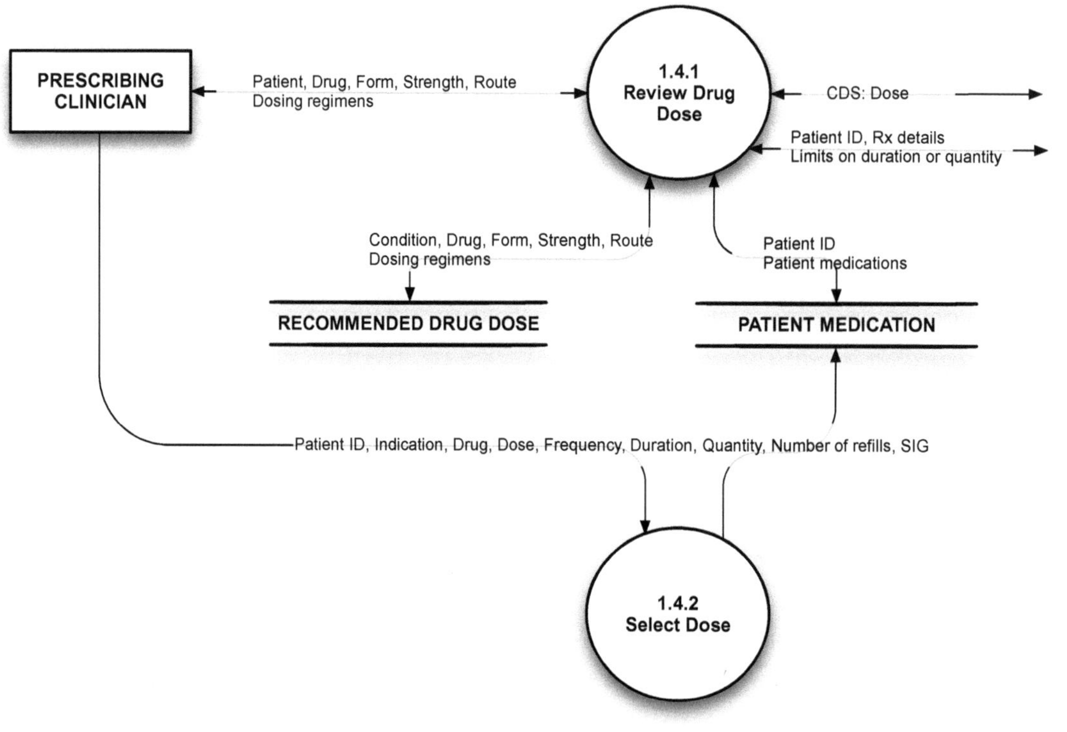

Figure 3.12 DFD 1.4: Select Dose.

There are two steps for setting a drug dose:

1.4.1 Review Drug Dose:
← Sends condition, drug, form, strength, route to *RECOMMENDED DRUG DOSE*
→ Receives from *RECOMMENDED DRUG DOSE* tentative dosing regimens for selected condition, drug, form, route, and strength
← Sends *Patient ID* to *PATIENT MEDICATIONS*
→ Receives patient medications
← Sends drug dose to *CDS*
→ Receives *CDS* recommendations regarding drug dose for specific patient
← Sends drug to *PBM*
→ Receives *PBM* information on quantity or duration limits for drug as per patient's insurance plan

1.4.2 Select Dose:
→ For a given patient, drug, form, strength, and route, system captures and stores in *PATIENT MEDICATION* clinician's selection of
Dose
Frequency
Quantity
Number of refills
SIG
Indication

Again, the clinician's decision is not final until *Signed*, an activity performed by process 1.6 *Sign Rx*.

DFD 1.5 Consider Formulary

We have already met process 1.5 *Consider Formulary* in relation to three other processes: 1.2 *Review Patient Data,* 1.3 *Select Drug,* and 1.4 *Select Dose*. Let us isolate process 1.5 *Consider Formulary* from DFD 1 (Figure 3.13).

The 1.2, 1.3, and 1.4 processes each expect slightly different responses from 1.5 *Consider Formulary*:

- General information on patient's eligibility under *PBM* plan
- Information on drug tier, copay, need for preapproval, and therapeutic alternatives on *PBM* formulary
- Limitations on duration or quantity covered by *PBM*

The processes of *1.5* are correspondingly (Figure 3.14)

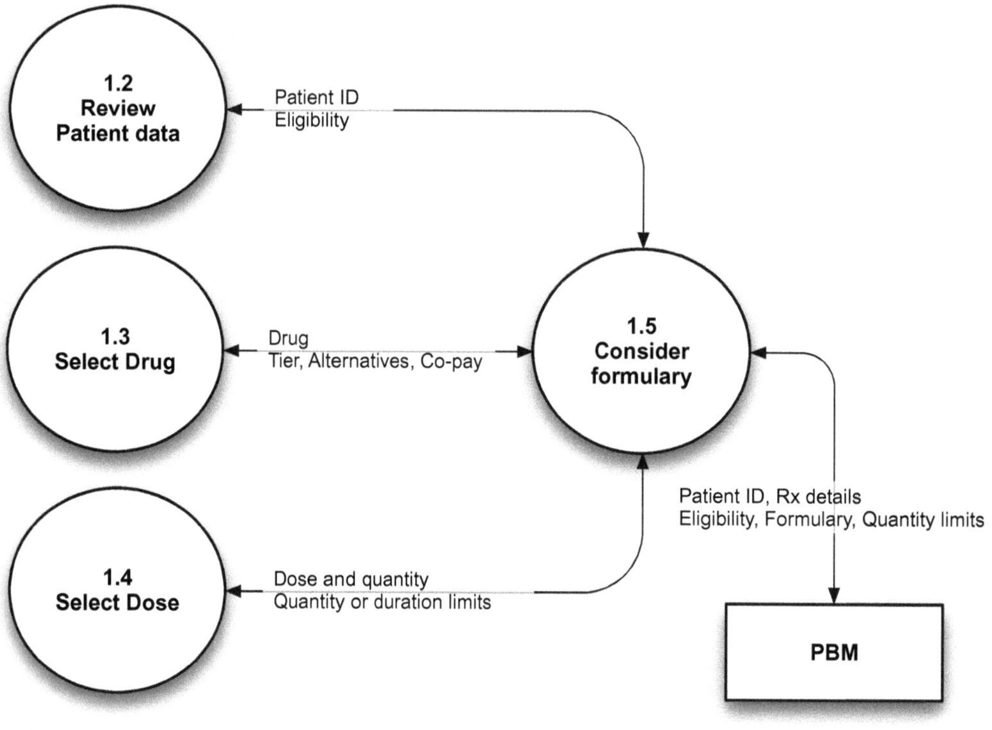

Figure 3.13 Consider Formulary process: Isolated from DFD 1.

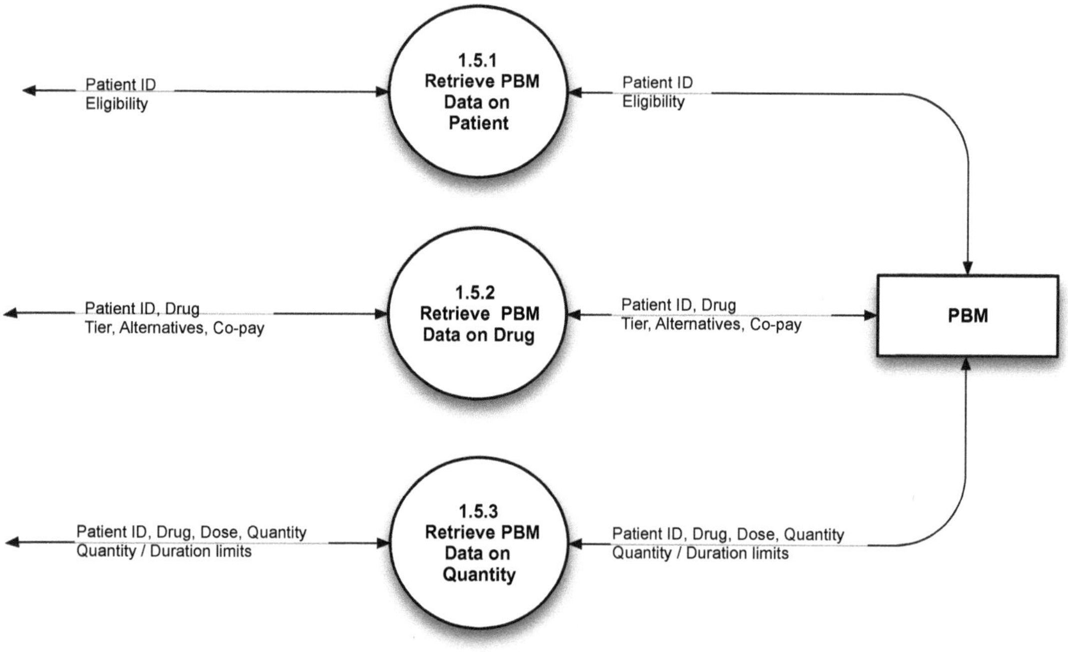

Figure 3.14 DFD 1.5: Consider Formulary.

1.5.1 Retrieve PBM Data on Patient
← Sends *Patient ID* to *PBM*
→ Receives information about patient's eligibility under *PBM* plan

1.5.2 Retrieve PBM Data on Drug
← Sends *Drug* to *PBM*
→ Receives information from *PBM* about drug tier on formulary

There are usually four tiers: tier 1, the least expensive, usually generics, to tier 4, the most expensive or specialty drugs.

In addition, process 1.5.2 receives information on therapeutic alternatives, patient's copay, and the need for preapproval

1.5.3 Retrieve PBM Data on Quantity
← Sends *Drug, Dose, Quantity* to *PBM*
→ Receives information about *Quantity* or *Duration* limitations under *PBM* plan

DFD 1.6 Sign Rx

The processes discussed are not final until *Signed*. Just before finalizing the *Prescribe* workflow, the clinician still has to pick a *Destination* for the prescription (Figure 3.15). Process 1.6 *Sign Rx* has two main subprocesses (Figure 3.16):

1.6.1 Select Rx Destination
← Sends *Patient ID* to *NON-DRUG REPOSITORY*
→ Receives list of patient's preferred pharmacies
→ Captures clinician's selection of *Rx output* (print, fax, electronic) and *Rx destination*

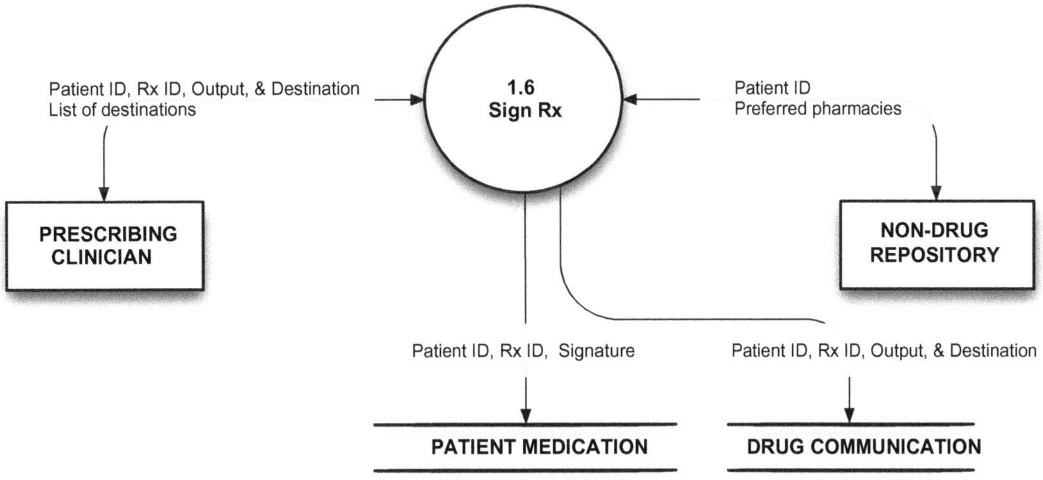

Figure 3.15 Sign Rx process: Isolated from DFD 1.

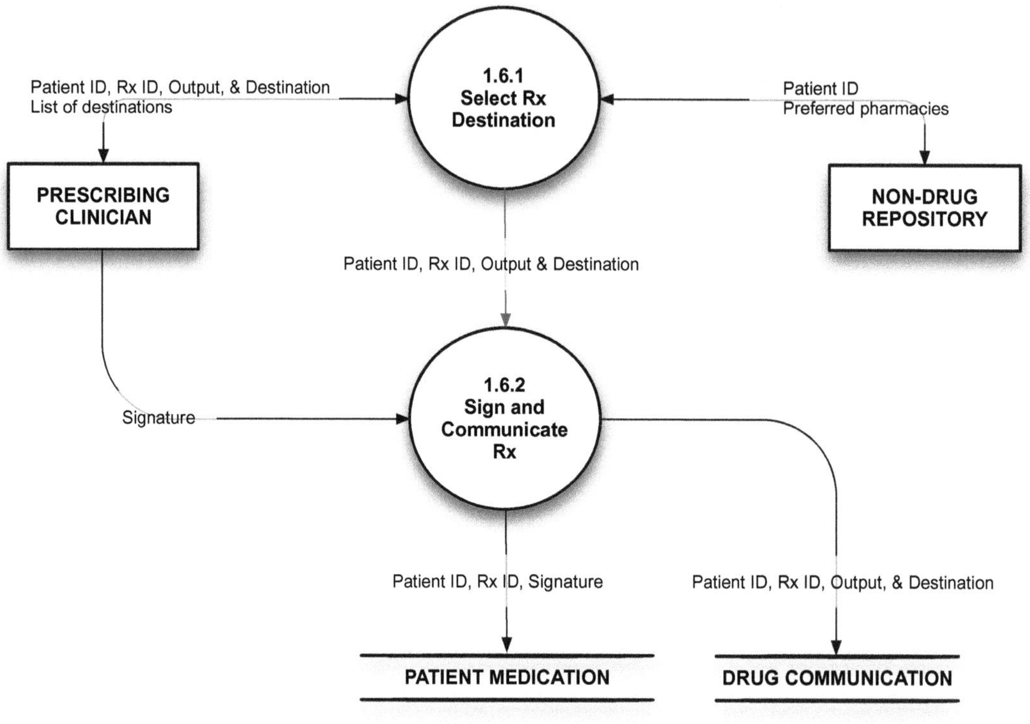

Figure 3.16 DFD 1.6: Sign Rx.

1.6.2 Sign & Communicate Drug
← For *Patient ID* and *Rx ID,* the system changes status to *Signed* in *PATIENT MEDICATION* and finalizes the *Prescribe* workflow
← Sends details to *DRUG COMMUNICATION*

Controlled Substances

Some of the medications prescribed may fall under the category of controlled substances—drugs with the potential of abuse that may cause physical or psychological dependence with limited or disputable medical benefits. The U.S. *Drug Enforcement Administration* (*DEA*) has classified substances in five schedules (Controlled Substances Act 1970): from schedule 1 with the highest potential for abuse and without any currently accepted use in medical practice to schedule 5 with the lowest abuse potential (4). Published March 2010 and effective June 2010, the DEA has ruled on the electronic prescription of controlled substances (5): A clinician may electronically prescribe controlled substances if the physician complies with the requirements defined in the *National Institute of Standards and Technology* (*NIST* Special Publication

800-63-1 Assurance Level 3). *NIST* has defined the need for two-factor authentication for the electronic prescription of controlled substances from the following list:

- Something the user knows (such as a password)
- Something the user has (a token—a cryptographic key stored on a hardware device "other than the user computer," such as a cellular phone or PDA [personal digital assistant])
- Something uniquely defining user (such as a biometric, e.g., fingerprint)

When signing a controlled substance in an electronic prescription, it is assumed user and prescribing software have been certified by the regulatory agencies defined in the DEA act for this specific activity.

We have finished the analysis of the processes in workflow 1 *Prescribe,* so we can focus on the other analytical path: the data structures and elements needed to support the processes just defined.

Data Elements

While developing the logical model, we use the conceptual model draft (Figure 2.12) as the starting point and

- Define the relationships between the entities with *PK* and *FK*
- Define the entities' attributes (the tables' column names)
- Modify the model while eliminating redundancy and any other potential data anomalies

Readjustments to the conceptual model are expected, such as adding a *DRUG COMMUNICATION* table, as discussed in this chapter.

Another readjustment becomes apparent from the analysis of the relationships between the concepts of *Patient, Clinician, Prescription,* and *Drug.* In the conceptual model draft, there is a direct relation between *Patient* and *Drug* in the form of a table, *PATIENT MEDICATION*: One patient may have many drugs prescribed. There is also a direct relation between *Clinician* and *PATIENT MEDICATION*: One clinician may prescribe many drugs. Using this model, the *Patient ID* and *Clinician ID* need to be stored, redundantly, with each drug in the *PATIENT MEDICATION* table. This representation may work, but because it suffers from redundancy, it should be considered an incorrect data model.

A better data model would have one patient—many prescriptions and one prescription—many drugs: an indirect relationship between *Patient* and *Drug.* From the clinician's point of view, the same indirect relationship should be modeled: one clinician—many prescriptions and one prescription—many drugs.

The *Rx ID* uniquely identifies the concept of *Prescription* in real-life communications with pharmacies and *PBM*. *Prescription* has its own attributes, such as the *Patient* and *Clinician* involved and the *Date & time*. *Prescription* may have one or more *Drugs* (1:*M* relationship with *PATIENT MEDICATION*). Reflexively, we could define a table *PRESCRIPTION* in between *PATIENT, CLINICIAN,* and *PATIENT MEDICATION*. If we think about the other workflows a patient may be part of, in lieu of medications—*Order, Dispensation, Administration*—then we better use a more generalized concept to cover all four possible *Patient-Clinician-Drug-Workflows*, and not only *Patient-Clinician-Drug-Prescriptions*. What are the *Patient-Clinician-Drug-Workflows* common characteristics? All four workflows have the following similar relationships:

- One patient may have many workflows (prescriptions/orders/dispensations/administrations).
- One workflow (prescription/order/dispensation/administration) may have many drugs.
- One workflow instance belongs to only one patient.
- One clinician may be involved in many workflow instances.
- One workflow instance has only one related clinician.

It seems the attributes of this entity are *Patient, Clinician, Type of workflow (Prescription, Order, Dispensation or Administration)*, and a *Date & time* stamp. Each instance in this entity can be uniquely identified and has its own attributes. There are enough reasons to justify the creation of a new entity, but the name *WORKFLOW* may be too ambiguous, so let us name the entity *PATIENT CLINICIAN* in between the concepts of *Patient, Clinician,* and *Drug*. *PATIENT CLINICIAN* groups information on *Prescriptions* (and *Orders, Dispensations, Administrations)* and is linked via a *FK* to the actual *Drugs* in the *PATIENT MEDICATION* table. *Patient ID, Clinician ID,* and *Date & Time* are stored within *PATIENT CLINICIAN*, eliminating the issue of storing it redundantly in *PATIENT MEDICATION*. Based on the conceptual model and the two additional tables mentioned, *DRUG COMMUNICATION* and *PATIENT CLINICIAN*, we can refine the conceptual model (Figure 3.17).

We define now the relationships between entities using *PK/FK* and identify their attributes, the data elements needed by the system's processes. Which data elements are expected to be part of the information exchange related to *Prescribe* workflow?

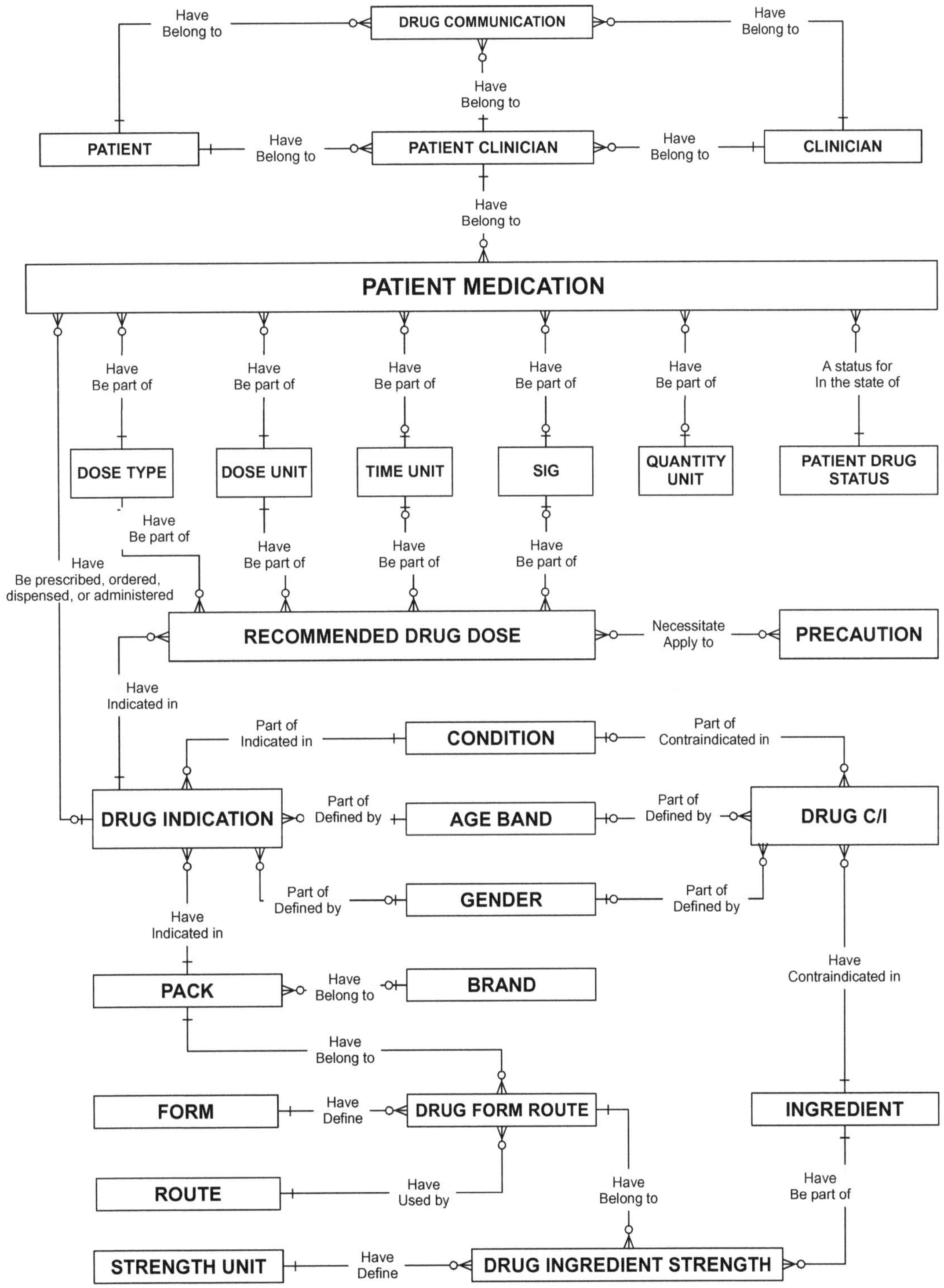

Figure 3.17 Conceptual model refined: Prescribe workflow.

To identify the data elements necessary to satisfy the processes mentioned, we review the dataflows on the DFDs 1.x:

DFD 1.1	Prescription-related communications
DFD 1.2	Patient medications Non-drug-related parameters *CDS* recommendations based on patient parameters *PBM* information on eligibility
DFD 1.3	List of drugs *CDS* recommendations on drug *PBM* information on drug
DFD 1.4	List of dosing regimens *CDS* recommendations on dose *PBM* information on quantity
DFD 1.5	PBM information: eligibility, formulary
DFD 1.6	Patient's preferred pharmacies DEA schedule

For each and every item, we should decide whether it is within the boundaries of the conceptual data model. Is it an item stored within the *Medications* system database, or is the data element supplied by an external actor?

Internal (medications system database):

- Prescription-related communications
- Patient medications
- List of drugs and dosing regimens
- CDS recommendations
- DEA schedule

External repositories:

- Patient non-drug related parameters
- PBM information
- Patient's preferred pharmacies

For the discrete data elements that are in scope, we need to define the tables that store them and their attributes (tables' columns). Using the conceptual model redefined in Figure 3.17, we draft a logical data model.

For the information items provided by external actors, we will not build any data model. Still, we need to identify the data elements involved in the interactions with the medications system. Since the external actors are beyond the system's boundaries and responsibilities, we treat them as black box entities: We define the input into and output from these actors, but not the processes these actors perform within their own systems.

Prescription-Related Communications

Communications related to prescriptions are within the scope of the medications system. The *DRUG COMMUNICATION* table, used by process *1.1 Communicate Prescription* has the minimum following attributes:

Patient ID: *5667870098712338 a.k.a. Patient J*
Workflow ID. *322455779021 (Rx ID)*
Request/Response: *Request*
Origin: *Pharmacist X*
Destination: *Dr. A*
Date & time: *11/15/2010 15:35*
Communication type: *Prescription refill request*
Content: *Please approve refill for attached prescription*
Priority: *High*
Status: *Refill approval requested*
Media: *email*

Let us use this list of attributes in the logic data model (Figure 3.18).

Patient ID and *Clinician ID* are the *PK*s of *PATIENT* and *CLINICIAN* tables, respectively. These unique identifiers are used to connect the two tables in an intersection entity such as the *DRUG COMMUNICATION* or *PATIENT CLINICIAN* tables. Note the *PK* in one table becomes the *FK* in the intersection table. Communication threads in the form of *Request/Response* between patient and clinicians are stored in *DRUG COMMUNICATION*. The table *PATIENT CLINICIAN* holds information about the *Patient, Clinician, Type of workflow* (*Prescribe, Order, Dispense* or *Administer*), and a *Date and time* stamp. Each instance (row) in *PATIENT CLINICIAN* may have only one patient, one clinician, and many related instances of drugs in the *PATIENT MEDICATIONS* table. A uniquely identified prescription may have only one *Patient ID* and one *Clinician ID* as *FK*. *Workflow ID* in our case is a synonym for *Rx ID*. One *Workflow ID* may relate to many *Patient Drug ID* in *PATIENT MEDICATION* table.

Patient Medications

Once a patient is uniquely identified with *Patient ID*, all information on the patient's medications may be retrieved, including immunizations, in-hospital, self-administered, over the counter, herbs, vitamins, and nutritional supplements. Given

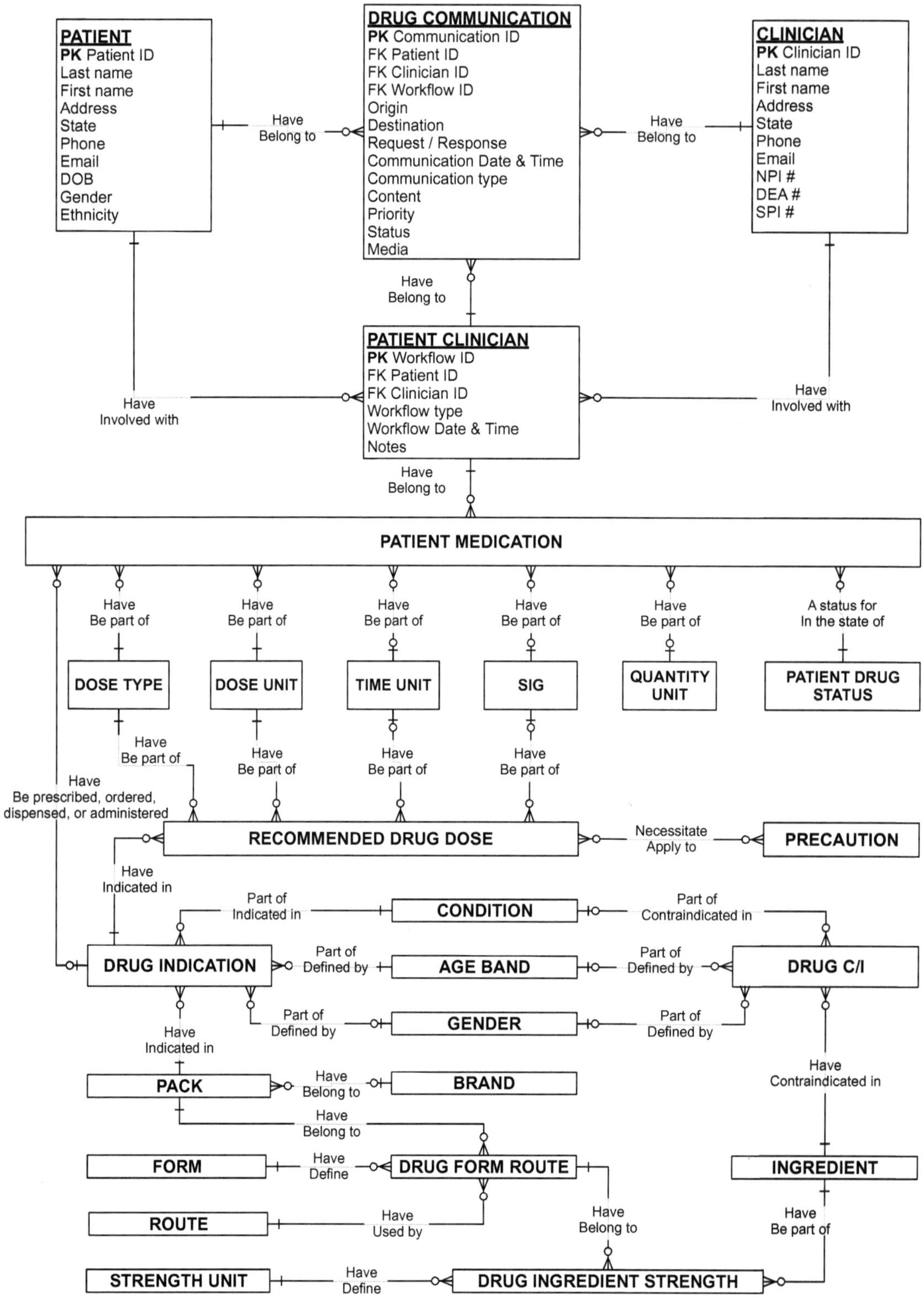

Figure 3.18 Logical data model: Patient, Clinician, and Communications.

a *Patient ID, PATIENT MEDICATION, PATIENT CLINICIAN,* and related tables, the system should be able to provide for prescribed drugs the following parameters:

Patient ID: *Patient K.*
Clinician ID: *Dr. W.*
Workflow ID *(Rx ID): 356788790081*
Workflow type: *Prescription*
Indication for drug: *Hypertension*
Drug name: *Inderal*
Form: *Tab*
Strength: *40*
Strength unit: *mg/tab*
Route: *PO*
Dose: *2*
Dose unit: *tab*
Frequency: *2*
Frequency unit: *day*
Duration: *1*
Duration unit: *year*
Quantity: *4*
Quantity unit: *refills*
Dosing regimen: *maintenance*
SIG: *SIG ID 18899282*
Dispense as written (DAW) or substitutes allowed: *Substitutes allowed*
Meds samples have been dispensed: *No*
Clinician comments: *Diuretics alone not sufficient to keep BP under control*
Drug status: *Signed—Pending dispensation*
Workflow Date & time: *2010.12.14 09:47*

Let us add the attributes of *PATIENT MEDICATION* and related entities to the logical model (Figure 3.19).

PK Dose type ID from the "source" table *DOSE TYPE* is used as an *FK* in the intersection entity *PATIENT MEDICATION*. The same mechanism is used connecting *PATIENT MEDICATION* with *QUANTITY UNIT, TIME UNIT,* and so on. An interesting aspect to note is that usually intersection entities hold most of the "knowledge" of a database. The small tables (around these large intersection entities) are analogous to dictionaries, lists of discrete data elements the system uses as building blocks for instances of information. These rows, meaningful combinations of the elementary building blocks, comprise the knowledge intersection tables.

From process 1.6 *Sign Rx* and the discussion on controlled substances, we need to add the DEA schedule at the level of "substance"—in our case, *INGREDIENT*. The attributes of *RECOMMENDED DRUG DOSE* and *INGREDIENT* are represented in the logical model (Figure 3.20).

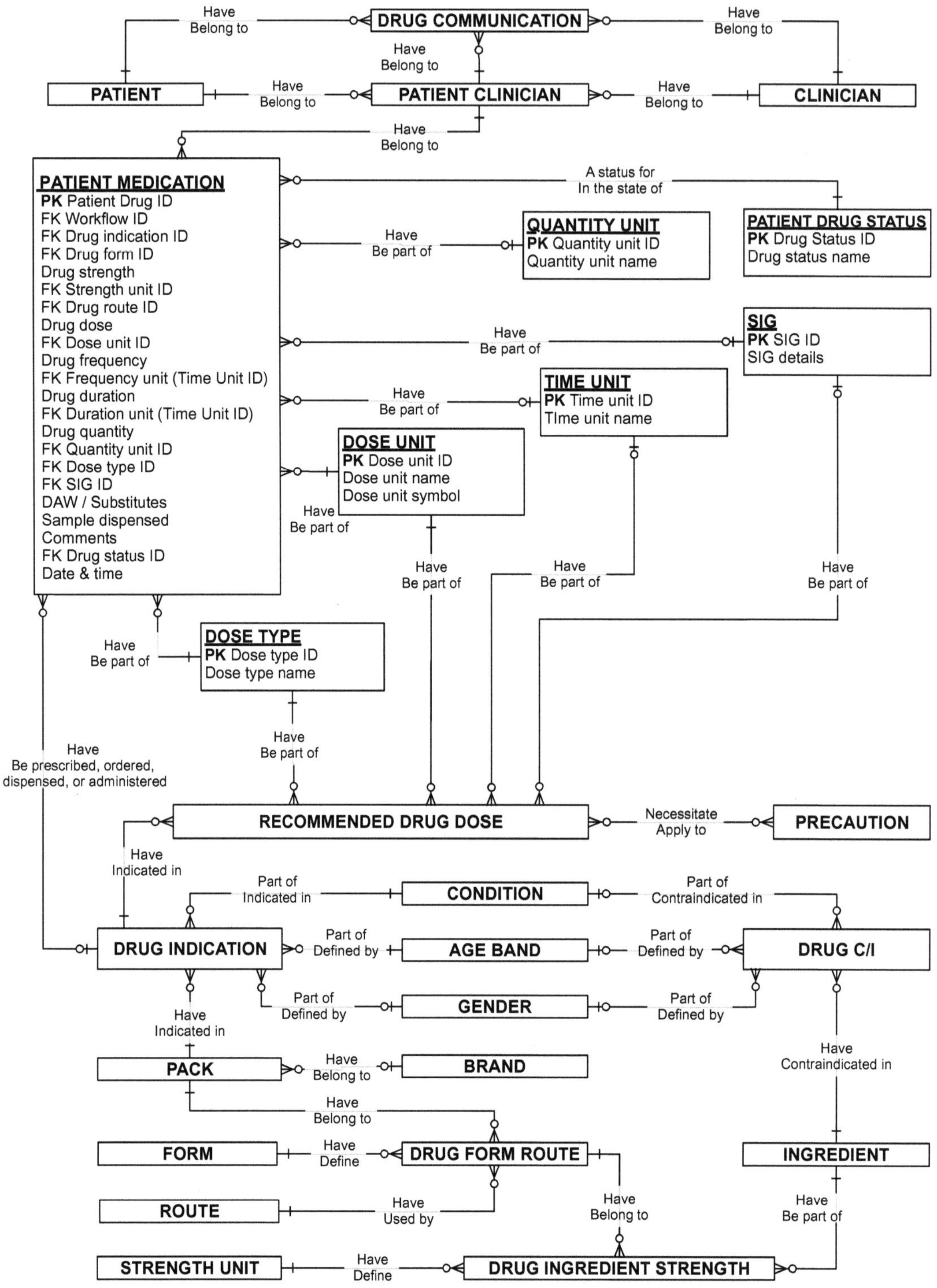

Figure 3.19 Logical data model: Patient's medications.

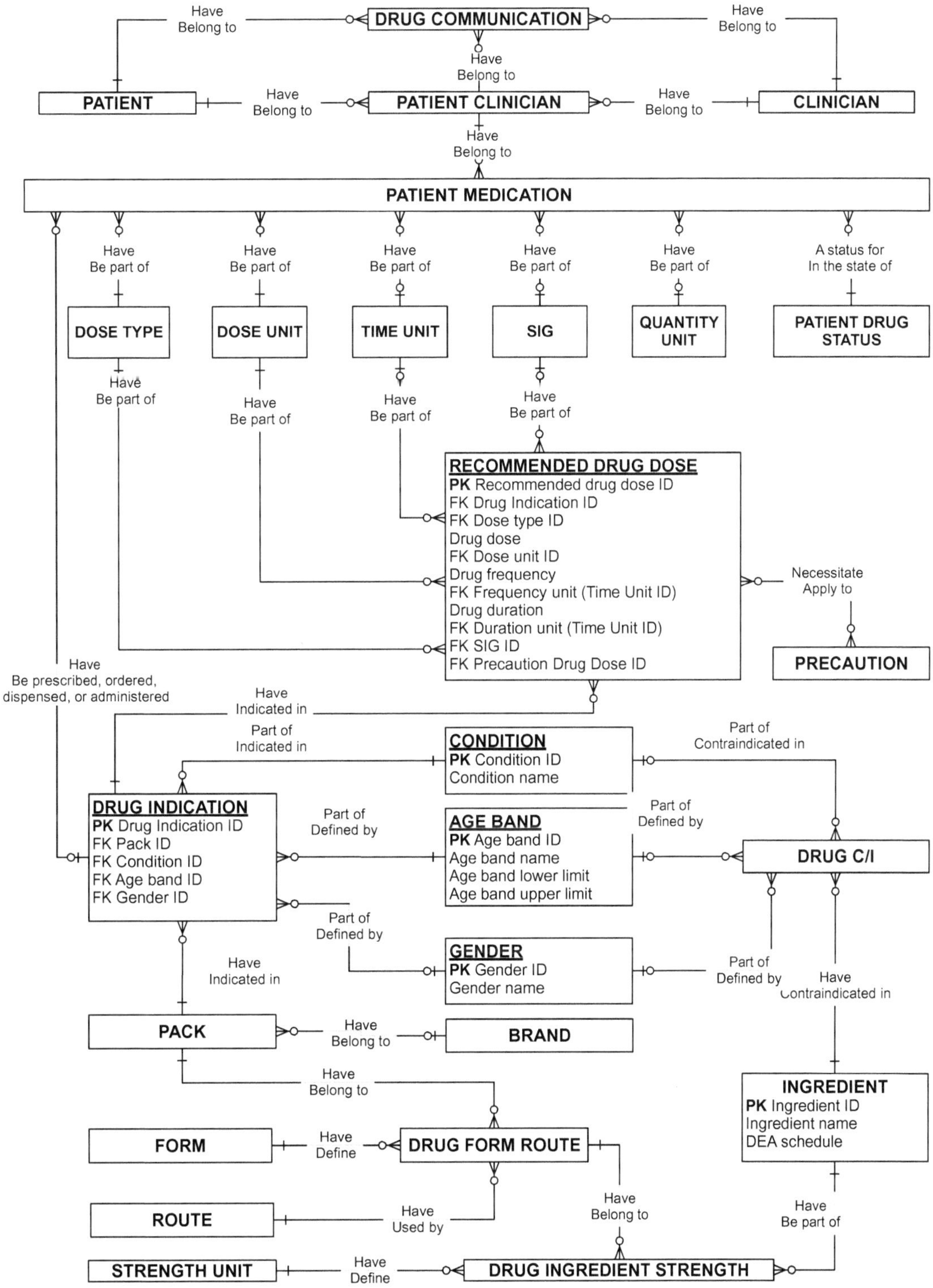

Figure 3.20 Logical data model: Recommended drug and dosing regimens.

Using the data model, the system can traverse between tables and retrieve the correct discrete data elements needed for system functions and workflows: *Patients, Clinicians, Medications, Form, Strength, Route, Dose, DEA schedule,* and so on.

Patient Non-Drug-Related Parameters

This information comes from an external entity, *NON-DRUG REPOSITORY,* a cluster of databases or services beyond the medications system boundaries. As such, we are not supposed to model or build any data stores for it. The medications system has no control over these people and systems. Still, we need to define the data elements we expect the *Medications* system to exchange with these external actors.

The Medications system sends *Patient ID* to *NON-DRUG REPOSITORY* and receives back information, via process *1.2.1 Retrieve Non-Drug Parameters.* Clinical information, non-drug related, may be classified as:

Reason(s) for visit, history of present illness: *Hypertension, Headaches, Sore throat*
Problem list, active and past diagnoses: *Diabetes since 1996, s/p Acute MI 2008.03.01*
Physical exams and vital signs: *HR 98/min regular, BP 190/100 mmHg—2010.11.28*
Allergies: *Penicillin, Iodine*
Past surgical procedures: *s/p TUR-P 1999.12.01, s/p Lap. Cholecystectomy 2001.11.13*
Lab results: *Hb A1C 8.3%, Blood Glucose 178 mg% 2010.11.02*
Imaging results: *Chest XR 2010.10.22, Chest XR 2003.11.01*
Consultations: *Cardiologist 2010.11.07, Urologist 2011.03.02*
Treatment/Care plans
Life style: *Smoking, Alcohol, OTC, Herbs, Dietary restrictions*

These data structures are extracts or summaries of other vast data stores; each one may represent many discrete data elements. Future information may include genomic data and possibly the DNA sequence of the patient. The standards and vocabularies used in this complex information exchange are detailed in the chapter on interoperability. Considerations on how to optimally present the user with huge amounts of information can be found in the chapter on the user interface.

CDS

CDS workflow number 5 is invisible in all the DFDs 1.x since it is a higher-level workflow. Dataflows from several *Prescribe* processes are being shown as two-headed arrows to and from *CDS*. These 1.x processes send data elements to *CDS*, which considers the information and sends back recommendations as its output. The actual processes are detailed in the chapter on *CDS*. Since from the 1.x perspective *CDS* is a higher-level process, we will treat *CDS* as a black box, similar to an external actor. From the *Prescribe* workflow perspective, *CDS* receives input and produces an output, but the processing inside the *CDS* workflow is detailed elsewhere. Considering the dataflows related to prescribing, *CDS* might be invoked at three different levels: *Patient, Drug,* and *Dose.*

Input being sent to *CDS* (request):
Gender, Age, Weight, Height, Vital signs, Diseases, Allergies, Surgical procedures (past and planned), Medications, Lab results and orders, Life style

Output expected from *CDS* (response):
Evidence-based medicine (EBM): *guidelines and best practices for treating community acquired pneumonia*
Drug—indication: *Esmolol for rapid atrial fibrillation*
Drug—contraindicated for age: *Quinolone—1 month old* baby
Drug-contraindicated for gender: *Leuprolide—females*
Allergies: *Proventil and peanut allergy cross-sensitivity*
Drug—contraindicated for patient's conditions: *Asthma and beta-blocker*
Drug-drug interactions: *Erythromycin and disopyramide*
Drug—vital sign: *Beta-blocker—bradycardic patient*
Lab affecting drug selection: *Antibiotic selection as per culture results and sensitivity*
Drug affecting lab result: *Promethazine causing false positive pregnancy test*
Labs that need to be monitored when using drug: *Liver function tests—Griseofulvin*
Duplicate therapy same drug: *Atenolol and Tenormin*; same class: *Atenolol and Propranolol;* same intent: *Heparin and Warfarin*
Drug—surgery: *Metformin 48 hours before surgery*
Drug—food: *St. John's wort and oral contraceptives*
Dose adjustment for *age, indication, weight, height, BMI, BSA, renal, hepatic function*
Special precautions: *Renal, hepatic, immunocompromised*
Adverse reactions and side effects: *Erythromycin causing diarrhea*
Pregnancy/lactation considerations: *Acitretin in pregnancy*
Pharmacologically safer alternatives: *Proventil—Ventolin*
Intravenous admixtures: *Calcium chloride and Sodium bicarbonate*

PBM

Process 1.5 *Consider Formulary* is visible on some of the 1.x DFDs and is invisible on the diagrams that are at a lower level. Still, there are dataflows to and from process 1.5, and this information exchange is related to *Patient, Drug,* and *Quantity:*

Input being sent to *PBM* (request):
  Patient ID
  Patient medications

Output expected from *PBM* (response):
  Patient eligibility under plan
  Medicare versus non-Medicare coverage
  Tier
  Preapproval needed
  Restrictions: *Indication, Duration, or Quantity*
  PBM percentage coverage
  Patient's copay
Therapeutically equivalent, less-expensive, formulary alternatives

Patient's Preferred Pharmacies

The list of patient's preferred pharmacies is needed for process 1.6 *Sign Rx*. Since it is not part of the conceptual model, we assume an external entity, the *NON-DRUG REPOSITORY,* is providing this list. Each pharmacy may need to list in addition to address, phone, and so on the *National Provider Identifier* (*NPI*), a 10-digit unique identification number issued by the *Centers for Medicare and Medicaid Services* (*CMS*). The system stores the selected *Rx destination* in *DRUG COMMUNICATION.*

Summary

In this chapter, the *Prescribe/eRx* workflow in an outpatient environment was analyzed from two perspectives:

- Processes—dataflow diagrams
- Data—conceptual and logical models

We identified six main processes and their subprocesses, external actors, and data stores related to *Prescribe*. Using the dataflows on DFDs 1.x, we uncovered the data elements necessary for the *eRx* workflow. We have refined the conceptual model, added a couple of entities, defined the entities' attributes, and set the relationships between entities with *PK/FK*.

Issues related to *CDS* are discussed in a separate chapter. The standards and vocabularies to facilitate the information sharing between disparate systems are covered in the chapter on interoperability, and the user interface considerations are discussed in Chapter 7. In the next chapter, we proceed with the structured analysis of workflow number 2: *Order/CPOE (*computerized physician order entry) in an in-hospital environment.

References

1. Van Ornum, M. 2009. *Electronic prescribing—A safety and implementation guide.* Sudbury, MA: Jones and Bartlett.
2. Anderson, P.O., S.M. McGuinness, and P.E. Bourne. 2010. *Pharmacy informatics.* Boca Raton, FL: CRC Press.
3. Goudrey-Smith, S. 2008. *Principles of electronic prescribing.* New York: Springer.
4. Wikipedia. Schedule V controlled substances. http://en.wikipedia.org/wiki/Controlled_Substances_Act#Schedule_V_controlled_substances (accessed March 5, 2011).
5. U.S. Department of Justice, Drug Enforcement Administration, Office of Diversion Control. Electronic prescriptions for controlled substances. March 31, 2010. http://www.deadiversion.usdoj.gov/ecomm/e_rx/faq/practitioners.htm (accessed March 5, 2011).

Review Questions

Find the errors in the following four diagrams:

1.

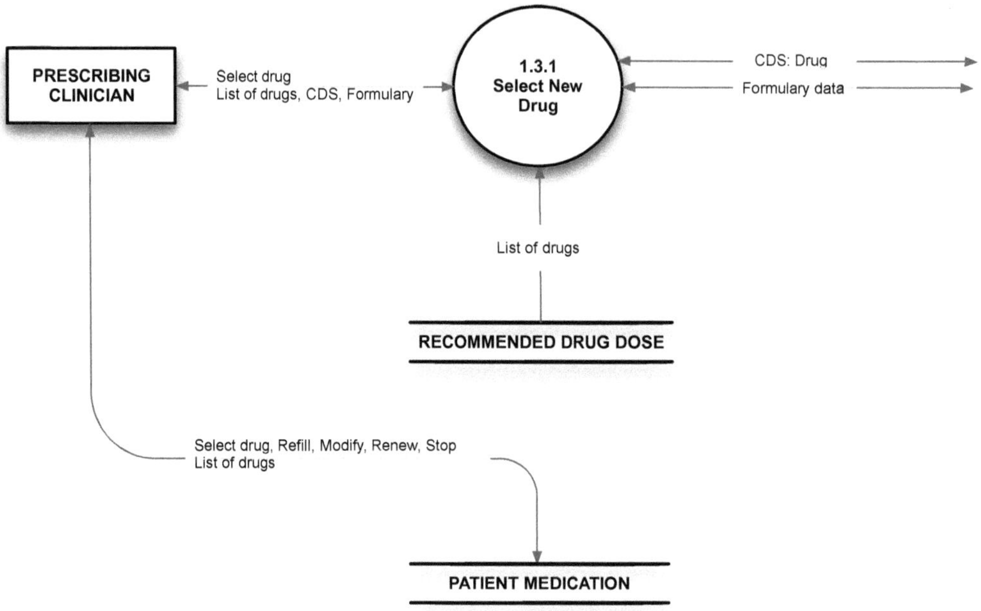

2.

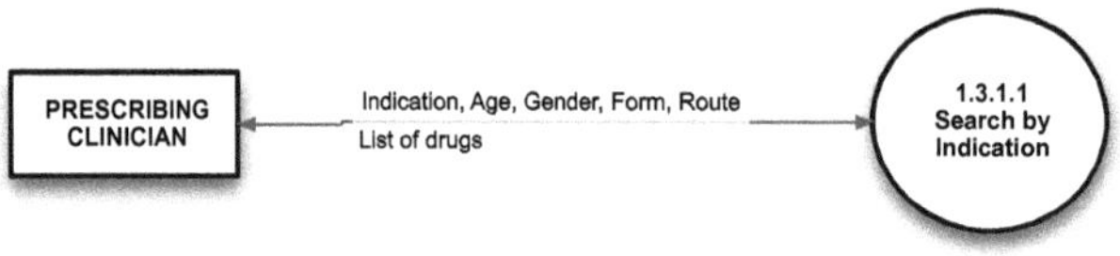

3.

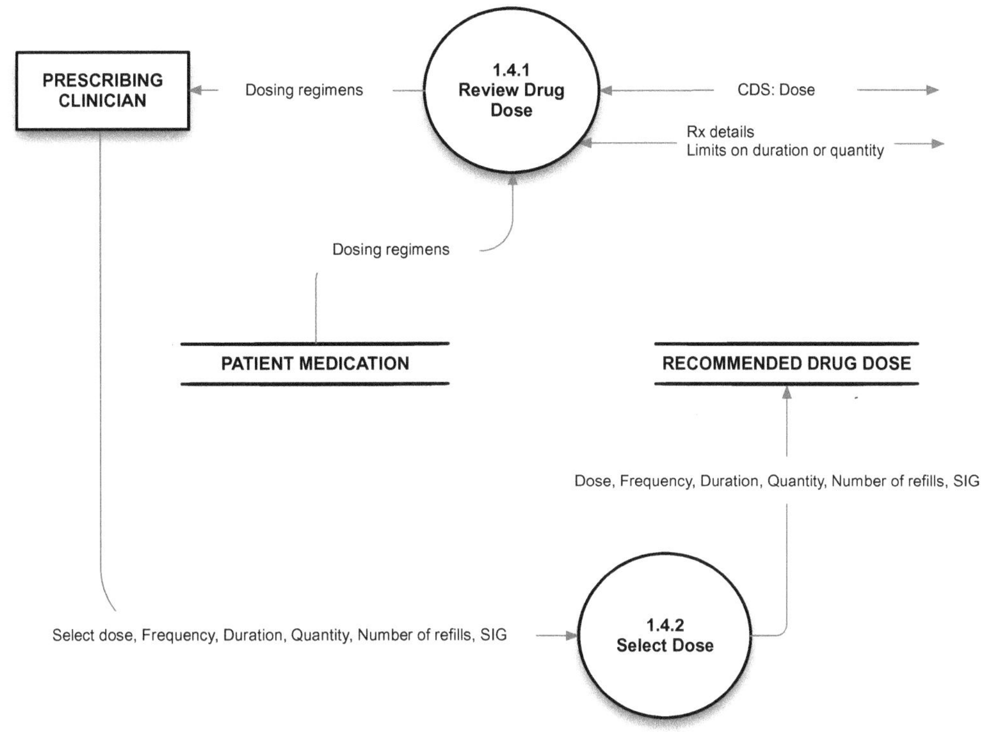
PRESCRIBING CLINICIAN
Dosing regimens
1.4.1 Review Drug Dose
CDS: Dose
Rx details
Limits on duration or quantity
Dosing regimens
PATIENT MEDICATION
RECOMMENDED DRUG DOSE
Dose, Frequency, Duration, Quantity, Number of refills, SIG
Select dose, Frequency, Duration, Quantity, Number of refills, SIG
1.4.2 Select Dose

4.

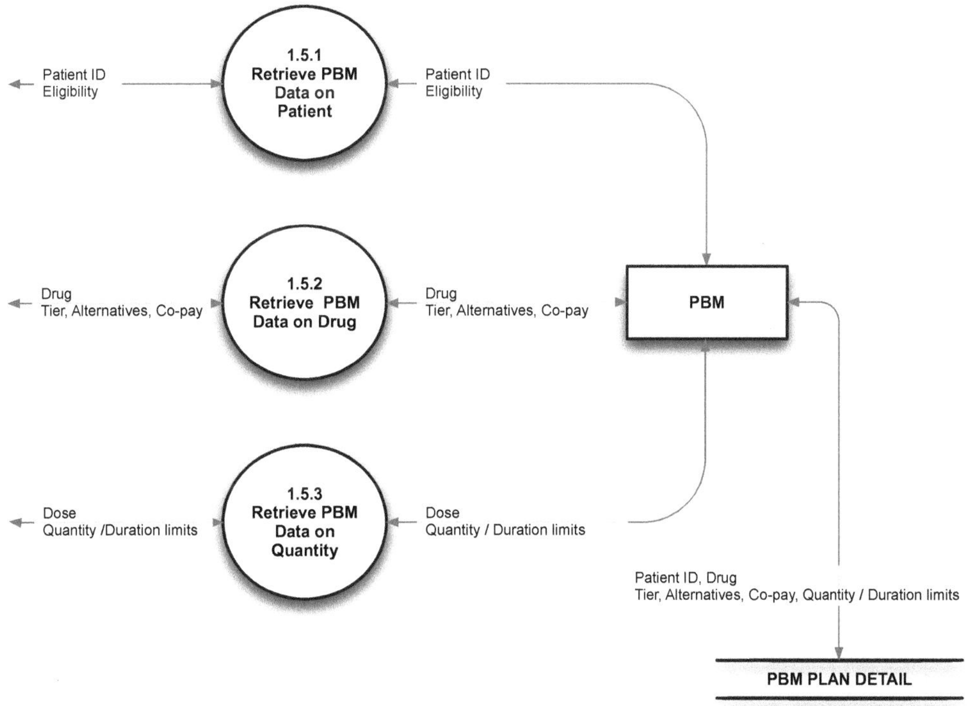
Patient ID
Eligibility
1.5.1 Retrieve PBM Data on Patient
Patient ID
Eligibility
Drug
Tier, Alternatives, Co-pay
1.5.2 Retrieve PBM Data on Drug
Drug
Tier, Alternatives, Co-pay
PBM
Dose
Quantity /Duration limits
1.5.3 Retrieve PBM Data on Quantity
Dose
Quantity / Duration limits
Patient ID, Drug
Tier, Alternatives, Co-pay, Quantity / Duration limits
PBM PLAN DETAIL

5. Explain why *CDS* does not appear in Figure 3.2.
6. Explain why process *1.5 Consider Formulary* does not appear in Figure 3.6.
7. Explain what happened to *CLINICAL REPOSITORY*: It appears in Figure 3.1 and does not appear in Figure 3.2.
8. Are the processes numbered correctly in the following figure?

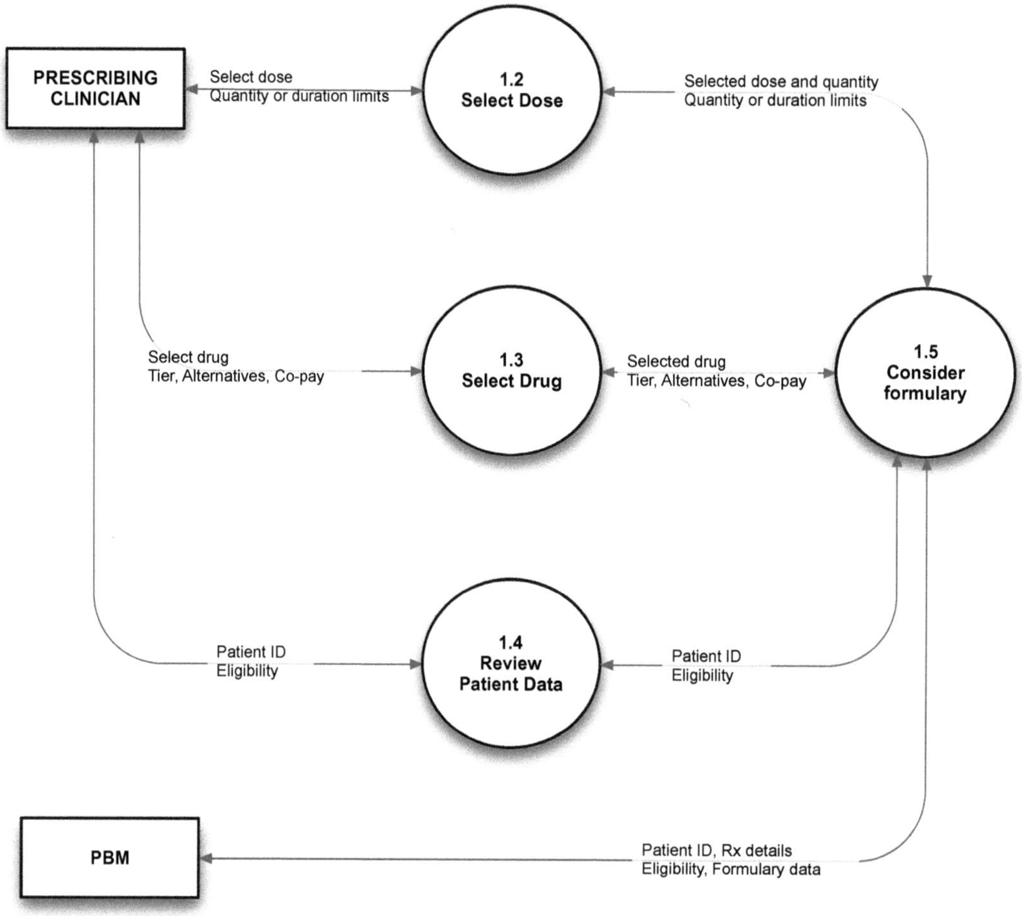

9. Which table provides the clinician with the patient's name, address, date of birth, and gender?
10. Which entity or actor provides the clinician with knowledge on safer therapeutic alternatives? Less-expensive alternatives?

Chapter 4

Order/CPOE

The power to command has never meant the power to remain mysterious.

Ferdinand Foch

This chapter analyzes workflow number 2: *Order/Computerized Physician Order Entry* (*CPOE*) from the same two perspectives as the previous *Prescribe/eRx* workflow: processes and data elements.

Initially, we isolate the *Order* workflow from the higher-level diagram (dataflow diagram [DFD] 0), and using top-down analysis we decompose it into lower-level processes: DFD 2.x.

We identify similarities with the *Prescribe* workflow, and this allows us to reuse, with some changes, most of the *Prescribe* processes in the *Order* workflow. Still, there are enough differences between *eRx* and *CPOE* to justify a two-chapter approach. The similarities and differences will help us define DFD 2 and its lower subprocesses and activities.

Using the dataflows from the DFD 2.x, we uncover the discrete data elements required by the *Order* workflow to perform its functions. Dataflows may relate to external actors or to the *Medications* database. We identify both the data structures involved in the interactions with the external actors and the data elements used by the evolving logical model.

Processes

DFD 2 Order Workflow

First, let us isolate the *Order* workflow from the *Medications* system:

Healthcare → Medications → Order

From DFD 0: Medications system (Figure 2.3), we cut and paste workflow number 2 with its external entities, data stores, and other processes attached (Figure 4.1).

The *Order* workflow is similar to *Prescribe* workflow in many aspects (1–3). In terms of processes, an ordering clinician may need to:

- Review patient data
- Select a new drug
- Select an existing drug (and *Modify, Stop, Hold,* or *Renew*)
- Select a dose
- Communicate with other external actors
- Consider *Formulary* information
- Consider *CDS* (clinical decision support) recommendations
- Sign and distribute order

The seven processes in the column in Figure 4.2 (*Communicate, Review Patient Data, Select Drug, Select Dose, Consider Formulary, CDS,* and *Sign*) with some minor differences are actually common to all the clinical workflows related to the medications domain—*Prescribe, Order, Dispense* and *Administer*—and we consider reusing processes from the previous chapter as much as possible. In

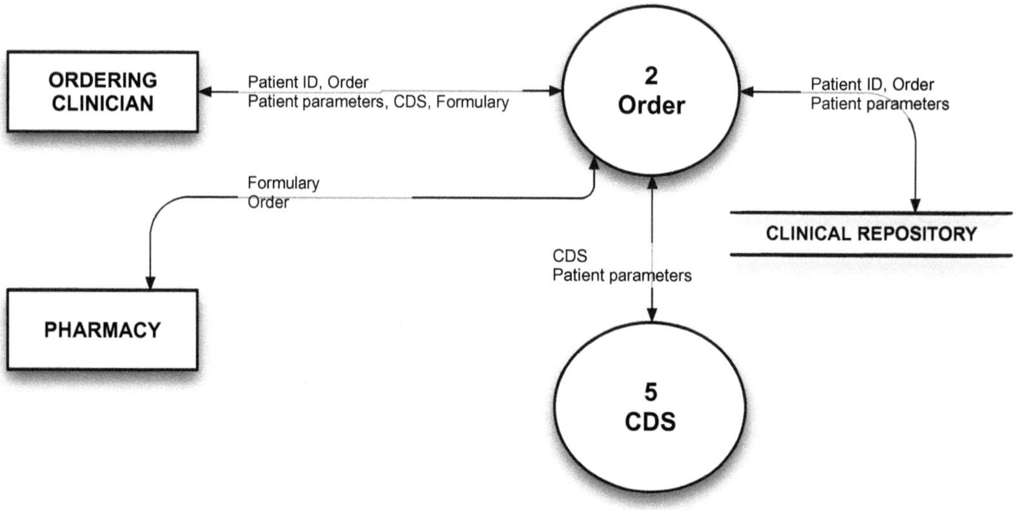

Figure 4.1 Order workflow: Isolated from DFD 0.

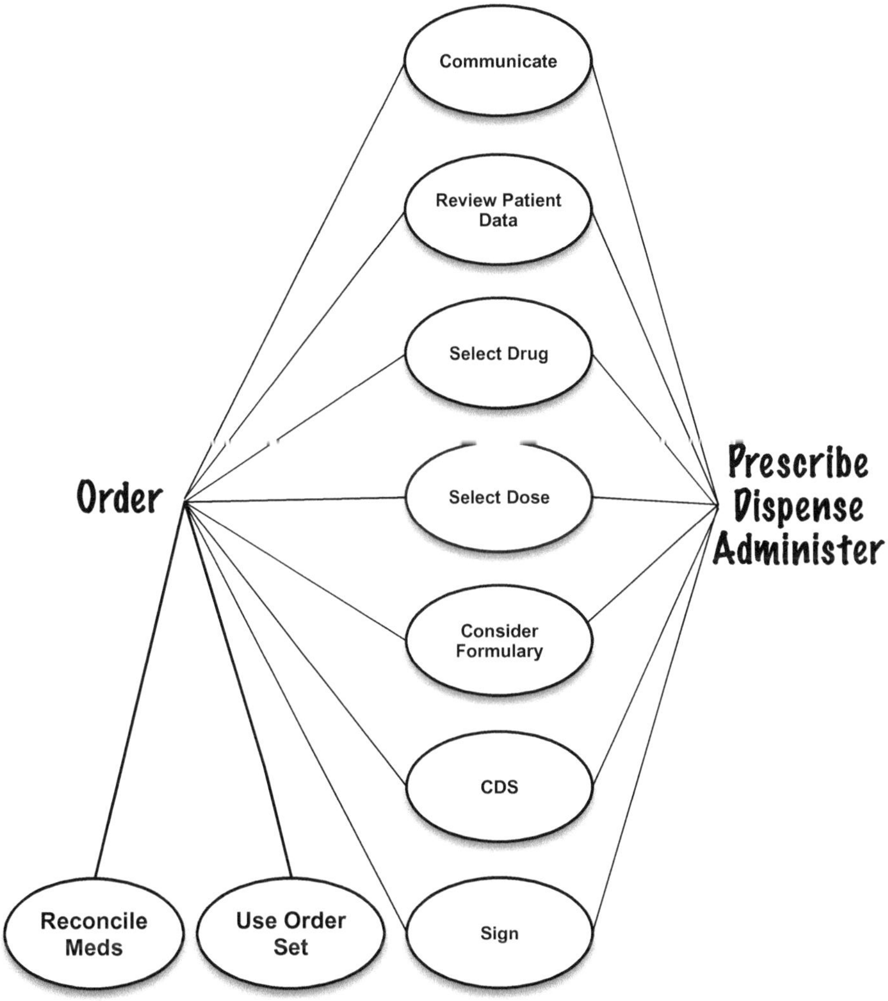

Figure 4.2 Common processes: Order, Prescribe, Dispense, and Administer.

addition, *the Order* workflow has two specific processes to be detailed further in this chapter:

- *Use Order Set*
- *Reconcile Meds*

Let us examine the differences between the *Prescribe* and *Order* workflows (Table 4.1).

There are a couple of preconditions to DFD 2 that must be satisfied before the system may proceed with the *Order* workflow:

1. Clinician is uniquely identified.
2. Clinician is authorized to use system in lieu of ordering medications.
3. Patient is uniquely identified.
4. All system's activities are logged.

Table 4.1 Differences between *Prescribe* and *Order* Workflows

	Prescribe/eRx	*Order/CPOE*
Setting	Private practice, outpatient	Inpatient, ambulatory
Pharmacy	Community, mail, independent	Hospital pharmacy
Formulary	PBM	Hospital pharmacy
Complexity	Simple	Complex
Order set	Rare	Common
Meds reconciliation	Rare	Common
Need to cosign	Rare	Common

The *Order* workflow is decomposed into eight main processes (bubbles):

2.1 Communicate Order
2.2 Review Patient Data
2.3 Select Drug
2.4 Select Dose
2.5 Consider Formulary
2.6 Sign Order
2.7 Use Order Set
2.8 Reconcile Meds

If we compare the *Order* workflow (Figure 4.3) with the *Prescribe* workflow from Chapter 3 (Figure 3.2), we can see several differences:

Patient and *PBM (*pharmacy benefit management)—external actors in *Prescribe*—are not participating in the *Order* workflow. The *Patient* is not planned to interact directly with the *Order* workflow.

The role of *PBM* in the *Prescribe* workflow is represented by the hospital *PHARMACY* in the *Order* workflow.

There are a couple of new actors (*ORDER SET REPOSITORY* and *RULES REPOSITORY)* and two new processes: 2.7 *Use Order Set* and 2.8 *Reconcile Meds*. The other six processes, 2.1–2.6, are reused, with some minor changes from DFD 1.x *Prescribe* into DFD 2.x *Order*.

Even though several processes mentioned use it, *CDS* does not appear on DFD 2.x. As workflow number 5, it should not appear on a lower DFD. Still, we define the dataflows to and from *CDS* related to *Order* workflow, as we will see that several *Order* processes use *CDS* on medications. Details about the *CDS* can be found in Chapter 8.

Let us proceed with the top-down analysis of the *Order* workflow and decompose DFD 2 into its children processes: the DFD 2.x.

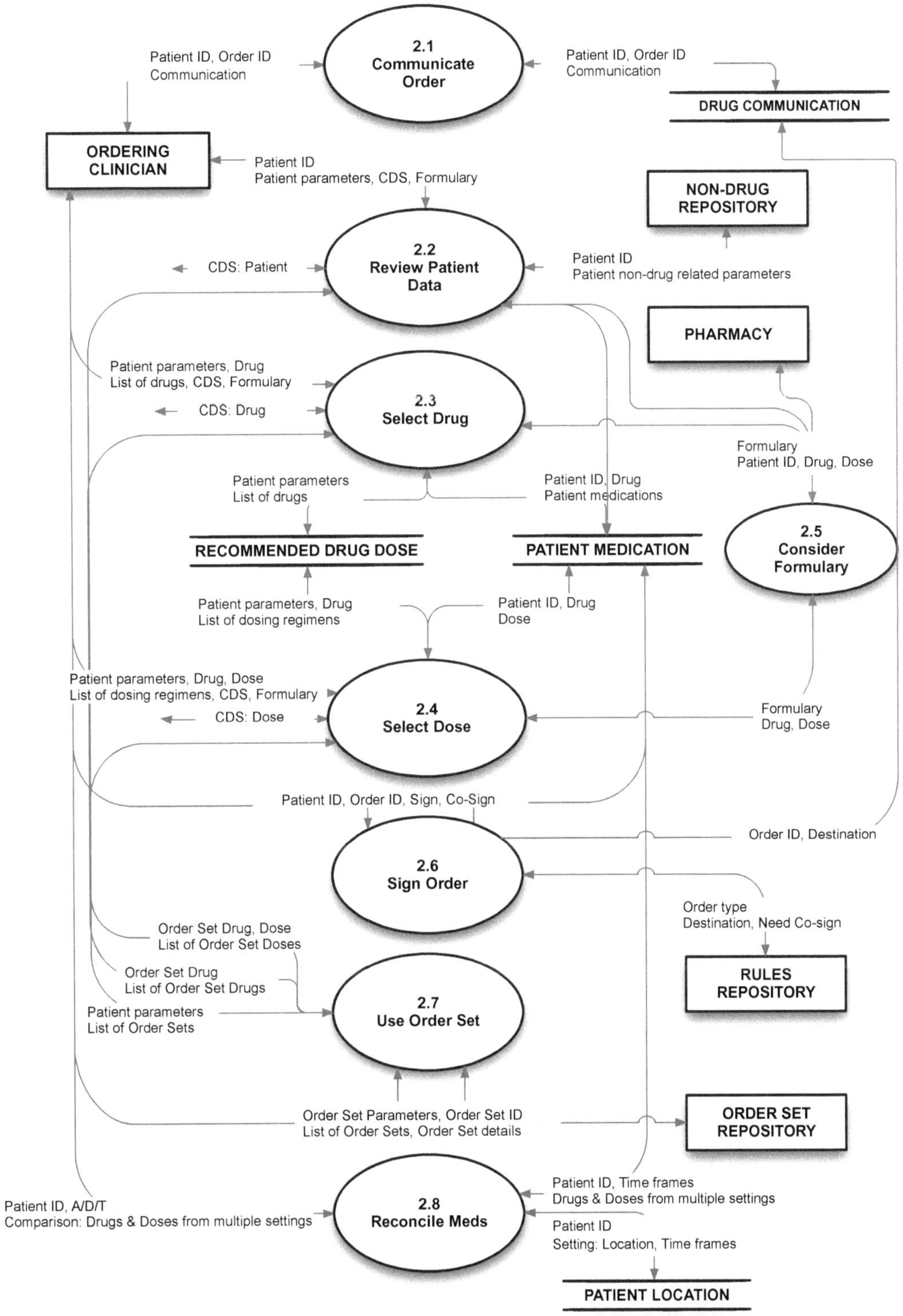

Figure 4.3 DFD 2: Order workflow.

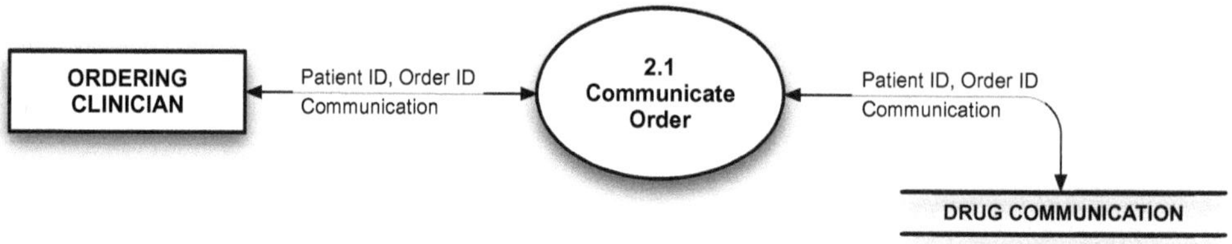

Figure 4.4 Communicate Order process: Isolated from DFD 2.

DFD 2.1 Communicate Order

Again, we isolate a process—2.1 *Communicate Order*—from its higher-level DFD while we keep its related actors, data stores, and (same-level) processes attached (Figure 4.4).

Only *ORDERING CLINICIAN* interacts with process 2.1. The *Order*-related communications are in the form of *Read/Write,* and the table to store these data elements is *DRUG COMMUNICATION.* The breakdown of the DFD 2.1 *Communicate Order* (Figure 4.5) is following the same line as DFD 1.1 *Communicate Prescription.*

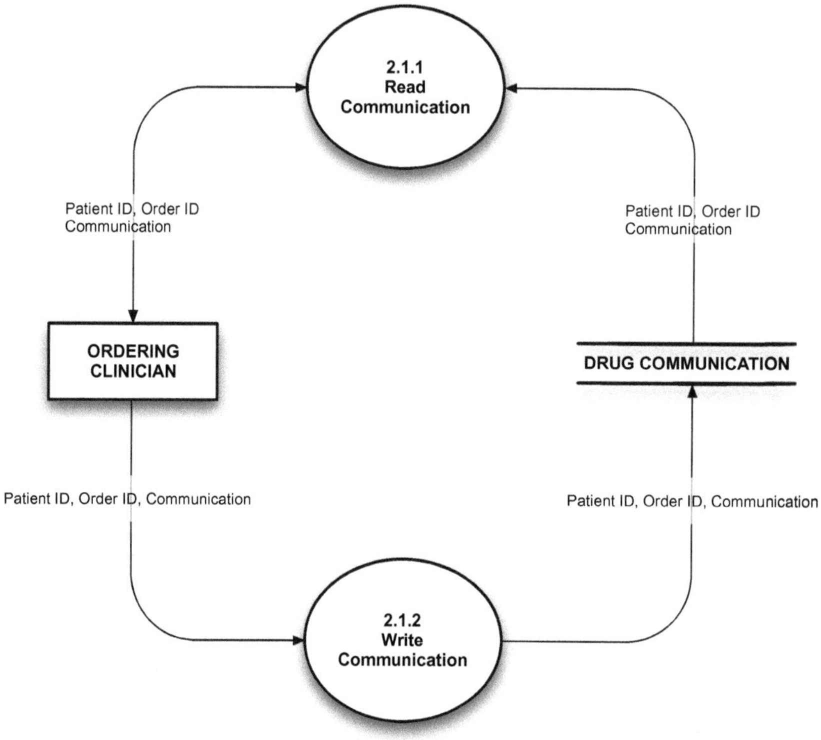

Figure 4.5 DFD 2.1: Communicate Order.

The communications processes of *DISPENSING* and *ADMINISTERING CLINICIAN* are detailed in the next chapters, and that is the reason the ORDERING CLINICIAN appears to be the only actor in DFD 2.1:

← Sends *Patient ID* and *Order ID* to *DRUG COMMUNICATION*
→ Receives communications related to patient
← Sends updated communications to *DRUG COMMUNICATION*

We can reuse process 1.1.x (Figures 3.3 and 3.4) with only some minor differences for process 2.1.x.

DFD 2.2 Review Patient Data

Isolating process 2.2 *Review Patient Data* from the parent DFD 2 (Figure 4.6).

Process 2.2 *Review Patient Data* uses the following subprocesses:

← Sends *Patient ID* to *PATIENT MEDICATION*
→ Receives information on patient medications
← Sends *Patient ID* to *NON-DRUG REPOSITORY*
→ Receives clinical, non-drug-related information
← Sends patient parameters to *CDS*
→ Receives *CDS* recommendations
← Sends patient parameters to *Consider Formulary*
→ Receives *Formulary* data on patient medications
← Sends patient parameters to Process *2.7 Use Order Set*
→ Receives a list of potentially relevant *Order Sets* for the specific patient parameters
← Sends patient (drug and non-drug-related) data, *CDS*, *Formulary*, and a list of relevant *Order Sets* to the user

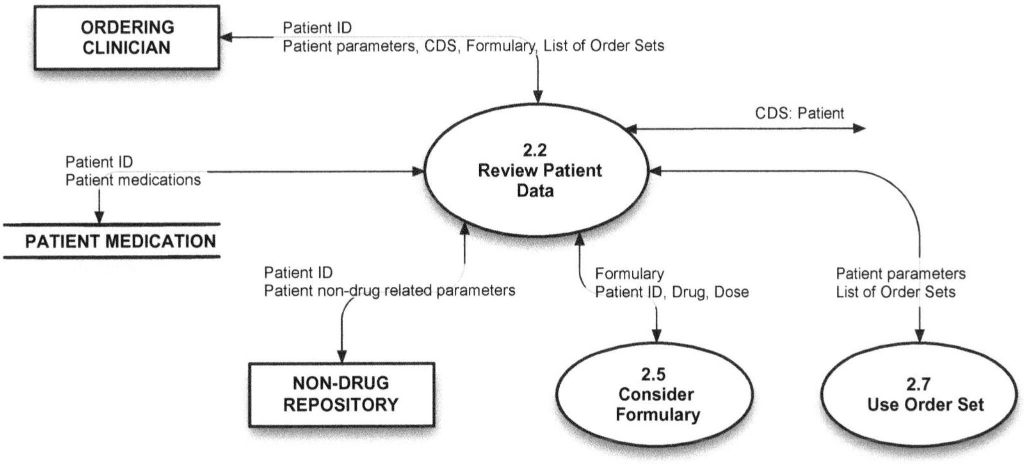

Figure 4.6 Review patient data process: Isolated from DFD 2.

The differences between DFD 1.2 and 2.2 are as follows:

- Process 2.7 *Use Order Set* may recommend an *Order Set* relevant to the patient parameters.
- Eligibility with *PBM* is usually not performed by a hospital pharmacy.

Thus, DFD 1.2.x (Figures 3.5 and 3.6) may be reused with these two changes as DFD 2.2.x.

DFD 2.3 Select Drug

Isolating process 2.3 *Select Drug* from the *Order* workflow (Figure 4.7).

Process 2.3 has the following subprocesses:

← Sends *Patient ID* to *PATIENT MEDICATION*
→ Receives the information on the medications patient is administered
← Sends patient parameters to *RECOMMENDED DRUG DOSE*
→ Receives list of drugs relevant to patient condition, age, gender, and so on
← Sends patient parameters to *CDS*
→ Receives *CDS* recommendations on drugs
← Sends drug parameters to 2.5 *Consider Formulary*
→ Receives information on drug tier and therapeutic alternatives
→ Receives drug options from 2.7 *Use Order Set*
→ Receives drug selected by user
← Sends drug selected to 2.7 *Use Order Set* (the actual selection of an *Order Set* is detailed further in this chapter)
← Sends drug to *CDS* and *Formulary* for final check
← Sends selected drug to *PATIENT MEDICATION* (setting the dose is detailed next)

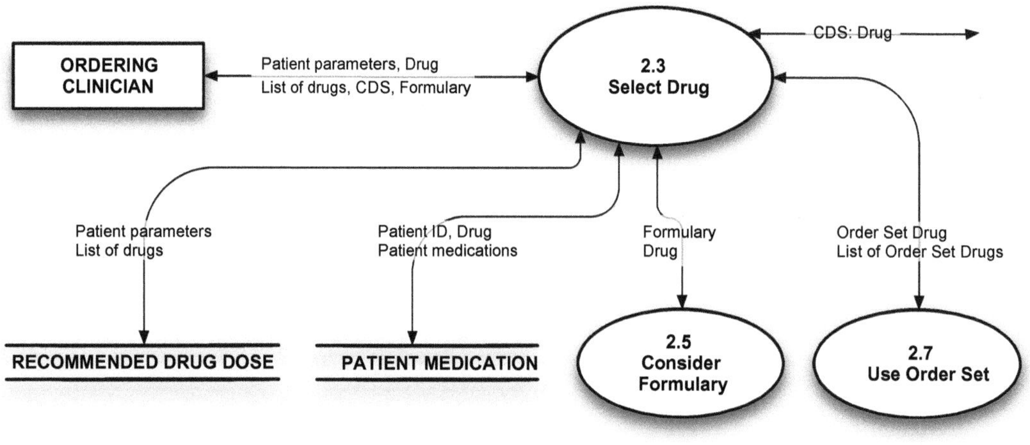

Figure 4.7 Select Drug process: Isolated from DFD 2.

Analyzing the process of *Select Drug* within the *Prescribe* context versus the *Order* context reveals only one difference: A drug may be selected as part of an *Order Set*. Once selected, an *Order Set Drug* will still have to pass the *CDS* and *Formulary* filters.

Select New Drug (*Search by Indication* and *Search by Drug*) and *Select Existing Drug* (*Refill, Modify, Stop,* and *Renew*) are the same processes as detailed in *Prescribe* workflow. DFD 1.3.x (Figures 3.7–3.10) may be reused as 2.3.x with only minor changes.

DFD 2.4 Select Dose

Once a drug (and *Strength, Form,* or *Route*) is selected, the next step is to set the dose (Figure 4.8). This is needed for a new drug as well as for an existing drug that may require a dose modification.

← Sends *Patient ID* and drug to *PATIENT MEDICATION*
→ Receives details on patient's medications and doses
← Sends patient parameters to *RECOMMENDED DRUG DOSE*
→ Receives details on doses
← Sends patient parameters and drug selected to *CDS*
→ Receives *CDS* recommendations on dose
← Sends patient drug to 2.5 *Consider Formulary*
→ Receives formulary information on quantity or duration limitations
← Sends drug to 2.7 *Use Order Set*
→ Receives dosing options from 2.7 *Use Order Set*
→ Receives dose selected by user
← Sends dose to *CDS* and *Formulary* for final check
← Sends selected dose to *PATIENT MEDICATION*

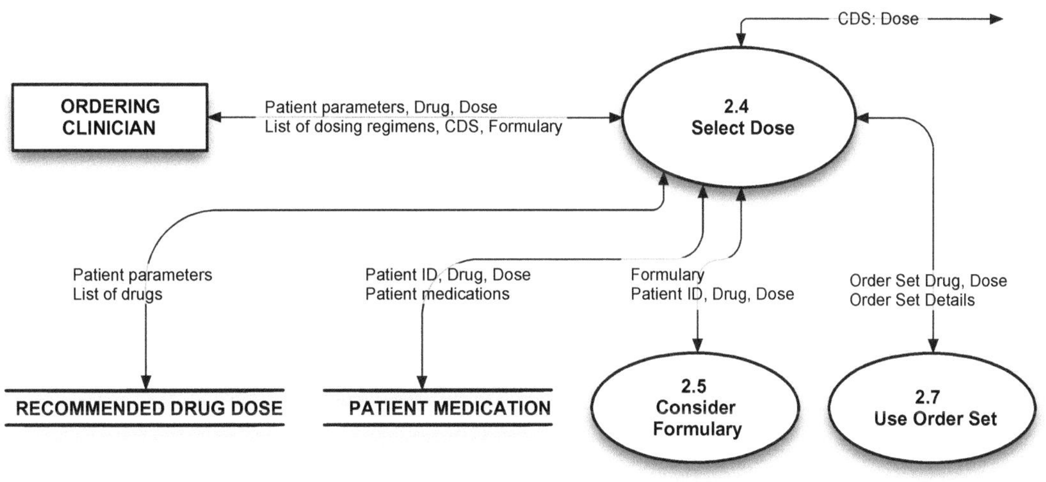

Figure 4.8 Select Dose process: Isolated from DFD 2.

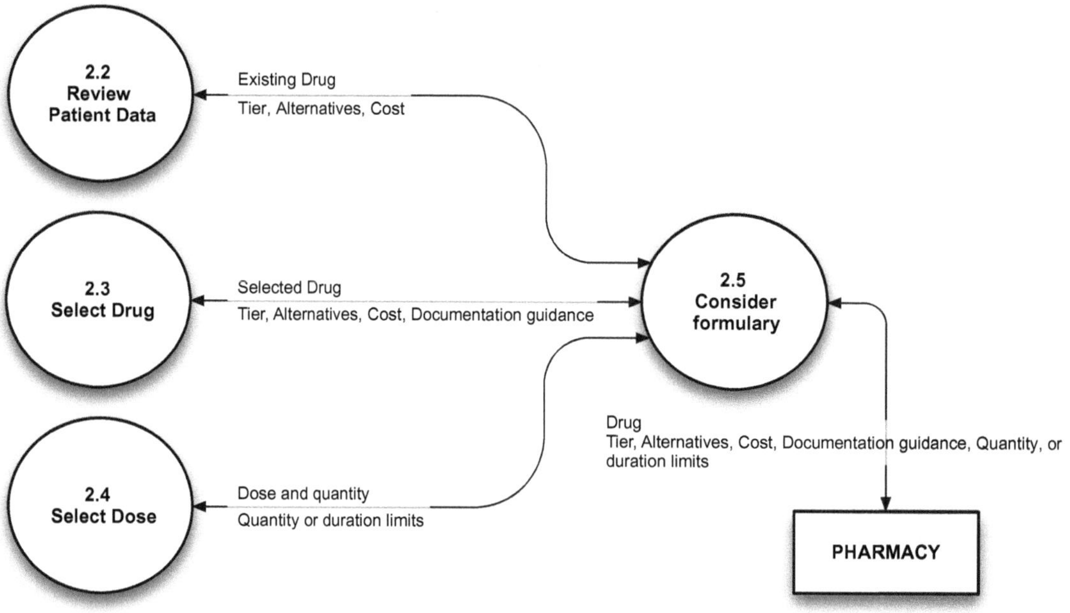

Figure 4.9 Consider Formulary process: Isolated from DFD 2.

In the *Order* workflow, 2.4 *Select Dose* is similar to process 1.4 *Select Dose* in the *Prescribe* workflow, with an additional potential source for dose information from an *Order Set.* Process 1.4 *Select Dose* (Figures 3.11 and 3.12) from *Prescribe* can be reused in the *Order* workflow with only this minor adjustment.

DFD 2.5 Consider Formulary

Figure 4.9 shows the process for isolating *2.5 Consider Formulary* from the *Order* workflow.

The actor providing *Formulary* information in an in-hospital environment is the hospital *PHARMACY* and not the *PBM.* Usually, the issue of patient eligibility, which was part of the *Prescribe* workflow, is not dealt with by the hospital *PHARMACY.* (This is not always true: Sometimes a hospital *PHARMACY* may dispense drugs to *Medicare* patients, and then the eligibility needs to be checked; in this case, *PHARMACY* workflow is *Prescribe* and not *Order.*) Other than this difference, processes 1.5 and 2.5—*Consider Formulary*—are similar, and the DFD from *Prescribe* workflow (Figures 3.13 and 3.14) can be reused in the *Order* workflow.

DFD 2.6 Sign Order

To finalize and communicate an order, the clinician needs to sign it, DFD 2.6 *Sign Order* (Figure 4.10). There are two main differences between signing an

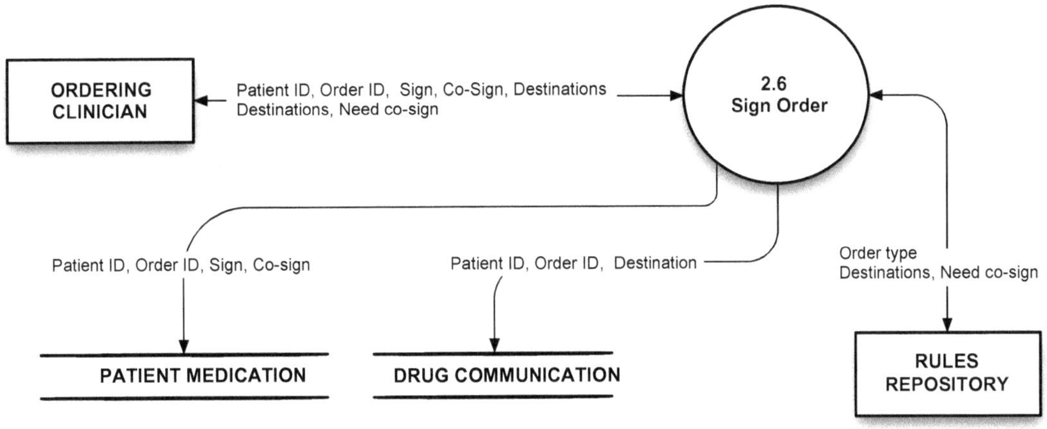

Figure 4.10 Sign Order process: Isolated from DFD2.

order versus signing a prescription:

- Need to cosign an order: For either a group of users (students, residents, or other clinicians in training) or for specific drugs or age bands, an order may require two clinicians to sign/cosign.
- Order destinations: While the *Prescribe* workflow used the patient's preferred pharmacies as tentative destinations for the prescription, the *Order* workflow will set the destinations according to the order type. For example, drug-related orders should be communicated to the pharmacy, attending physician, medication administration record, data warehouse for quality indicators, and so on. Orders for an imaging study such as computed tomography (CT) or magnetic resonance imaging (MRI) would normally be communicated to the radiology department with an update to the ordering party and so on.

Where is the information (on the need to cosign an order and its destinations) actually stored? The *RULES REPOSITORY* actor is the source for this information: Given an order type, the *RULES REPOSITORY* will respond with the order destinations and whether there is a need to cosign the order.

Rules may be related to the clinician, patient, or drug parameters. A rule may be simple, such as *"all orders signed by a first-year resident need the attending cosignature,"* or complex, such as *"chemotherapy drugs to children under 6 years of age need a board certified oncologist signature."*

There are two subprocesses (Figure 4.11):

2.6.1 *Select Order Destination:* for reviewing and possibly modifying the list of destinations for an order to be communicated and determining the need to cosign an order

2.6.2 *Sign and Communicate Order:* user is signing the order (or cosigning), which finalizes the order process, documents it in the *PATIENT MEDICATION* and the *DRUG COMMUNICATION* tables

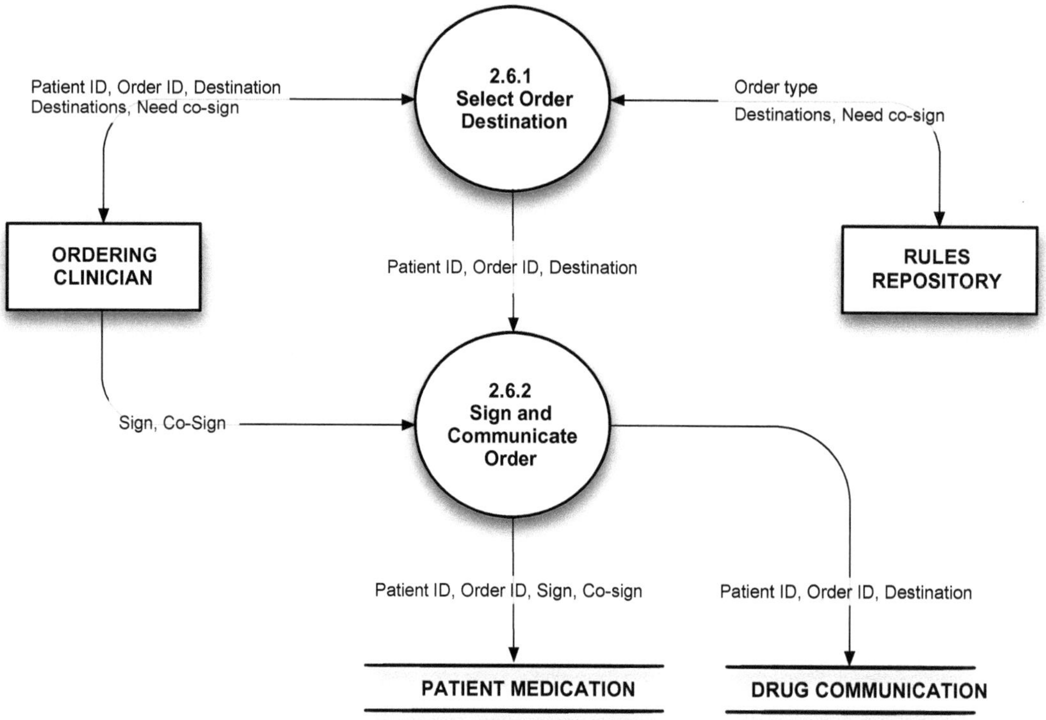

Figure 4.11 DFD 2.6: Sign order.

Reuse Considerations

The six processes mentioned (*Communicate, Review Patient Data, Select Drug, Select Dose, Consider Formulary,* and *Sign*) are common to both *Prescribe* and *Order* workflows. This is an opportunity to reuse documents, diagrams, models, software components, and code in more than one workflow. Still, there are some minor differences between these processes one should pay attention to when considering reuse (Table 4.2).

We continue the analysis of two new processes that are used by the *Order* workflow: 2.7 *Use Order Set* and 2.8 *Reconcile Meds.* These processes were not present in the *Prescribe* workflow.

DFD 2.7 Use Order Set

An *order set* is a collection of predefined orders and goals related to a specific condition, either a disease (such as *pneumonia*) or a procedure (such as *cholecystectomy*) (4).

Order sets may exist on paper in an organization, but for our discussion we refer to the digital format, used by *CPOE* during *Order* workflow.

These order sets may be developed by one physician, a group of clinicians or a department in a hospital, or a healthcare organization, or they may be purchased from a vendor that authors order sets.

Table 4.2 ***Prescribe* and *Order* Reuse Considerations**

Process	*Prescribe/eRx*	*Order/CPOE*
x.1 Communicate	1.1 Patient is an actor	2.1 Patient is not an actor
x.2 Review Patient Data	1.2 Eligibility check	2.2 No eligibility check Order sets matching patient data
x.3 Select Drug	1.3 Order Sets is not an actor	2.3 Order Sets is a source for drugs
x.4 Select Dose	1.4 Order Sets is not an actor	2.4 Order Sets is a source for dose
x.5 Consider Formulary	1.5 PBM provides formulary Eligibility check	2.5 Hospital pharmacy provides formulary No eligibility check
x.6 Sign	1.6 No need for co-sign Destinations by patient	2.6 Need for co-sign Destinations by order type

An order set may have multiple categories, and each category may have a number of data elements (5). The medications are usually only one of many parts of an order set (Table 4.3). The advantages and disadvantages of using *Order Sets* are summarized in Table 4.4.

An *Order Set*, like any other data item, has a unique identifier. In addition, an order set has a unique title, which is usually a concatenation of

- Condition (*Community Acquired Pneumonia*)
- Population age band (*Adult*)
- A/D/T (admission, discharge, transfer) setting (*Admission*)

The title of an *Order Set* may be *Community Acquired Pneumonia—Adult—Admission. The* system will try to match an order set to these specific patient "triggers" within the *ORDER SET REPOSITORY*. Focusing on the medications part of an order set, we isolate process 2.7 *Use Order Set* from the DFD 2 (Figure 4.12).

Using the parameters of *Condition* or *Procedure, Age Band,* and *A/D/T setting*, a query is sent to *ORDER SET REPOSITORY*. (Of course, one can search for matching *Order Sets* with additional parameters such as *Gender*, but for the moment the three parameters mentioned cover most cases.)

Once selected, an *Order Set* may provide relevant drugs to be considered (either for ordering or for modifying, stopping/holding) as well as the corresponding dosing regimens. Thus, an order set is another potential, structured source for drugs and doses, similar to *PATIENT MEDICATION, RECOMMENDED DRUG DOSE, CDS,* and *FORMULARY*.

Table 4.3 *Order Set* Categories and Data Elements

Category	*Data Elements*
Patient details	ID, Name, Age, Gender Diagnosis Medications list, Allergies Weight, Height, BMI, BSA Current location (*ICU bed # 5*)
Order Set condition/ procedure	Diagnosis (*Community Acquired Pneumonia*) or Procedure (*Coronary Artery Bypass Graft*)
A/D/T	Admission (*internal medicine 3N*) or Discharge (*home*) or Transfer (*NICU*)
Vitals	*Temperature* *Blood pressure* *Oxygen saturation* *Heart rate*
Activity	*Bed rest*
Respiratory/oxygen therapy	*40% oxygen mask*
Intravenous access and fluids	*Peripheral IV lock* *Glucose 5% Normal Saline 500 ml—next 8 hours* *Fluids balance every shift*
Nutrition and diet	*Low sodium—less than 2 Grams/day* *Lactose free* *Gluten free*
Lab	*Blood basic chemistry panel* *CBC*
Medications	*Tab. Digoxin 0.125 mg PO Once a day* *Sol. Cefotaxime 1 gram IV 2 times a day* *Regular Insulin 1 unit/hour IV*
Blood	*Red Blood Cells one unit during next 4 hours* *2 units of Fresh Frozen Plasma*
Imaging	*Chest X-Ray* *Abdominal US* *Brain MRI*
Consults	*Cardiology for setting the pacemaker on fixed rate pre-surgery*
Nursing	*Surgical incision care* *Bed sores prevention protocol*
Patient education	*Smoking cessation*

Table 4.4 *Order Sets* Pros and Cons

Advantages	*Disadvantages*
Improved adherence with guidelines, best practices and evidence-based medicine	High-cost, time-consuming efforts in configuring, building, maintaining, and tweaking order sets
Increased speed of documentation	Push-back from physicians regarding standardized order sets as "cookbook" medicine versus the "art" being practiced
Improved quality by reducing variability of care processes (standardization)	Technological barriers integrating order sets with other EHR components
Reduced errors while ordering	
Reduced costs while eliminating unnecessary tests or treatments	

The user interacts with 2.7 *Use Order Set* via processes 2.2, 2.3, and 2.4:

2.2 *Review Patient Data*
2.3 *Select Drug*
2.4 *Select Dose*

These three processes provide *CDS* to 2.7 while filtering the *Order Set Drugs* and *Doses* as detailed in Chapter 8.

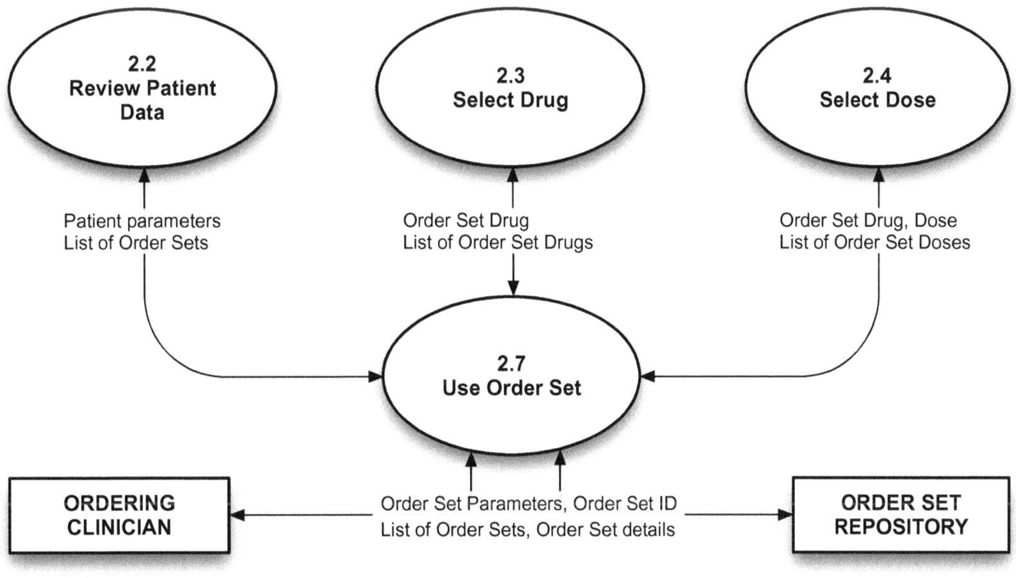

Figure 4.12 Use order set process: Isolated from DFD 2.

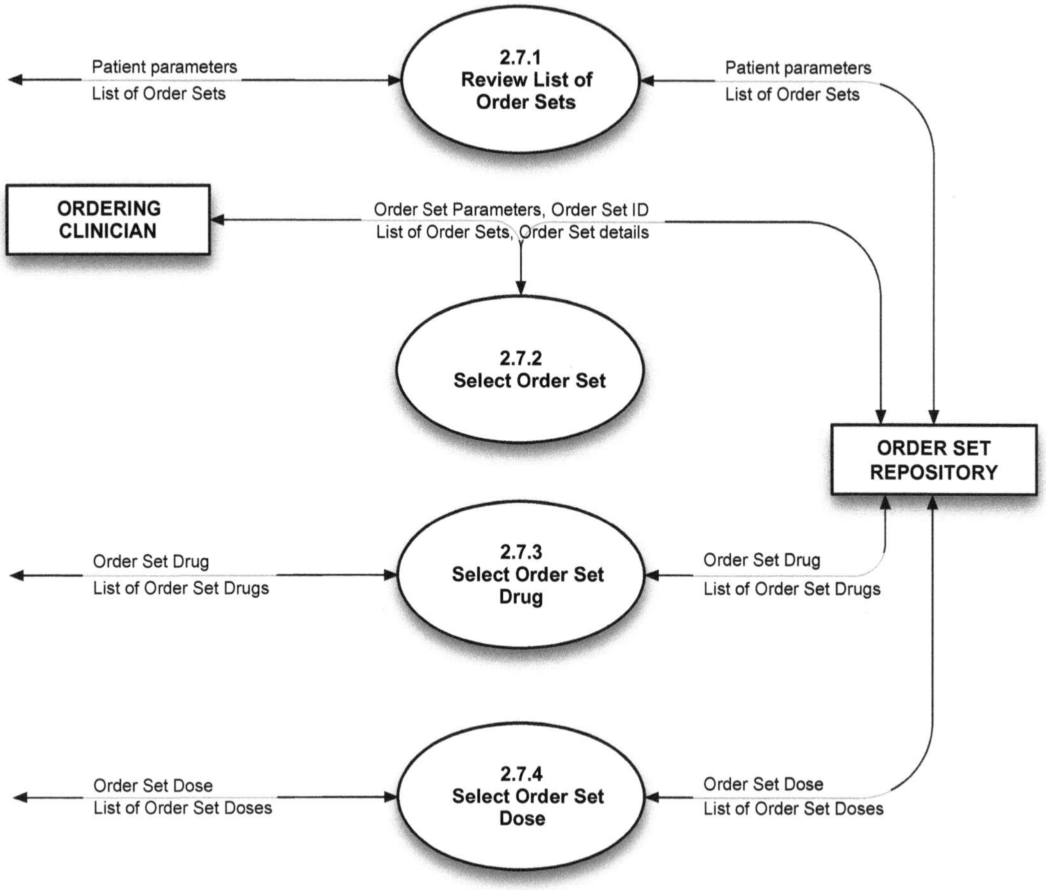

Figure 4.13 DFD 2.7: Use Order Set.

Decomposing process 2.7 further is shown in Figure 4.13:

2.7.1 *Review List of Order Sets* receives from process 2.2 *Review Patient Data* (not shown in Figure 4.13), the patient parameters, such as *Condition/Procedure, Age Band,* and *A/D/T setting,* and it returns a list of relevant *Order Sets.*

The user may select an order set and review the order set details in subprocess 2.7.2 *Select Order Set.*

During process 2.3 *Select Drug* (not shown in Figure 4.13), the user may choose a drug from an *Order Set* using subprocess 2.7.3 *Select Order Set Drug.*

Finally, if a drug was selected from an order set, the user may continue choosing a corresponding dose via process 2.4 *Select Dose* (not shown in Figure 4.13), using subprocess 2.7.4 *Select Order Set Dose.*

The reason processes 2.2, 2.3, and 2.4 are not shown in DFD 2.7 (Figure 4.13) is that processes do not appear in a lower-level DFD: 2.7.3 *Select Order Set Drug* interacts with 2.3 *Select Drug,* but the correct DFD will not show process 2.3 in a DFD depicting processes three levels deep, such as 2.7.x. This is one of the rules to prevent clutter on DFDs.

DFD 2.8 Reconcile Meds

Medication reconciliation is a process of addressing discrepancies between a patient's medications from two (or more) different time frames and settings to avoid errors, adverse drug events, and ultimately patient harm (6,7,8).

Transitions of care happen with a change in the patient's environment, level of care, service, or practitioners, usually during A/D/T:

- *Admission:* reconciliation between medications taken at home versus drugs to be administered in the hospital
- *Transfer*: reconciliation between drugs administered at one level of service, such as pediatric ward, versus drugs to be administered in another setting, such as an intensive care unit (ICU)
- *Discharge*: reconciliation between drugs administered in the hospital versus medications to be taken at home

The following is a typical patient journey through different transitions of care:

Home → Emergency Department → *(A)* → Ward → *(T)* → ICU → *(T)* → Ward → *(D)* → Home

At each transition of care, there exists a real, significant potential to err: Drugs may be omitted or ordered incorrectly, appear as duplicate therapy, be under- or overdosed, and so on (9,10).

From the *Order* workflow perspective, the system should be able to present a user with drugs and doses from two epochs to compare. During this process of comparison, the user may order a new drug and dose, or the user may select a drug and then modify its parameters, discontinue (stop/cancel or hold), renew, or refill it. The main goal for 2.8 *Reconcile Meds* is to enable the comparison of drugs and doses from two (or more) settings: locations and time frames.

Isolating 2.8 *Reconcile Meds* from DFD 2 is shown in Figure 4.14.

Decomposing the process of medication reconciliation further, we identify two subprocesses (Figure 4.15):

2.8.1 *Select Reconcile Settings*

← Sends *Patient ID* and the transition of care *(A/D/T)* settings to *PATIENT LOCATION* table
→ Receives the time frames and locations relevant to the reconciliation process
← Sends the time frames to *PATIENT MEDICATION*
→ Receives the lists of drugs relevant to the time frames

The 2.8.2 *Compare Drugs from Multiple Settings* consolidates the results from 2.8.1 and presents them to the user for comparison (see the chapter on the user interface).

If a patient, for example, is to be discharged home, the user selects *Discharge* as the transition of care; the system queries the *PATIENT LOCATION* table

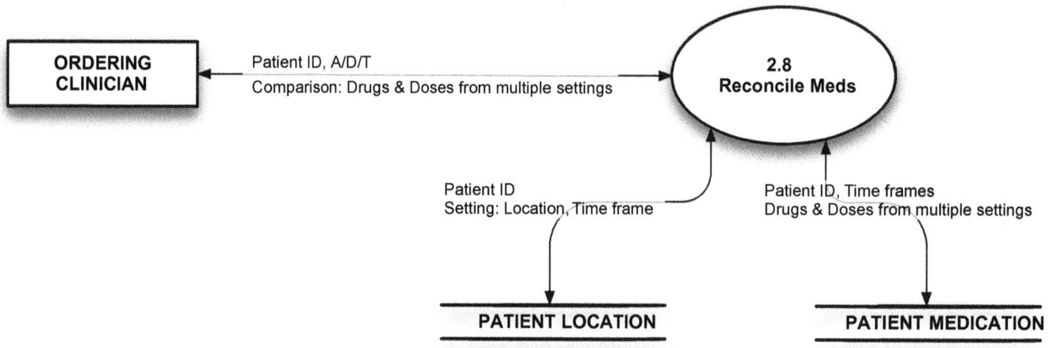

Figure 4.14 Reconcile meds: Isolated from DFD 2.

and retrieves the time frame the patient was in the hospital versus at home. Using these time frames, the system then retrieves the following from *PATIENT MEDICATION* and presents it to the user:

1. A list of drugs and doses patient was taking at *home preadmission* (which may have been modified when patient was admitted)
2. Drugs and doses administered in the *hospital*
3. A new list of drugs and doses for patient to take *home after discharge*

The user compares as a starting point two lists (1 and 2) and creates a third, new list of drugs for the next setting (3). The selection and modification of drug parameters in list 3 is done using the previously mentioned processes 2.3 *Select Drug* and 2.4 *Select Dose.*

We have finished the analysis of the *Order* workflow from the perspective of the functional processes: data in motion. Next, we focus on the data at rest, the databases and the conceptual and logical models.

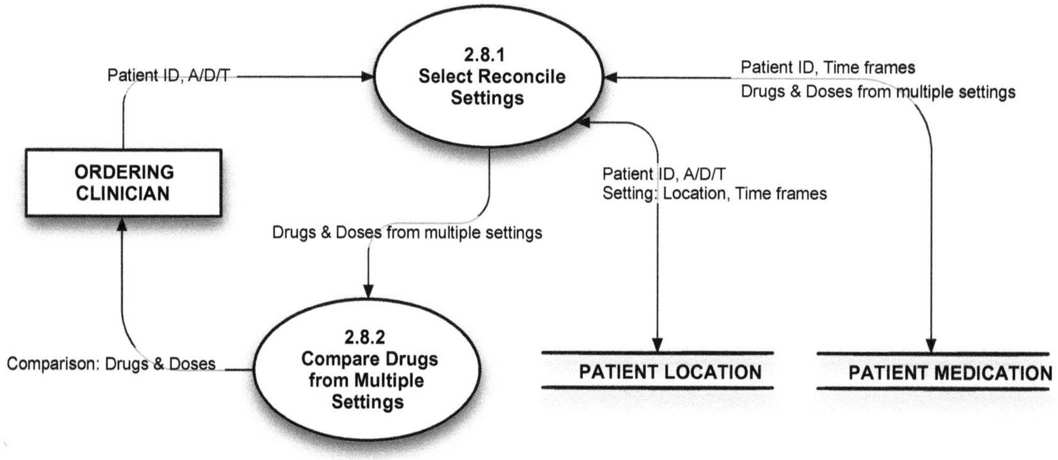

Figure 4.15 DFD 2.8: Reconcile meds.

Data Elements

In this section, we continue to develop the conceptual and logical model we have worked on in the previous chapter using Figures 3.17–3.20. There is no need to restate the discrete data elements that are common to both *Prescribe* and *Order* workflows, such as pack, generic name, ingredient, form, strength, recommended dose, and so on. Nor do we need to redefine the patient and clinician attributes or their respective tables and relationships. We focus only on those data structures that either are new to the *Order* workflow or need to be changed following our functional analysis of the processes of *Order*.

Repeating a table in a logical model—such as one *PATIENT MEDICATION* table for *Prescribe* and another *PATIENT MEDICATION* table for *Order*—is considered bad modeling practice, a sure path to redundancy, and redundancy breeds inconsistency and other data anomalies.

Thus, we use the logical model developed for the *Prescribe* workflow as a basis for further analysis related to the data structures needed by the *Order* workflow. In a similar fashion to the data analysis of *Prescribe*, we have to decide which data elements are inherent to our data model and thus belong to the system database versus the data elements that are part of the information exchange with external actors. For the latter, we are not supposed to model the entities and relationships since by definition it is beyond our control; the external actors are just that—external to the medication system—and we have no influence over their database model or design. Still, we need to identify the data elements the medications system needs to exchange with these external actors.

Analyzing the *Order* dataflows reveals the following new or changed data structures when compared to *Prescribe*:

DFD 2.1	Multiple destinations for *Drug Communication*
DFD 2.2	Multiple *Physical Measurements* *BMI/BSA (Body Mass Index/Body Surface Area)* List of *Order Sets* matching patient's parameters
DFD 2.3	Conditional and sequential complex orders *Drug from Order Set*
DFD 2.4	Dose from *Order Set*
DFD 2.5	Formulary data elements—no change
DFD 2.6	Need for cosign *Destinations by Order* type
DFD 2.7	Selection of an *Order Set* (matching patient parameters)
DFD 2.8	Patient *A/D/T* settings: location and time frames

CDS and *Formulary* data elements have been defined previously in the chapter on *Prescribe/eRx*. Comparing the *Order* workflow with the *Prescribe* workflow, there are no significant changes as far as *CDS* and *Formulary* data structures are concerned, so this information is not repeated here.

Which data structures are within the medications logical model scope, and which are related to the external actors information exchange?

Internal Data Store

- Multiple destinations for one *Drug Communication*
- Multiple *Physical Measurements*
- *BMI/BSA*
- Conditional and sequential complex orders
- Patient *A/D/T* settings: location and time frames

External Actors

- List of *Order Sets* matching patient's parameters
- Selection of an *Order Set*
- Drug from *Order Set*
- Dose from *Order Set*
- Need for cosign
- Destinations for *Order* type

Multiple Destinations for One Drug Communication

The previously defined logical model (Figure 3.20) allows a communication to be sent to only one destination. This model assumption is wrong for a prescription and even more so for an order. We need the system to be able to send a drug communication to multiple destinations. This requirement should translate into an entity *DESTINATION*, which lists the possible destinations for a drug-related communication, be it a prescription, order, dispensation, or administration of a drug. In addition, one destination (such as *PHARMACY* or *Dr. A*) may receive one or more communications sent to them. Therefore, between *DRUG COMMUNICATION* and *DESTINATION* entities there is a many-to-many (*M:N*) relationship, which necessitates an intersection table in between: *DRUG COMMUNICATION DESTINATION*. Let us change the logical model to accommodate this requirement, as shown in Figure 4.16.

Multiple Physical Measurements

A patient's weight and height are obviously expected to change during the patient's lifetime; thus, the system needs to store multiple values in a longitudinal electronic health record (EHR). Several processes in the medications system, such

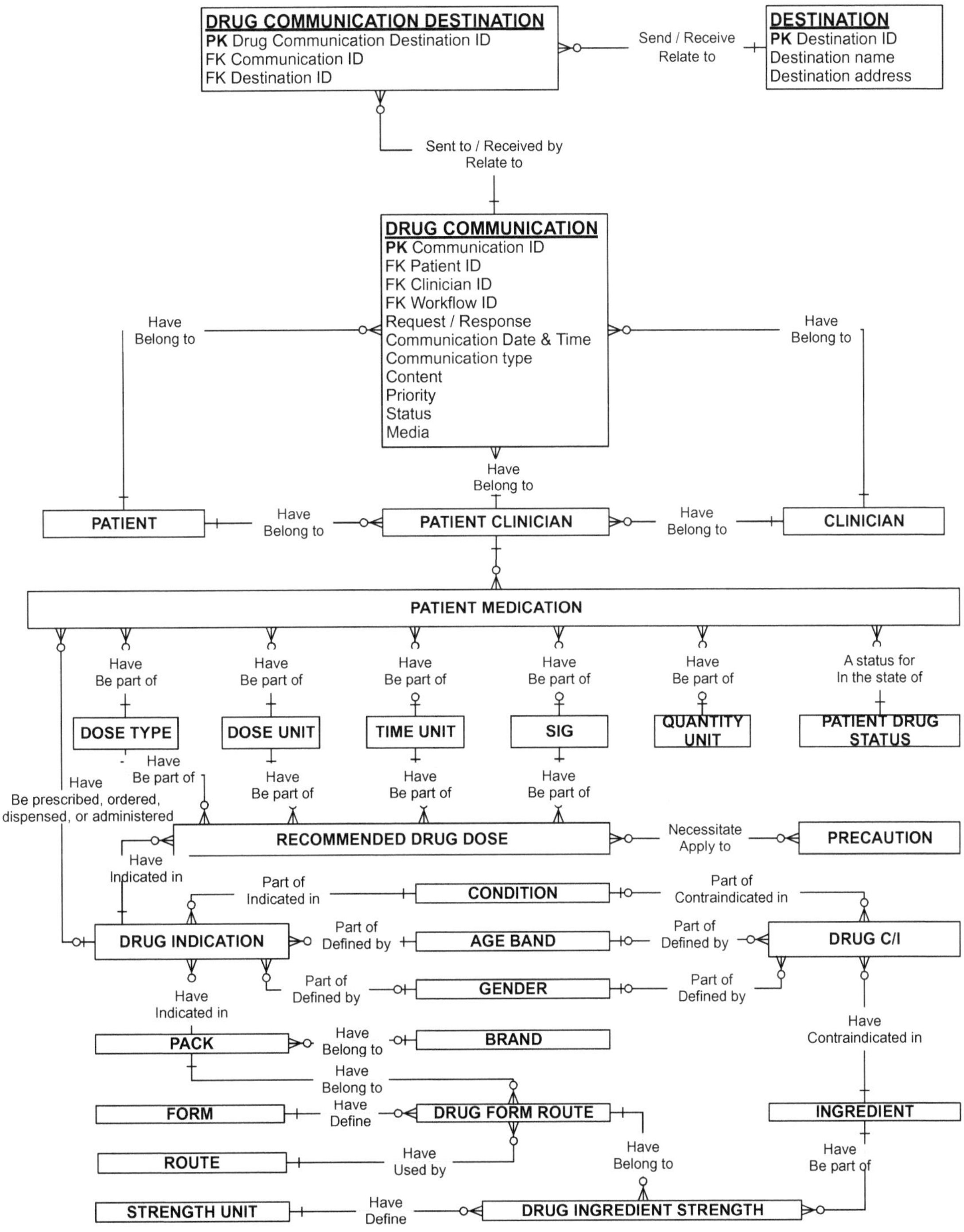

Figure 4.16 Logical model refined: Communication destinations.

as those related to dosing, may use weight and height. Still our logical model (Figure 3.20) allows only one value to be stored. This error in the model means we cannot keep a history of the weight and height values. In addition to storing multiple values for one patient, the system needs to know about a measurement:

- Date and time stamp
- Unit of measurement (*kg, pound, inch, meter*)
- Source of the information (*patient/relative/nurse/monitor*)

We need to refine the model to allow historical data: One *PATIENT* may have multiple instances in *PHYSICAL MEASUREMENT.* A *PHYSICAL MEASUREMENT* type (*weight, height*) may be recorded for many *PATIENTs,* and it should have a unit of measurement and a source attached. Thus, between PATIENT and PHYSICAL MEASUREMENT there exists a *M:N* relationship that needs a resolution with an intersection table. The logical model adapted to this requirement is shown in Figure 4.17.

The refined model, with the new tables of *PHYSICAL MEASUREMENT, PHYSICAL MEASUREMENT SOURCE,* and *PATIENT MEASUREMENT,* can be used for storing the patients' vital signs as well. Blood pressure, heart rate, oxygen saturation, and other vital signs change at a much faster pace than weight or height. Heart rate literally changes with each and every heartbeat, but for most clinical purposes, even in an intensive care environment, systems will average the heart rate over 15–60 seconds before presenting and storing it. Just because the rectified data model allows the storage of vital signs does not necessarily mean the *Medications* system database should actually store this vast amount of information. A prudent design decision may restrict access only to slowly changing dimensions, such as weight and height, to this table and leave the high-volume vital sign storage responsibility to others, such as the *NON-DRUG REPOSITORY.*

BMI and BSA

Calculated values such as *BMI* or *BSA* have the weight and height parameters in their formulae:

$$BMI = \text{Weight in kilograms}/(\text{Height in meters})^2$$

BSA: There are several applicable formulae (11).

BMI and *BSA* are sometimes used for dosing calculations, such as for chemotherapeutic agents. Why aren't *BMI* and *BSA* represented in the logical model? The reason *BMI* and *BSA* are not included in the database is that calculated values, as a rule, should not be stored since they may become out of sync and redundant and thus a source of data anomalies. *BMI* and *BSA* should be calculated on the fly, each and every time they are needed, using the relevant values of

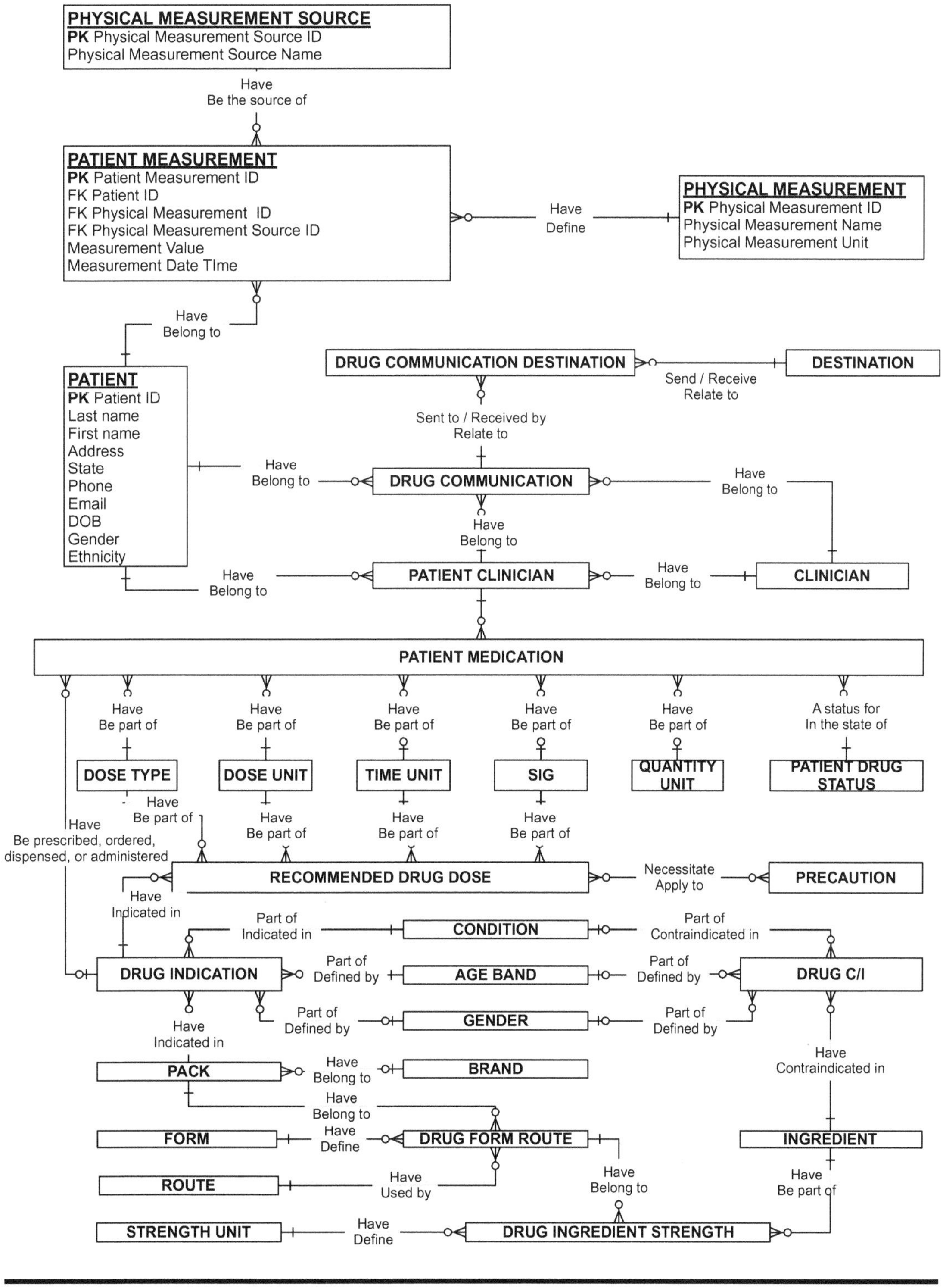

Figure 4.17 Logical model refined: Physical measurements.

weight and height from the database. The data model will store historical values of weight and height with the correct units of measure and source of information, so *BMI* and *BSA* can be recalculated, ad hoc and in the correct historical context.

Conditional and Sequential Complex Orders

Orders can be quite complex: conditional, sequential, with different flavors of SIG (instructions on how to take/administer the drug).

Conditional:

"*if* urine is less than 1 cc/kg/hour, *then* increase rate to 10 mcg/kg/min"
"*if* blood glucose is more than 400 mg%, *then* increase rate to 5 unit/hour"

Sequential:

"administer drug X 60 minutes *before* drug Y"
"administer drug Y 20 minutes *after* drug Z"
"administer drug A together *with* drug B"
"*do not* administer drug K *with* drug G"

Conditions may be nested and with multiple logic branching. Combinations of conditional, nested, and sequential orders may exist, and it is far from trivial to translate them to software logic.

CDS and *Order Sets* are two potential sources for these types of complex orders. Once a complex order is selected, the medications part of it is processed and stored as discrete data elements in the *PATIENT MEDICATION* table.

Patient A/D/T Settings Location and Time Frames

A patient may move from one setting or location to another during time: from home admitted to the hospital, within the hospital transferred from one ward to another level of care such as an ICU, back to a different ward, and eventually discharged from the hospital to a long-term facility or back home. At each and every location, there are start/end date and time stamps defining the time frame. These discrete data elements, patient locations and time frames, as far as the medications system is concerned, are important for process 2.8 *Reconcile Meds* to retrieve information from two different time frames for the user to compare and contrast.

Since a *PATIENT* may move to and from multiple *LOCATIONs* and one *LOCATION* may have many *PATIENTs*, we have a *M:N* relationship between these two entities; therefore, there is an intersection entity in between: *PATIENT LOCATION*. The logical model redefined with this requirement in mind is shown in Figure 4.18.

Now, let us identify the data structures exchanged with the newly added, two external actors: *ORDER SET REPOSITORY* and *RULES REPOSITORY*.

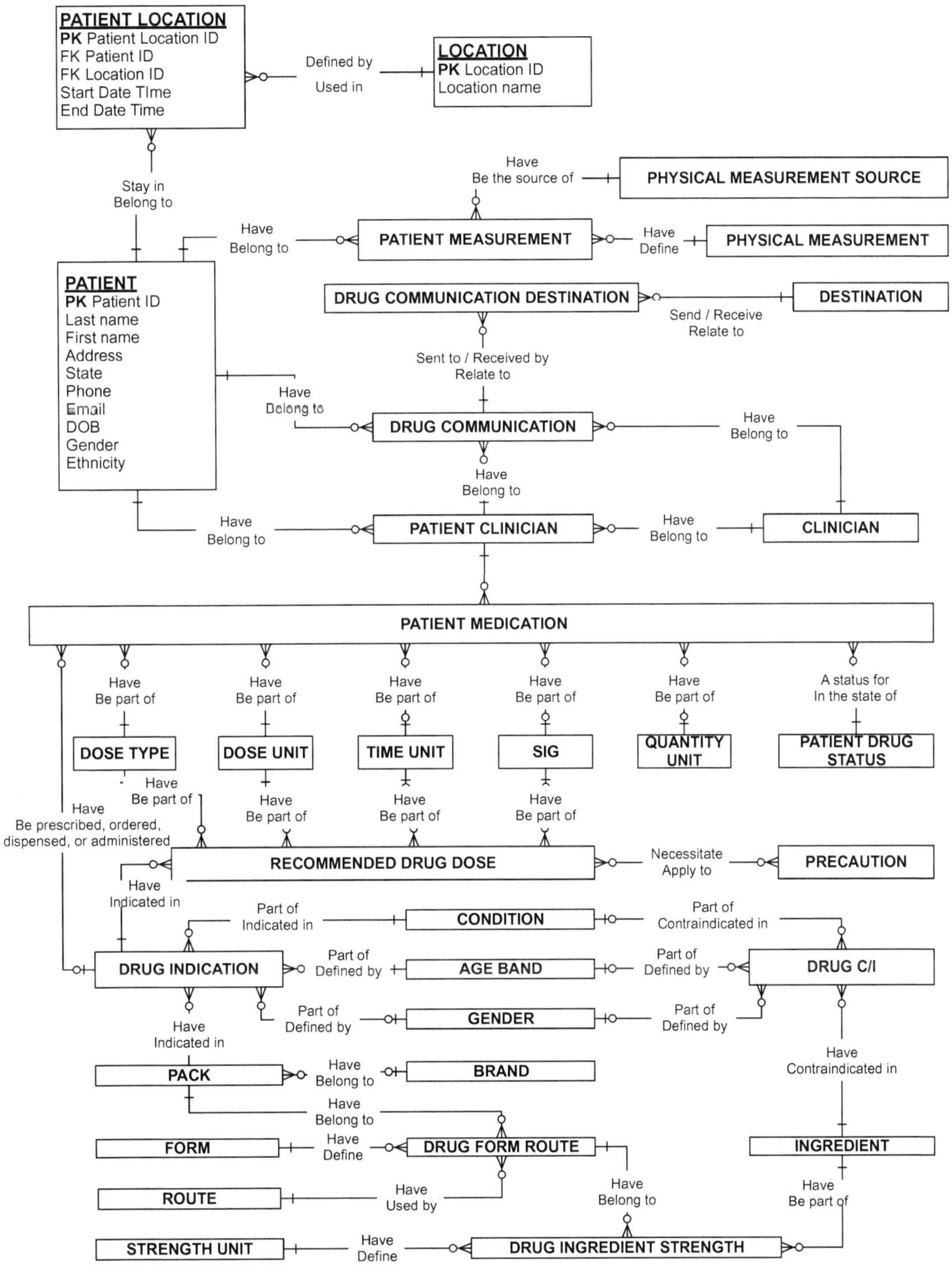

Figure 4.18 Logical model refined: Patient locations.

Data Exchange with External Actors

ORDER SET REPOSITORY is the source for all the data elements listed in Table 4.3, but our focus is on the medications domain:

← System sends patient parameters (age, gender, weight, condition, procedure, etc.)
→ *ORDER SET REPOSITORY* responds with a list of *Order Sets* matching patient's parameters
→ User selects an *Order Set*
← System presents list of drugs from an *Order Set* for user to select; list originates from the *ORDER SET REPOSITORY*
→ User selects drug from *Order Set*
← System presents a list of doses for specific drug
→ User selects dose from *Order Set*

RULES REPOSITORY: Organizations such as hospitals, after investing significant resources in a medications system, would like to enforce certain business rules, for example, the need to cosign certain orders; these rules must be created, stored, and maintained somewhere. In the *Medications* system, these responsibilities are within the scope of the *RULES REPOSITORY.*

Which drugs, ingredients, patient-related parameters, and so on are defined as "triggers" for a rule, such as *need to cosign,* to fire?

The system may be expected to send both patient and drug-related parameters to the *RULES REPOSITORY: Age Band, Gender, Weight, Condition/Indication, Drug, Dose,* and so on, and the *RULES REPOSITORY* will respond whether there is a *need for cosign.*

The destinations for orders are another list of data elements provided by the *RULES REPOSITORY.* Specific orders, defined by *Order* type (*Lab, Imaging, Meds*) may need specific routing and mailing, such as pharmacy, automatic dispensers, medication administration records, and other clinicians. Since each and every institution prefers to decide how to tweak its own rules, the *RULES REPOSITORY* is an external actor and not part of the medications system.

Summary

In this chapter, the *Order/CPOE* workflow was analyzed and compared to the *Prescribe/eRx* workflow from the perspectives of processes and data.

Similarities have been identified, and the principle of reuse of a deliverable (diagram, model, module, etc.) was introduced together with a summary of the main differences between the *Prescribe* and *Order* workflows.

Two new processes are used by the *Order* workflow and are analyzed with DFD: *Use Order Set* and *Reconcile Meds*.

The logical model went through another cycle of refinement necessitated by the dataflows in DFD 2.x, and some modeling adjustments were done to the previous ERD iteration.

References

1. Van Ornum, M. 2009 *Electronic prescribing—A safety and implementation guide*. Sudbury, MA: Jones and Bartlett.
2. Goudrey-Smith, S. 2008. *Principles of electronic prescribing*. New York: Springer-Verlag.
3. U.S. Department of Health and Human Services/Office of the National Coordinator for Health Information Technology. Medication management detailed use case. http://www.hhs.gov/healthit/documents/UseCaseMM.pdf (accessed March 8, 2011).
4. McClay, J. 2006. Structuring order sets for interoperable distribution. *AMIA Symposium Proceedings* 549–553.
5. Kangarlu, S. 2010. *Hospitalist admission order sets: CPOE complete*. Bloomington, IN: AuthorHouse.
6. NHIN trial implementations. Medication management use case requirements document V(2).
7. U.S. Department of Health and Human Services/Agency for Healthcare Research and Quality. Medication reconciliation. http://www.psnet.ahrq.gov/primer.aspx?primerID=1 (accessed March 8, 2011).
8. Al-Ubaydli, M., M. van Wagenberg, and E. Johnson with artwork by S. Salvatierra. *Electronic medication reconciliation. Practices for streamlining information transfer*. 2007. Washington, DC: Advisory Board.
9. Test Procedure for §170.302 (j) Medication Reconciliation. http://healthcare.nist.gov/docs/170.302.j_%20MedicationReconciliation_v1.1.pdf (accessed March 12, 2011).
10. U.S. Department of Health and Human Services/Agency for Healthcare Research and Quality. Medications at Transitions and Clinical Handoffs (MATCH): Multi-disciplinary team approach to medication reconciliation. http://www.ahrq.gov/about/annualmtg07/0927slides/gleason/Gleason-contents.html (accessed March 9, 2011)
11. BSA formulae. http://www.halls.md/body-surface-area/refs.htm (accessed March 11, 2011).

Review Questions

1. Identify the possible sources of information for an order drug and dose.
2. Which workflows or processes may alter, filter, or categorize the information on drugs and doses included in an order?
3. Which processes from *Prescribe* workflow are good candidates for reuse in *Order* workflow?
4. What are the main differences between *Prescribe* and *Order* workflows?
5. During *Order* workflow: Who is the source of *Formulary* information? Which process is responsible for retrieving the *Formulary* information?
6. What are the main categories of information in a typical *Order Set*?
7. What are the benefits of using *Order Sets*?
8. In Figure 4.16, a communication on a drug may be communicated to many destinations, but all with the same priority level. How would you change the model to accept for each destination a possible different priority? For example: *Pharmacy—High Priority, Dr. X—Medium Priority, MAR—Low priority.*
9. Calculate the number of rows in the table *PATIENT MEASUREMENT* after 1 year, assuming the following parameters: 100 patients/year, 10 vital signs parameters monitored continuously for each patient, and each value stored every 1 minute.
10. Show in the logical model the following data elements:
    Patient X, 20110403 22:39:12, Weight 2.4 kg, Source: nurse weighed patient.
    Patient Y, 20110413 10:03:09, Weight 180 pounds, Height 5 feet 6 inches, Source: patient.

Chapter 5

Dispense/ePharmacy

Water, air, and cleanness are the chief articles in my pharmacy.

Napoleon Bonaparte

Between *Prescribe* or *Order* and the actual administration of a drug is the *Dispense/ePharmacy* workflow. Thus, pharmacy has the unique opportunity to become an important safety checkpoint in the medication life cycle. The focus of this chapter is on the integration of the *Dispense* workflow with the *EHR* (electronic health record) in general and the *Medications* system specifically—so pharmacy may have all the information necessary to make the right dispensing decision.

Small, community-based as well as large hospital pharmacies need automation to dispense drugs safely and efficiently. This automation comes in many forms, and it is introduced in this chapter: robots counting tablets or preparing intravenous solutions and injections, bar code labeling, automated dispensing cabinets, and so on. Realizing that currently most pharmacies use a mix of manual and automated processes in their operations, the structured analysis can be applied to a manual as well as a highly automated pharmacy environment.

First, we analyze the *Dispense* workflow with a DFD (dataflow diagram) and then identify the data elements needed to support the *Dispense* processes.

Dispense Workflow

There are many types of pharmacies: community based, independent or chain related, mail order, hospital based, specialty oriented (e.g., for chemotherapy), and so on. While each type of pharmacy may use its own flavor of an operational model, our goal is to identify the recurring theme, the similarities of various *Dispense* workflows, used by different types of pharmacy models.

Pharmacies need to safely and efficiently procure, stock, and dispense drugs. This chapter focus is on the *Dispense* workflow: The pharmacist coordinates the medications therapy of patients, who are often treated by many clinicians.

Let us follow a drug during its life cycle and identify processes and entities that may be involved in the *Dispense* workflow (Figure 5.1).

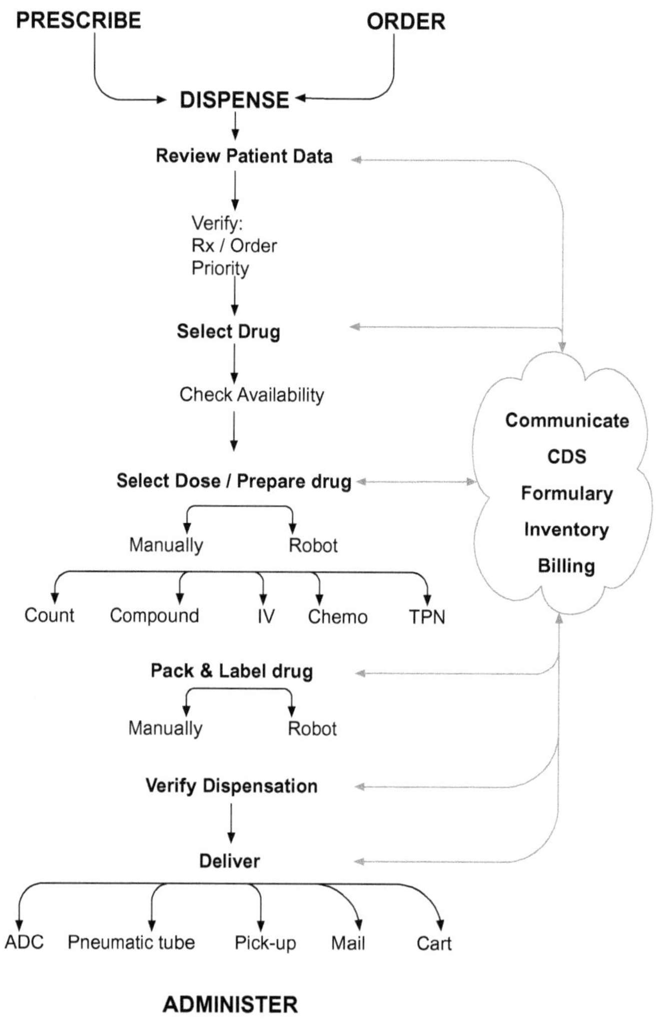

Figure 5.1 Drug life cycle: Dispense workflow.

Pharmacies have similar requirements related to the *Dispense* workflow (1,2):

- Review patient data
- Verify prescription (Rx)/order
- Check availability in stock
- Select drug and dose
- Pack and label
- Verify dispensation
- Deliver

During this workflow, the pharmacy communicates with others (prescribing clinician, patient, nurse, billing, etc.) and considers the *CDS* (clinical decision support) and formulary recommendations. Even with automated workflows and robots to help, safety risks and concerns during the *Dispense* workflow still lurk in two areas (2):

- Drug preparation
- Drug labeling

A pharmacy with millions of dispensations per year and a low error rate may still suffer from too many dispensing errors (3).

The error-prone process of manually counting tablets was the first to be automated: Pharmacy "robots" capable of counting tablets and capsules arrived in pharmacies in the 1980s (4). Later, these counting machines were connected to the inventory systems. The next step in the *ePharmacy* evolution was the automatic process of labeling medications with bar codes at the *unit dose* level. The concept of *Unit Dose* emerged as the amount of a medication to be administered as a single dose: uniquely identifying a combination of *Patient, Drug, Route, Dose,* and *Time* and connecting it, preferably via bar code, to an actual drug dispensed (5,6). The safety benefits of the uniquely identified, bar-coded drugs from the pharmacy were then extended into a *Bar Code Medication Administration* (*BCMA*) system to ensure the right patient, drug, route, dose, and time are correct during the *Dispense* as well as the *Administer* workflow. The concepts of *BCMA* and *Unit Dose* are further discussed in the next chapter.

In addition to the process of counting tablets and capsules, other areas of pharmacy automation now include the robotic preparation of *intravenous* (*IV*) solutions, injections, and *total parenteral nutrition* (*TPN*), automatic packing, bar code labeling, and drug distribution.

The *Dispense* workflow in Figure 5.1 covers most of the spectrum of pharmacy operating models: from small, independent, community based and mostly manual to large, highly automated pharmacy operations of academic medical centers or multihospital healthcare organizations.

Automated Dispensing Cabinet

Once a drug is ready to be dispensed in the pharmacy, there are two main delivery paradigms (2,3):

- Central pharmacy, where drugs are "pulled" by a courier, delivered in a cart, via pneumatic tube, FedEx, mail, and so on
- Decentralized *automated dispensing cabinets* (*ADCs*): geographically dispersed drug storage units from which one may remove drugs with the explicit permission of the pharmacy

For the drugs stored in an *ADC* nearer to the point of care, the act of dispensation becomes one of delivering a data package—the pharmacy's permission—instead of an actual delivery of a physical package of medications.

The pharmacy periodically restocks the *ADCs,* which function as "minipharmacies" located throughout a hospital, at the ward/floor level. *ADCs* may contain emergency drugs and usually have procedures in place to allow pulling these medications during an emergency with or without an order or pharmacy dispensation, to be reconciled at a later time.

The *ADC* should be regarded as a buffer between the central pharmacy and the location where medications are administered. The *ADC* offers an opportunity for the pharmacy to improve the safety of the medications life cycle since the pharmacy can intervene before the drug is available for administration. The *ADC* usually provides a faster turnaround time, a more secure method of drug distribution, better inventory management, as well as improved charge capture and billing (7,8). So, it should come as no surprise that approximately 80% of U.S. hospitals used *ADC* in 2007, and presently it is the primary method of drug delivery for a U.S. hospital pharmacy (2). The *Institute for Safe Medication Practices* (*ISMP*) provides an interdisciplinary approach to the implementation and use of *ADCs* (9).

In this book, *ADC* is represented by the generic *DRUG STORAGE* actor; it appears also in the *Administer* workflow in the next chapter as the location where a nurse pulls medications in a hospital environment.

The features of a modern pharmacy are (6,10)

- Automated *Unit Dose* preparation and packaging
- Automated bar code labeling of *Unit Doses* as an integral part of a *BCMA* system
- *ADCs*

Let us analyze the *Dispense* workflow as it integrates with the *EHR* in general and the *Medications* system specifically.

Processes

DFD 3 Dispense Workflow

Before the *Dispense* workflow may proceed, a set of preconditions has to be fulfilled:

1. Clinician is uniquely identified.
2. Clinician is authorized to use the system in lieu of dispensing medications.
3. Patient is uniquely identified.
4. All system activities are logged.

The *Dispense* workflow isolated from DFD 0: *Medications* system (Figure 2.3) is shown in Figure 5.2. At this level, the only entities interacting with the *Dispense* workflow are two external actors (*PHARMACY* and *FORMULARY*), one other workflow (*CDS*), and the *CLINICAL REPOSITORY*. The top-down analysis of this workflow, along the lines sketched in Figure 5.1, reveals additional external actors, data stores and their interactions with *Dispense* (Figure 5.3).

The steps in a drug life cycle identified in Figure 5.1 are represented as pharmacy processes in DFD 3 (Figure 5.3):

- Communicate
- Review patient data
- Select drug
- Select dose/prepare (including pack and label)
- Consider formulary
- Dispense/deliver

Let us continue the functional decomposition of the 3.x processes.

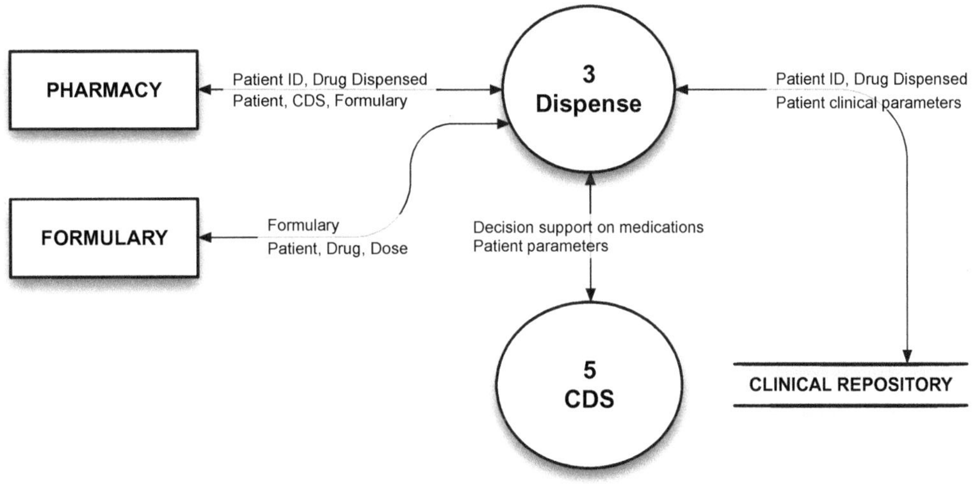

Figure 5.2 Dispense workflow: Isolated from DFD 0.

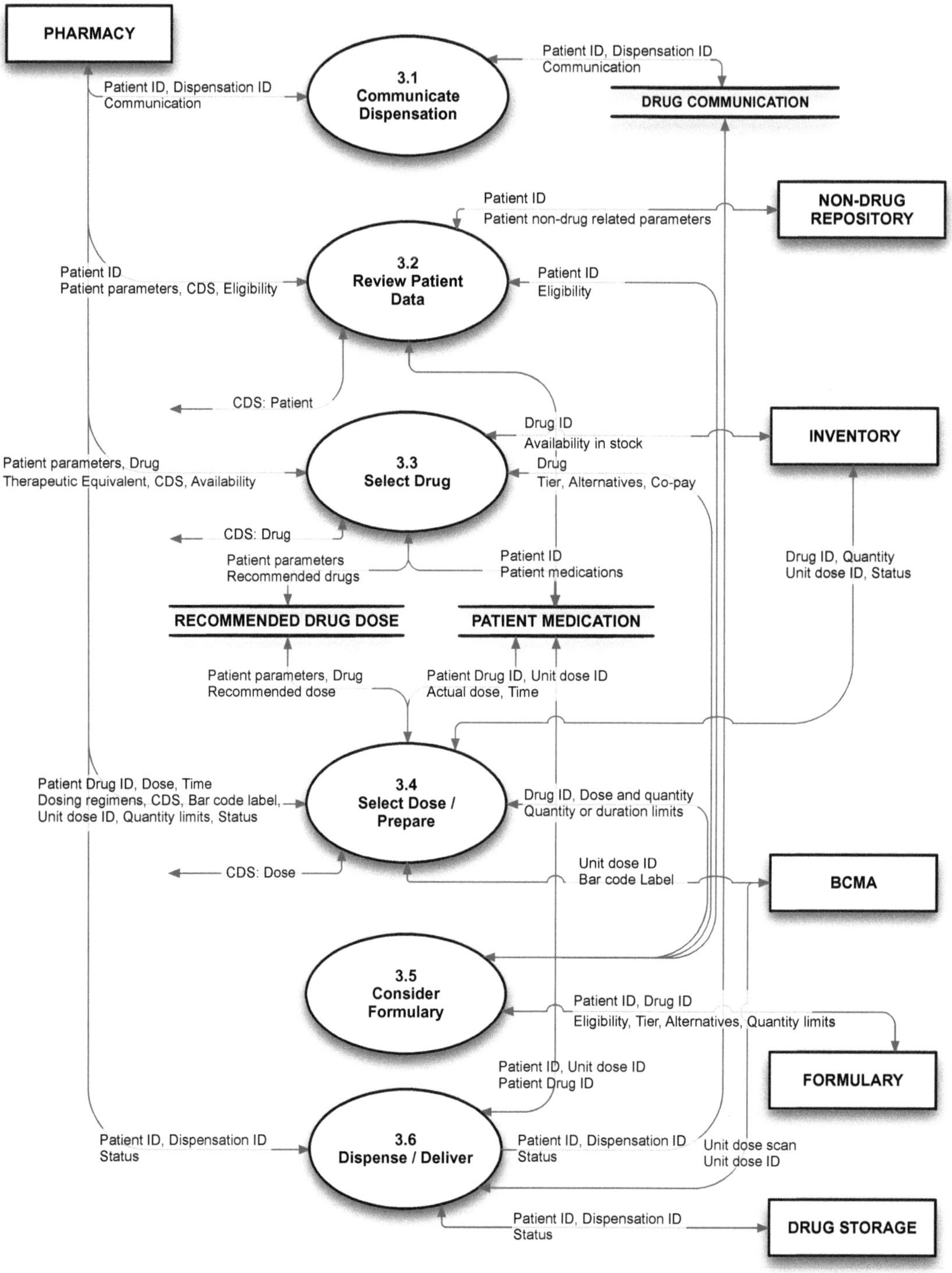

Figure 5.3 DFD 3: Dispense workflow.

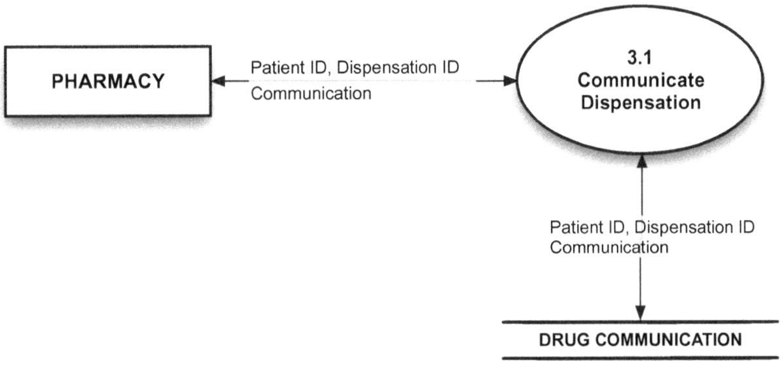

Figure 5.4 Communicate Dispensation process: Isolated from DFD 3.

DFD 3.1 Communicate Dispensation

The pharmacy is usually communicating with many entities: prescribing/ordering physicians, administering nurses, *PBM* (pharmacy benefit management), billing, and so on. These actors' communication processes, in the form of request/response or read/write, either were defined in other chapters (Figures 3.3, 3.4, 4.4, 4.5, 6.4, 6.5) or are not within the *Medications* system boundaries. Thus, *PHARMACY* appears as the only actor in DFD 3.1 *Communicate dispensation* (Figure 5.4).

PHARMACY receives communications from (read) and replies (write) to a data store introduced previously: *DRUG COMMUNICATION*. It is the same database table the other actors are using for their communication needs in DFD 1.1 during *Prescribe*, in DFD 2.1 during *Order*, and in DFD 4.1 during *Administer*. These DFDs can be reused with minor modifications as DFD 3.1.

DFD 3.2 Review Patient Data

To decide on the safest and most efficient dispensation, *PHARMACY* needs information about the patient. The sources for this information are the *PATIENT MEDICATION* table and the *NON-DRUG REPOSITORY*, both introduced previously. In addition to the patient clinical parameters (age, conditions, allergies, other drugs, lab results, etc.), *PHARMACY* may need to consider *CDS* and *FORMULARY* recommendations. *CDS*, as a higher workflow, is not shown on a lower-level DFD such as 3.2 *Review patient data*; *FORMULARY* is an external actor (Figure 5.5).

Process 3.2:

→ Receives Patient ID
← Sends Patient ID to an external actor *NON-DRUG REPOSITORY*
→ Receives clinical parameters, not related to medications, such as diagnosis, lab results
← Sends Patient ID to the *PATIENT MEDICATION* database table

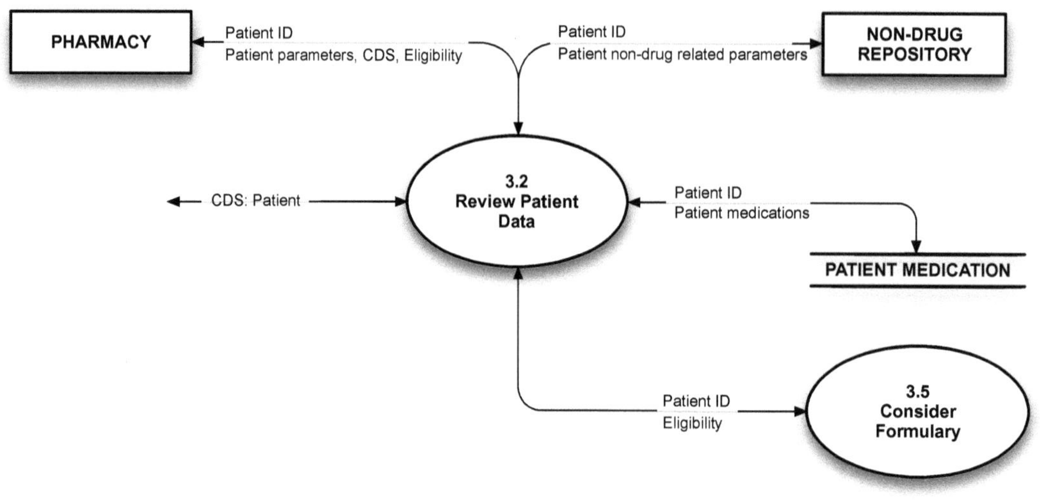

Figure 5.5 Review Patient Data process: Isolated from DFD 3.

→ Receives information on the medications prescribed, ordered, dispensed. and administered to patient
← Sends patient parameters to *CDS*
→ Receives *CDS* recommendations
← Sends patient parameters to *FORMULARY*
→ Receives information on patient's eligibility
← Consolidates this information and presents information to user

The *PHARMACY* needs for relevant clinical information during process 3.2 *Review Patient Data* are similar to the needs of other clinicians as depicted in Figures 3.5, 3.6, 4.6, 6.6, and 6.7.

DFDs 1.2, 2.2, and 4.2 can thus be reused for DFD 3.2. Minor modifications to be expected are the need for *PHARMACY* to check priority and proper documentation of an order or verify eligibility with *PBM* (an external actor) when dispensing a prescription out of the hospital.

DFD 3.3 Select Drug

Isolating process 3.3 *Select Drug* from DFD 3 is shown in Figure 5.6. Similar to other clinicians, *PHARMACY* receives information from a couple of sources and, using *CDS* and *FORMULARY,* decides on the drug to be dispensed. *PHARMACY* may dispense exactly the drug prescribed/ordered (dispense as written, DAW), or it may choose a substitute, a therapeutic alternative, because of safety, formulary, or availability issues.

Process 5.3:

→ Receives patient parameters
← Sends patient parameters to *PATIENT MEDICATION*

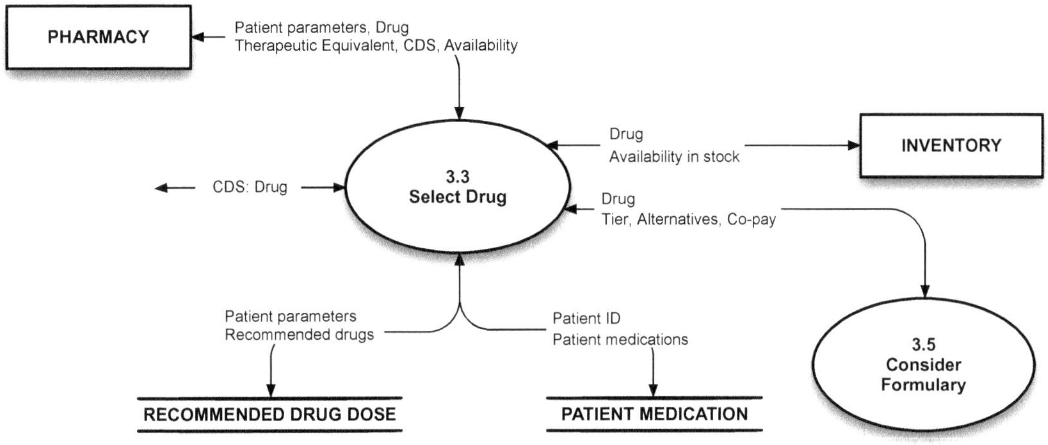

Figure 5.6 Select Drug process: Isolated from DFD 3.

→ Receives details on patient medications
← Sends patient parameters to *RECOMMENDED DRUG DOSE*
→ Receives recommendations on drugs
← Sends patient parameters to *CDS*
→ Receives *CDS* recommendations
← Sends patient parameters to *FORMULARY*
→ Receives *FORMULARY* considerations related to the drugs patient is already taking

At this point, the user may select a new (for the patient) drug based on an indication, a condition, or a disease or by searching for a drug by ingredient, class, and so on. Alternatively, the user may opt to select an existing drug—a drug the patient is already taking. Then, the user may stop, put on hold, or modify the dose, strength, form, and so on of an existing drug.

Before continuing with the dispensation, checking the drug availability in stock would be a good idea:

→ Receives the *Drug ID* user has selected
← Sends *Drug ID* to *INVENTORY*
→ Receives availability from *INVENTORY*

Once a specific drug is selected and it is available in stock, process 1.3 interacts again with *CDS* and *FORMULARY*:

← Sends selected *Drug ID* to *CDS*
→ Receives *CDS* recommendations specific to *Drug ID*
← Sends *Drug ID* to *FORMULARY*
→ Receives *FORMULARY* information regarding the drug tier, therapeutic alternatives, and (if in outpatient settings) the copay for the patient for the drug selected

The processes related to the selection of a drug have been detailed in DFD 1.3 as part of *Prescribe*, DFD 2.3 *Order* and DFD 4.3 *Administer* (see Figures 3.7–3.10, 4.7, 6.9), and with some minor changes these DFD can be reused as DFD 3.3.

DFD 3.4 Select Dose/Prepare

Once a drug is selected, *PHARMACY* can proceed with the dose selection and then prepare, pack, and label the drug. During the process, *CDS* and *FORMULARY* provide recommendations (Figure 5.7).

Process 3.4 associates the unique instance of a *Patient Drug ID* in the *PATIENT MEDICATION* table with the unique instance of a *Unit Dose ID*—the medication to be dispensed accordingly. From the process perspective,

→ Receives the *Drug* selected by user
← Sends *Drug* to *PATIENT MEDICATION*: If it is a new drug that *PHARMACY* decided to substitute for the one actually *Prescribed/Ordered*, then a new instance (and a new *Patient Drug ID*) would be created.
→ If it is an existing drug (one the patient is already taking), then its *Patient Drug ID* is fetched from *PATIENT MEDICATION*.
← Sends patient and drug parameters to *CDS*.
→ Receives recommendations related to drug dosing regimens.
← Sends patient and drug parameters to *FORMULARY*.
→ Receives information on quantity or duration limitations applying to given drug.

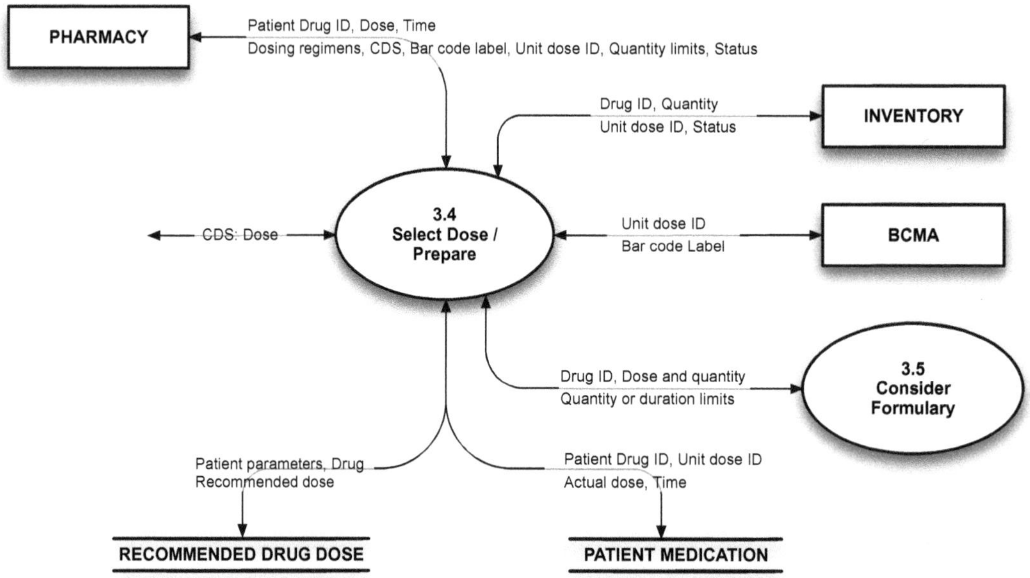

Figure 5.7 Select Dose/Prepare process: Isolated from DFD 3.

← Sends preparation details to *INVENTORY*, the production stage: manual, semiautomated, or fully automated preparation, packaging, and labeling of the drug to be dispensed.
→ Receives *Unit Dose ID* and status from *INVENTORY*.

Once the drug is ready for dispensation, it is packed and labeled/bar coded:

← Sends *Patient Drug ID* or *Unit Dose ID* to *BCMA*
→ Receives from *BCMA* the drug package bar code label as acknowledgment

The process of selecting the correct dose for a given drug is common to other workflows: DFD 1.4 *Prescribe* (Figures 3.11, 3.12) and DFD 2.4 *Order* (Figure 4.8). A similar process of selecting the dose and preparing a drug to be administered is presented in the next chapter as DFD 4.4 *Administer* (Figures 6.10, 6.11).

DFD 3.5 Consider Formulary

Using an external actor, *FORMULARY*, and depending on the settings (in hospital or community pharmacy), process 3.5 provides information on the formulary as a service to three other processes: 3.2 *Review Patient Data*, 3.3 *Select Drug*, and 3.4 *Select Dose/Prepare* (Figure 5.8).

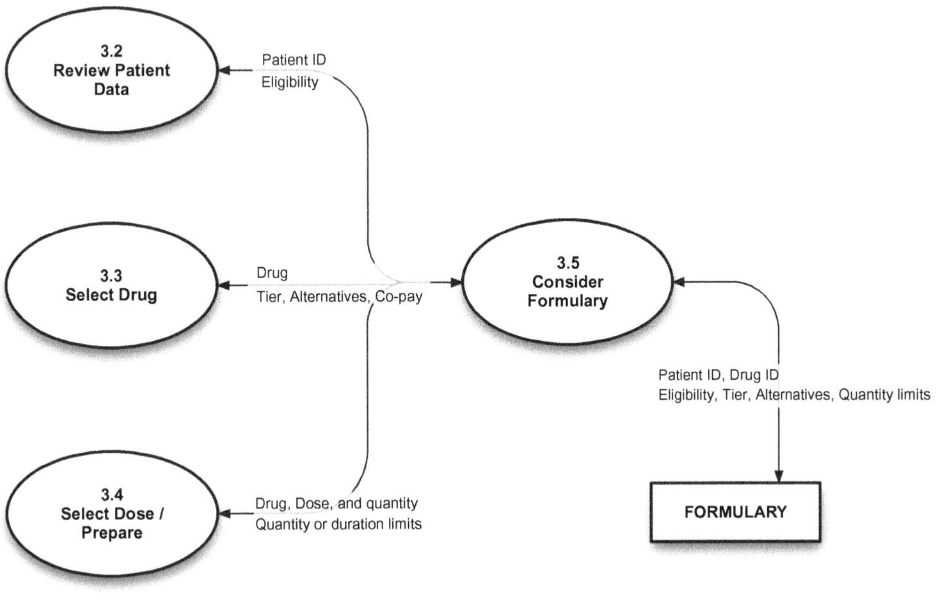

Figure 5.8 Consider Formulary process: Isolated from DFD 3.

In an outpatient setting, *PHARMACY* may check patient eligibility as part of process 3.5:

→ Receives *Patient ID*
← Sends *Patient ID* to *PBM* (represented by *FORMULARY*)
→ Receives information on patient's eligibility
← Sends information on eligibility to process 3.2

In both an in-hospital and a community-based pharmacy, once a drug is selected, process 3.5

→ Receives drug identifier from process 3.3
← Sends drug to *FORMULARY*
→ Receives information on drug formulary tier, therapeutic alternatives, and (in an out-of-hospital settings) the copay expected
← Sends information to process 3.3

Once the dose of a drug is selected, process 3.5

← Sends drug and dose/duration/quantity to *FORMULARY*
→ Receives information on quantity or duration limits
← Sends the formulary-related information on quantity or duration to process 3.4

DFD 3.6 Dispense/Deliver

Now, the medication is ready, packed, and labeled; just before delivery, *PHARMACY* verifies the prescription/order is correctly associated with the right patient, drug, route, dose, and timing, using process 3.6 *Dispense/Deliver* (Figure 5.9):

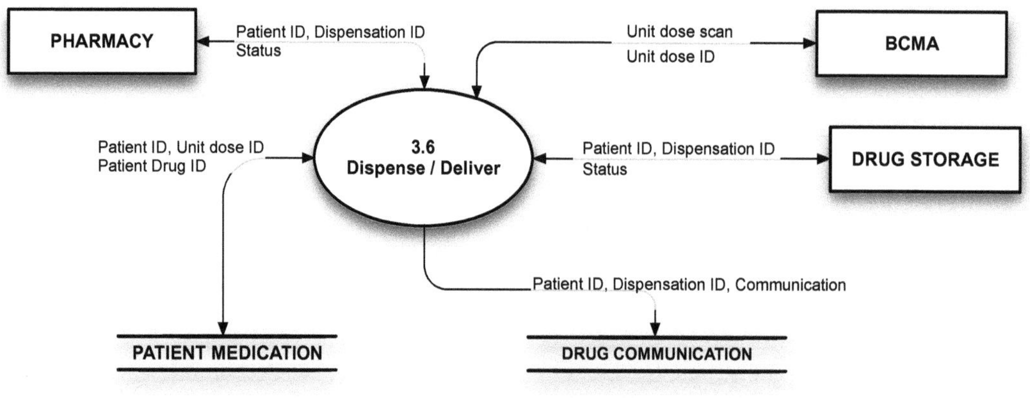

Figure 5.9 Dispense/Deliver process: Isolated from DFD 3.

Process 3.6:

→ Receives *Patient ID*
→ Receives drug label scan
← Sends scan to *BCMA*
→ Receives *Unit Dose ID*
← Sends *Patient ID* to *DRUG STORAGE*
→ Receives *Dispensation ID*, which uniquely identifies the dispensation instance

Process 3.6 then compares *Unit Dose ID* in *BCMA* with *Unit Dose ID* in *PATIENT MEDICATION*; *Patient ID*, *Dispensation ID* in *PATIENT MEDICATION* with their counterparts in *DRUG STORAGE* and, if all are correct, finalizes the workflow by changing the *Patient Drug Status* and

← Sends to *DRUG COMMUNICATION* the fact drug is available for pickup/delivery

During the last step in the *Dispense* workflow, a pharmacy may have to deal with several situations, such as partial dispensation, out-of-stock medications, multiple orders for the same patient, medications being returned or recalled, various delivery options (*ADC*, FedEx, mail, cart, pneumatic tube, etc.).

Data Elements

Analyzing the dataflows in the DFD mentioned reveals the data elements necessary to support the *Dispense* processes (Table 5.1).

The *PATIENT DRUG STATUS* is the data entity responsible for monitoring the status of a drug, its current location, and the next step planned in the workflow. The concepts of drug status as well as *Unit Dose*, *BCMA*, and *ADC/DRUG STORAGE* are further detailed in the next chapter.

Table 5.1 Data Elements Dispense Workflow

DFD 3.x Data Elements	*ERD Table*	*External Actor*
Patient ID	*PATIENT*	*BCMA* *DRUG STORAGE*
Clinician ID	*CLINICIAN*	
Dispensation ID	*PATIENT CLINICIAN*	
Communication	*DRUG COMMUNICATION*	*PHARMACY*
Patient parameters	*PATIENT MEDICATION*	*NON-DRUG REPOSITORY*

(continued)

Table 5.1 Data Elements Dispense Workflow (Continued)

DFD 3.x Data Elements	*ERD Table*	*External Actor*
Patient medications	*PATIENT MEDICATION* *PATIENT DRUG STATUS* *DRUG INDICATION* *PATIENT CLINICIAN* *DOSE TYPE* *DOSE UNIT* *TIME UNIT* *QUANTITY UNIT* *SIG*	*CDS* *BCMA* *FORMULARY*
Drug	*BRAND* *PACK* *INGREDIENT* *DRUG INGREDIENT STRENGTH* *FORM* *ROUTE* *DRUG FORM ROUTE* *DRUG INDICATION* *PATIENT MEDICATION*	*CDS* *FORMULARY* *BCMA* *DRUG STORAGE*
Recommended drugs	*RECOMMENDED DRUG DOSE*	*CDS* *FORMULARY*
Patient Drug ID	*PATIENT MEDICATION*	
Dose	*PATIENT MEDICATION* *PATIENT DRUG STATUS* *RECOMMENDED DRUG DOSE* *DOSE TYPE* *DOSE UNIT* *TIME UNIT* *QUANTITY UNIT* *SIG*	*CDS* *FORMULARY* *BCMA*
Unit dose ID	*PATIENT MEDICATION*	*BCMA* *DRUG STORAGE*
Eligibility		*FORMULARY (PBM)*
Tier		*FORMULARY*
Therapeutic alternatives		*FORMULARY*
Quantity limitations		*FORMULARY*

Summary

Located between the *Prescribe/Order* and the actual drug *Administer* workflows, the *Dispense* workflow has the unique opportunity to become an important layer of safety in a medication's life cycle. To dispense medications safely, a pharmacy needs information about the patient's conditions, allergies, other drugs taken, lab values, and so on. The role of the *Dispense* workflow is to provide the pharmacy with this information, *CDS* and formulary considerations on eligibility, therapeutic alternatives, and limitations on duration or amount of medication dispensed.

To dispense medications efficiently, a modern pharmacy needs automation; robots can help with counting tablets, preparing injectables, packing, labeling, bar coding, and dispensing medications. The integration of these technologies with the *EHR* in general and the *Medications* system specifically was introduced in this chapter together with the data elements needed to support the *Dispense* workflow.

References

1. Lee, J. 2010. Hospital information systems. In *Pharmacy informatics,* ed. P.O. Anderson, S.M. McGuinness, and P.E. Bourne, 77–93. Boca Raton, FL: CRC Press.
2. Boggie, D.T., J.J. Howard, and A.I. Simonian, A.I. 2010. Pharmacy information systems. In *Pharmacy informatics,* ed. P.O. Anderson, S.M. McGuinness, and P.E. Bourne, 93–107. Boca Raton, FL: CRC Press.
3. Cina, J.L., T.K. Gandhi, W. Churchill, et al. 2006. How many hospital pharmacy medication dispensing errors go undetected? *Joint Commission Journal on Quality and Patient Safety* 32(2): 73–80.
4. Wikipedia. Pharmacy automation. http://en.wikipedia.org/wiki/Pharmacy_automation (accessed August 15, 2011).
5. American Society of Hospital Pharmacies. 1989. ASHP Statement on Unit Dose Drug Distribution. http://www.ashp.org/DocLibrary/BestPractices/DistribStUnitDose.aspx (accessed August 14, 2011).
6. Murray, M.D., and K.J. Shojania. 2001. Unit-dose drug distribution systems. In *Making health care safer: A critical analysis of patient safety practices,* Evidence Report/Technology Assessment, No. 43. http://archive.ahrq.gov/clinic/ptsafety/chap10.htm (accessed August 15, 2011).
7. Pedersen, C.A., P.I. Schneider, and D.J. Scheckelhoff. 2009. ASHP National Survey of Pharmacy Practice in Hospital Settings: Dispensing and administration—2008. *American Journal of Health-System Pharmacy* 66: 926–946.
8. Franklin, B.D., K. O'Grady, P. Donyai, et al. 2007. The impact of a closed-loop electronic prescribing and administration system on prescribing errors, administration errors and staff time: A before-and-after study. *Quality and Safety in Healthcare* 16: 279–284.

9. Institute for Safe Medication Practices. Institute for Safe Medication Practices (ISMP) guidance on the interdisciplinary safe use of automated dispensing cabinets. http://www.ismp.org/tools/guidelines/ADC/default.asp (accessed August 14, 2011).
10. Murray, M.D. 2001. Automated medication dispensing devices. In *Making health care safer: A critical analysis of patient safety practices,* Evidence Report/ Technology Assessment, No. 43. http://archive.ahrq.gov/clinic/ptsafety/chap11.htm (accessed August 16, 2011).

Review Questions

1. Sketch a DFD with the *Patient, Prescribing, Ordering, Dispensing,* and *Administering Clinicians* communication processes
2. *Billing* as an entity is not shown in DFD 3. Which entity may provide information on drugs dispensed for *Billing* to capture charges?
3. The pharmacist needs to review the patient's renal and hepatic functions as reflected in the values of a laboratory analysis. Which entity is the source for this information?
4. Where does a pharmacy receive information on a patient's eligibility?
5. A medication is out of stock. Which process checks drug availability? Which entity provides this information?
6. A therapeutic equivalent is being considered. Which process recommends equivalent drugs? Which entity provides the information?
7. A robot is loaded with the wrong drug. Later, the robot dispenses the drug and labels it with a bar code. Which pharmacy process may prevent the wrong drug reach the patient?
8. Which process/entity provides information on limits on a drug quantity or treatment duration?
9. A patient's condition has recently changed. Which process will alert the pharmacist to the new condition, related contraindications, and any precautions that need to be considered?
10. A pharmacy sends a SMS (Short Message Service) text to patients' cellular phones, letting them know their prescription is ready for pickup. Which data store is the source for this information?

Chapter 6

Administer/eMAR

You take the red pill—you stay in Wonderland and I show you how deep the rabbit-hole goes.

Morpheus,
The Matrix

After *Prescribe/Order* and *Dispense,* the next workflow is *Administer/electronic Medication Administration Record* (*eMAR*).

Being the last line of defense before reaching the patient, the *Administer* workflow approach to safety is based on the *5 Rights*: A safe drug administration needs to validate the five parameters—*Patient, Drug, Dose, Route,* and *Time*—are all correct.

Using dataflow diagrams (DFDs), we decompose the *Administer* workflow into processes, making sure the *5 Rights* requirements are satisfied. During the analysis, we will identify concepts, such as *Automated Dispensing Cabinets, Bar Code Medications Administration* (*BCMA*) systems, and *Smart pumps* and their interactions with the *Medications* system.

Following the analysis of the processes, we focus on the data needed to support the *Administer* workflow: discrete data elements that are part of the evolving *Medications* ERD (entity-relationship diagram) and data elements exchanged with the actors.

A set of simple formulae clarifies the relationships between the various concepts and units of measurement that are used during the *Administer* workflow of a *Continuous drip,* such as *Volume Rate, Concentration, Dose, Drop Rate,* and so on.

The chapter ends with a high-level view of an automated process for calculating dose parameters.

5 Rights

Prescribe, Order, and *Dispense* are workflows covered in previous chapters. Following a prescription or an order and the dispensation of a drug (including over-the-counter drugs), the next workflow is *Administer.* It is the last step in a drug life cycle (Figure 6.1).

The recommendation is to adhere to the traditional *5 Rights* during the *Administer* workflow and make sure only the *Right*

1. *Patient*
2. *Drug*
3. *Dose*
4. *Route*
5. *Time*

parameters are involved in a drug administration (1). Sometimes, a sixth right is mentioned: the *Right Clinician.* We discuss several methods and systems to help make the *Administer* workflow safer, more efficient, while strictly adhering to the *5–6 Rights.*

The actual drug administration may necessitate first pulling the medication from a storage area. Next, there may be a need to further prepare the drug, such as diluting its strength (2). The *5 Rights* being addressed, a drug may be directly administered or pass through a *CDS* (clinical decision support) filter. *Bar code* and *Smart pump* systems, discussed further in this chapter, are technologies to automate the validation process of the *5 Rights.* Following actual administration

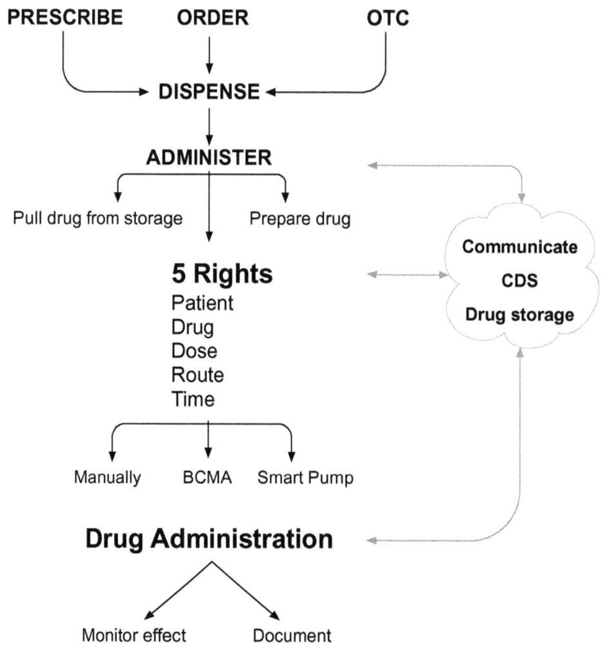

Figure 6.1 Drug life cycle: Administer workflow.

of a drug, there are two remaining tasks: *Monitor effect* and *Document* the administration (3).

In the following decomposition of the *Administer* workflow into its main processes (DFD), we have to satisfy the *5 Rights* requirements while integrating the actors and processes mentioned.

Processes

DFD 4 Administer Workflow

We first isolate workflow number 4 *Administer/eMAR* from the *Medications* system (Figure 6.2), which is represented as DFD 0 (Figure 2.3).

Besides the *ADMINISTERING CLINICIAN,* who is usually an actor in an inpatient environment, there is one other actor active in the *Administer* workflow: the *PATIENT,* who may use the *Medications* system in general and the *Administer* workflow specifically as his or her *Personal Health Record* (*PHR*), by keeping a log file up to date with the medications self-administered at home.

The preconditions to DFD 4, which must be satisfied before system may proceed with the *Administer* workflow, are as follows:

1. Clinician is uniquely identified.
2. Clinician is authorized to use system in lieu of administering medications.
3. Patient is uniquely identified.
4. Patient is authorized to use system in lieu of self-administering medications.
5. All the system activities are logged.

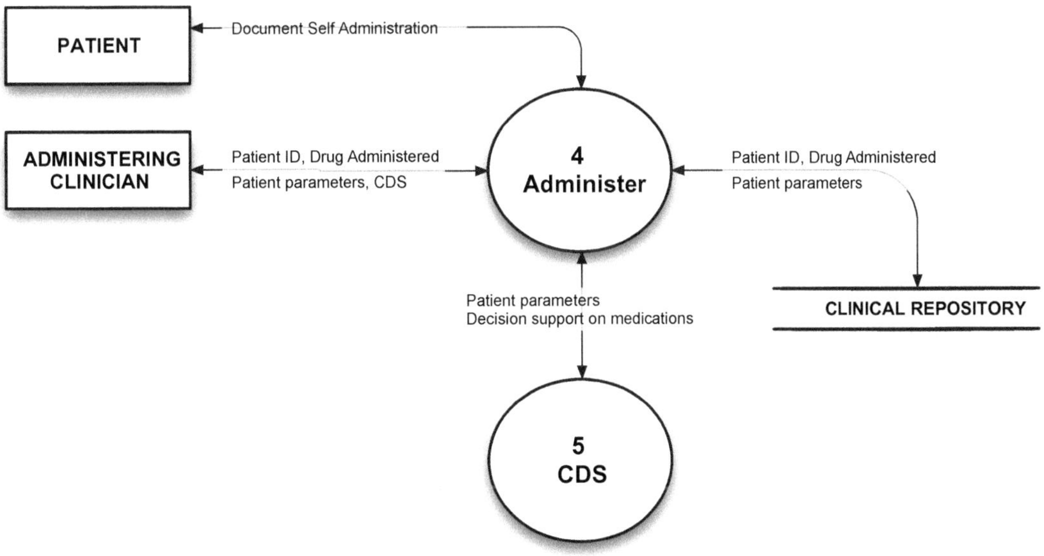

Figure 6.2 Administer workflow: Isolated from DFD 0.

Let us examine DFD 4 (Figure 6.3). The processes of the *Administer* workflow are

4.1 Communicate Administration
4.2 Review Patient Data (validating the *Right Patient,* enabling *Monitor effect*)
4.3 Select Drug (validating the *Right Drug* and the *Right Route*)
4.4 Select Dose (validating the *Right Dose* and the *Right Time*)
4.5 Interact with Storage
4.6 Administer/Sign

There are three actors that need an introduction: *BCMA, SMART PUMP,* and *DRUG STORAGE.*

Bar Code Medication Administration

Approximately a third of the errors related to medications occur at the time of administration (4). A significant reduction of errors during the drug administration process can be achieved with a *BCMA* system (5).

A *BCMA* system automates the validation of the *5 Rights* mentioned, based on bar codes, which in turn uniquely identify the *Patient, Drug, Dose, Route,* and *Time.*

Actually, a *BCMA* can validate the *6th Right* as well—the *ADMINISTERING CLINICIAN*—which may scan the clinician's badge bar code to be uniquely identified by the *BCMA* and *Medications* systems.

In addition to the potential for reducing errors, *BCMA* systems can improve inventory control, charge capture and billing, as well as save time both at the pharmacy and at the bedside. Still, only 25% of US hospitals used a *BCMA* in 2008 (6). Some of the obstacles to the widespread implementation of *BCMA* technology are the hefty price tag ($5–$10 million/hospital) and the whole flora and fauna of sociocultural change management issues. Lack of a single standard—there are more than 10 standards in the *BCMA* domain—may be another obstacle and is discussed in the chapter on standards.

The prerequisites to a successful *BCMA* implementation are as follows:

1. Integration with the *EHR/Medications* system
2. The capability to uniquely identify and bar code
   - Actors (patient via wristband, clinician via badge, pump bar code)
   - A large number of drugs at the *unit dose*

Unit Dose is the amount of a medication to be administered as a single dose, such as two tablets or one ampoule. The pharmaceutical and drug-distributing companies as well as pharmacies implement packaging and bar code systems to uniquely identify many very small items repeatedly and without errors (7).

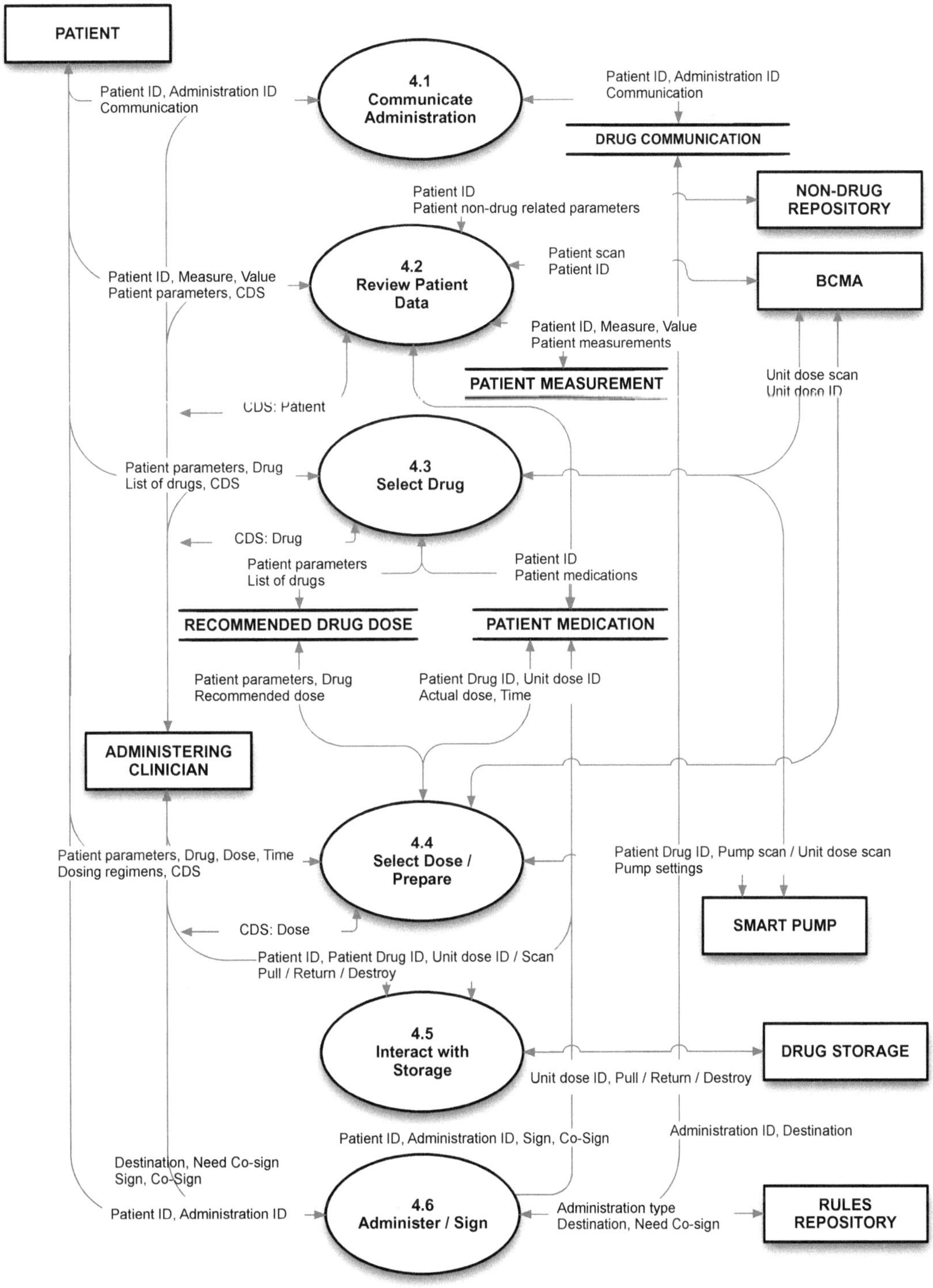

Figure 6.3 DFD 4: Administer workflow.

Once dispensed as a *Unit Dose Bar-Coded* (*UDBC*) item and just before its administration, the *BCMA* validates the *5 Rights* using a request/response data-flow detailed in this chapter.

A BCMA is an external system/actor, and since we have no control over its workflows, we are not detailing its processes, such as creating a bar code for a new patient or daily printing bar codes for thousands of *UDBC* medications.

Smart Pumps

Initially, pumps were introduced to administer medications in the form of fluids/solutions at a specified rate, usually *intravenously, epidurally,* or via a *nasogastric tube.* In a critical care environment, it is not unheard of for a single patient to be treated simultaneously with many pumps. Thus, the need for a pump to be uniquely identified is paramount, and it can be achieved with a bar code system similar to the *BCMA* mentioned. For example, scanning the *pump bar code* uniquely identifies the pump, the drug, and the current pump settings.

As shown in the following, using a couple of equations one can translate the parameters of quantity, concentration, volume, and dose into the parameter of volume rate to be programmed into the pump. The next natural step in the pump evolution was the capability to store multiple programs for many drugs and dosing regimens.

Once the old-fashioned infusion pump became a *computer-based infusion device,* it acquired the nickname of *Smart pump.* These devices now have libraries of drugs and dosing regimens as well as sets of rules and even their own *CDS* (8,9). For example, a *Smart pump* may have a *soft stop*: The pump alerts the user on exceeding the maximal recommended dose to be administered but allows an override or a *hard stop*, by which the pump refuses to administer the dose, and it cannot be overridden.

In my opinion, the IQ of a *Smart pump* should not be measured by the size of the drug library it has or the quality of the *CDS* it can provide. A really smart *Smart pump* should be able to communicate bidirectionally with the *EHR/Medications* system: receive details of an order and send information to the *MAR* (*Medication Administration Record*) chart. Thus, a truly *Smart pump* needs to accommodate two main scenarios:

→ Receive drug administration details

The pump accepts programming directly from the *EHR/Medications* system. A valid order can and should be transmitted directly from the *EHR* into the pump. This will eliminate the need to manually program the pump, a process notoriously prone to errors (9).

It will also eliminate the need to maintain a vast library of drugs, doses, and rules in two locations—*Medications/CDS* and *Smart pumps.* Creating and maintaining redundant data stores and application logic is time consuming, costly and usually breeds inconsistency.

If parameters for an order cannot be downloaded from the *EHR/Medications* system into the pump, a relatively *Smart pump* will at least be able to participate in the process of validating the *5 Rights* by comparing the programming parameters of the pump to those of the *EHR/Medications* system.

← Send pump settings to *MAR* chart

As information is flowing from the *EHR/Medications* system into the pump, it is only natural to expect a *Smart pump* to provide information back in the form of regular updates on administration data elements and contribute to the *MAR* chart. For example, a *Smart pump* should be able to send to the *MAR* once per minute the drug, strength, and dose rate being administered. This will automate the process of charting drug administration, which is traditionally the nurse's responsibility. If the *EHR* also collects the patient's vital signs, such as blood pressure, temperature, heart rate, and so on, then together with the drugs being administered it can automate a major part of the charting process of the nurse. Having information about *what the patient is doing* and *what is being done to the patient* helps in turn create a comprehensive user interface (see Chapter 7).

Drug Storage

This entity, introduced in the previous chapter, is actually a medications buffer between the pharmacy and the ward or unit. *DRUG STORAGE* keeps a 24-hour supply of drugs for the patients on the roster.

If an *Automated Dispensed Cabinet* (*ADC*) or *BCMA* is employed, then the pharmacist may provide an additional layer of safety to the process by reviewing the *Unit Doses* and the *Patient, Drug, Dose, Route,* and *Time* parameters before dispensing the drugs. Automated decentralized drug storage may help reduce errors, turnover time, and costs; it is discussed in Chapter 5.

DFD 4.1 Communicate Administration

Isolated from DFD 4, the process 4.1 *Communicate Administration* serves two actors (Figure 6.4). *PATIENT* may communicate self-administration and the effects of a medication. An *ADMINISTERING CLINICIAN,* usually a nurse, may request *PHARMACY* to restock the ward drug storage with a particular drug, reschedule the infusion administration for a certain patient, and so on. The communications are read from/written to a data store introduced previously: *DRUG COMMUNICATION* (Figure 6.5).

There are several actors apparently missing from Figures 6.4 and 6.5: *PHARMACY* is not shown as a participating actor on this diagram, even

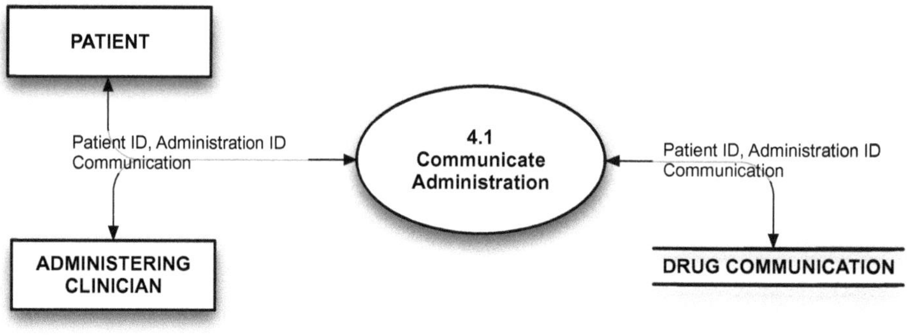

Figure 6.4 Communicate administration process: Isolated from DFD 4.

though it has to communicate with other actors participating in the *Administer* workflow. The *PHARMACY* communications process was detailed in Chapter 5 as DFD 3.1 (Figure 5.4). In a similar fashion, the *PRESCRIBING CLINICIAN* and *ORDERING CLINICIAN* do not appear in DFD 4.1 since their communications with the other actors are detailed in DFD 1.1 (Figures 3.3 and 3.4) and DFD 2.1 (Figure 4.4), respectively. Figure QA (in the Answers to Review Questions) Section 5.1 shows all actors communicating.

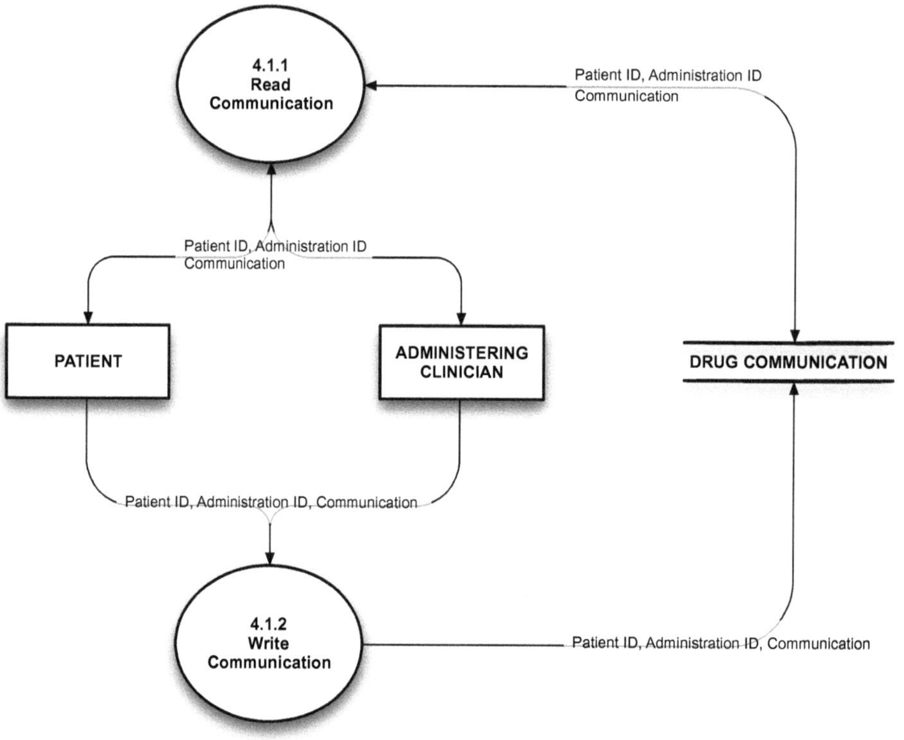

Figure 6.5 DFD 4.1: Communicate administration.

DFD 4.2 Review Patient Data

Isolating process 4.2 *Review Patient Data* from DFD 4, we can identify similarities with previous workflows as well as new actors and processes (Figure 6.6): Similar to the *Prescribe, Order, Dispense* workflow, the *Administer* workflow needs to retrieve data from both drug-related tables (*PATIENT MEDICATION*) and non-drug-related external actors (*NON-DRUG REPOSITORY*), consider *CDS*, and create a comprehensive clinical summary of the patient and present it to the *PATIENT* or the *ADMINISTERING CLINICIAN*. During this process, *BCMA* may be used for validating the *Right Patient*.

PATIENT MEASUREMENT is a table introduced in Chapter 4 and used by process 4.2 to retrieve and store discrete data elements, documented as part of DFD 4.2.5 *Monitor effect*.

Let us continue the analysis of process 4.2 *Review Patient Data* (Figure 6.7). The first three processes are similar to those detailed in the *Prescribe, Order,* and *Dispense* workflow (see Figures 3.5, 3.6, 4.6, 5.5):

4.2.1 *Retrieve Non-Drug Parameters* from an external actor
4.2.2 *Retrieve Patient Drugs* from a database table
4.2.3 *Review Clinical Parameters* receives *CDS* and integrates information

Since in this scenario the patient is one of the potential users, the system should enforce policies stating, for example, what information the patient is allowed to review or which drugs may the patient select and document self-administration.

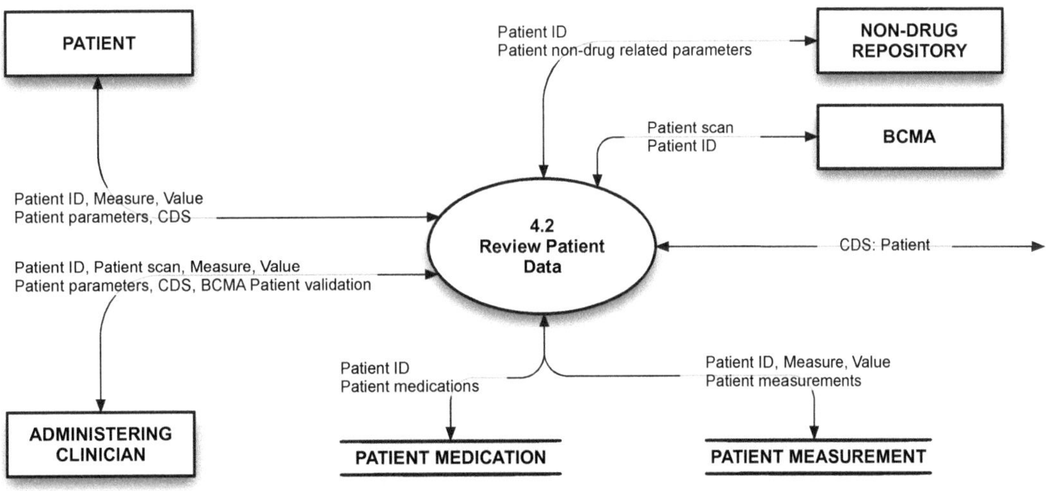

Figure 6.6 Review patient data process: Isolated from DFD 4.

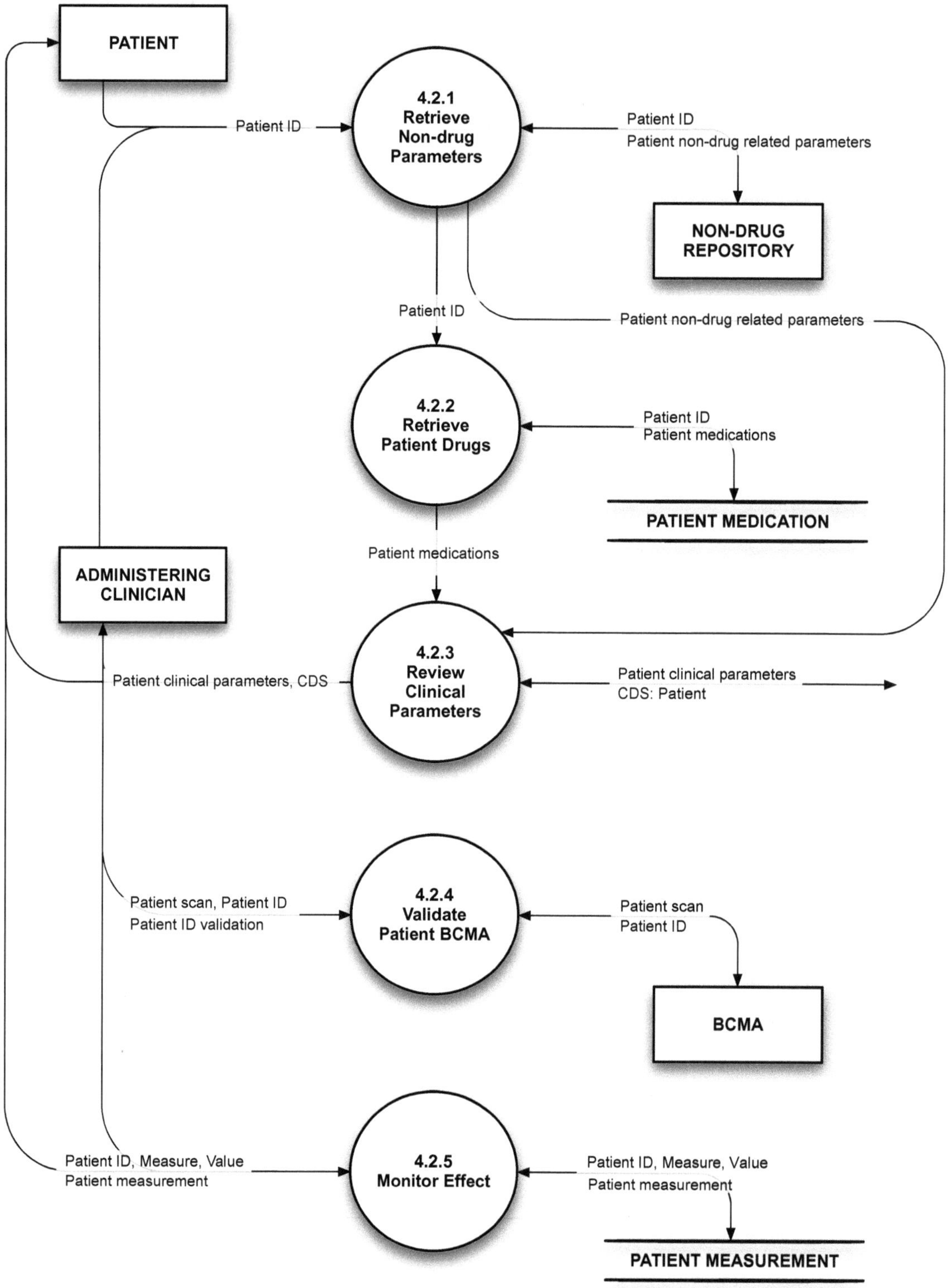

Figure 6.7 DFD 4.2: Review patient data.

Using the bar code system mentioned, process 4.2.4 *Validate Patient with BCMA* validates the *Right Patient:*

→ *BCMA* receives the scan of a patient's wristband
← *BCMA* returns a patient's unique identifier

Obviously, any discrepancy between the *BCMA* and the *EHR/Medications* systems should alert the user for a potential error. Process 4.2.4 *Validate Patient with BCMA* reconciles two systems regarding the patient identity:

- *Patient ID* in the *EHR/Medications* system
- *Patient ID* in the *BCMA* system as defined by the patient's wristband bar code

When challenged with a patient's wristband scan, process 4.2.4 decides whether

✓ Patients match: The patient in the *BCMA* system is the same as the patient in the *EHR/Medications* system.
× Discrepancy found (Alarm!): the patient in *BCMA* is different from the patient identified in *EHR/Medications.*

The *PATIENT* at home or the *ADMINISTERING CLINICIAN* in a hospital uses process 4.2.5 *Monitor Effect* to document and review the effects of medications. The process of monitoring drug effects may include a *before* and *after* set of values, for example, monitoring the respiratory rate, oxygen saturation, heart rate, blood pressure, and pain scale *before* administering a narcotic pain killer and *after* the administration, at certain time intervals. Process 4.2.5 *Monitor Effect* enables the selection of a measure, such as *blood pressure,* and then stores/retrieves its value and unit: *120/80 mmHg.*

The table *PATIENT MEASUREMENT* introduced in Chapter 4 is the data store for the values related to 4.2.5 *Monitor Effect* (Figure 4.17). Enabling multiple parameters to be captured repeatedly during time, 4.2.5 satisfies the requirement for monitoring drug effects *before* and *after* administration.

DFD 4.3 Select Drug

Process 4.3 *Select Drug* allows both the *PATIENT* and the *ADMINISTERING CLINICIAN* to select a new (to the patient) or an existing (the patient is already taking) drug (Figure 6.8). It may then be administered, self-administered, modified, stopped, pulled from or returned to storage, and so on.

The information sources for these drugs are the *RECOMMENDED DRUG DOSE* and the *PATIENT MEDICATION* tables. *CDS* is considered a part of the process of selecting a drug, in a similar fashion to *CDS* in the *Prescribe, Order,* or *Dispense* workflows.

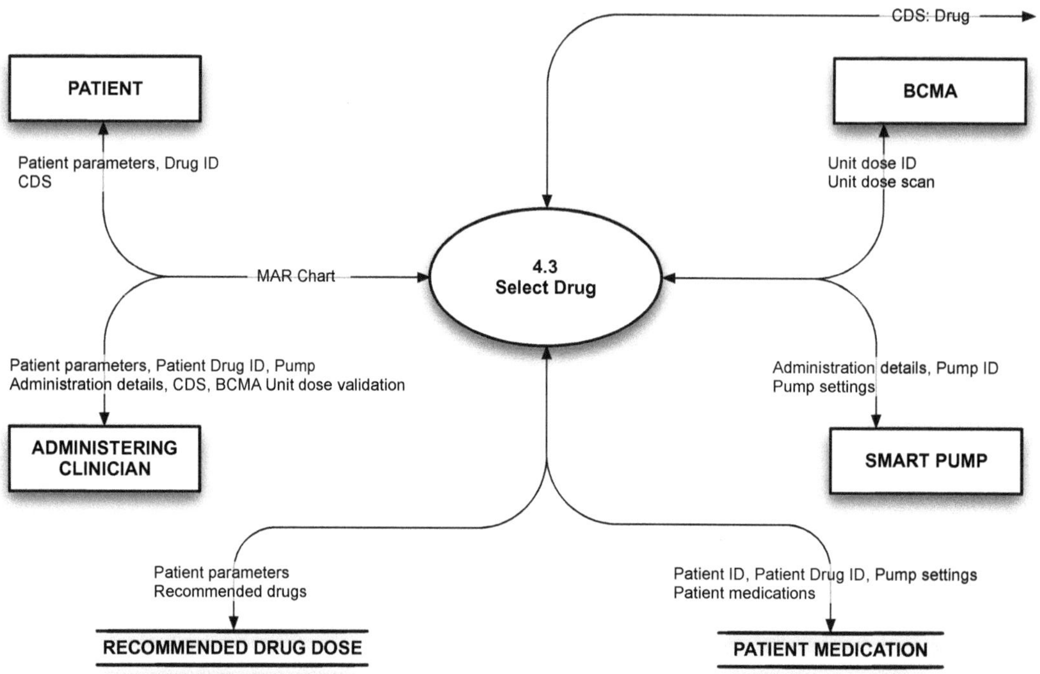

Figure 6.8 Select drug process: Isolated from DFD 4.

The *MAR* chart is a view, a result of a query, of the *PATIENT MEDICATION* table; it charts the drugs a patient is taking or being administered.

BCMA and *Smart pump* are also shown as having a role in the validation of the *Right Drug* and *Right Route*. Let us examine process 4.3 further (Figure 6.9):

Processes 4.3.1 *Select New Drug* and 4.3.2 *Select Existing Drug*, similar to previous workflows, allow both the *PATIENT* and the *ADMINISTERING CLINICIAN* to select a medication, modify its parameters, stop the administration, consider relevant *CDS*, and so on.

The patient choices should be restricted to *Over-the-Counter* (*OTC*) drugs and the medications a patient has been prescribed, ordered, and dispensed.

Select New Drug and *Select Existing Drug* have been detailed in previous chapters (see Figures 3.7–3.10) and with minor changes, DFD 1.3, 2.3, and 3.3 can be reused in the context of the *Administer* workflow as DFD 4.3.

Process 4.3.3 *Validate Drug BCMA* validates parameters with the *EHR/Medications* system and decides on the *Right Drug* and the *Right Route*.

→ *BCMA* receives the scan of a *Unit Dose* drug
← *BCMA* returns *the Unit Dose ID*

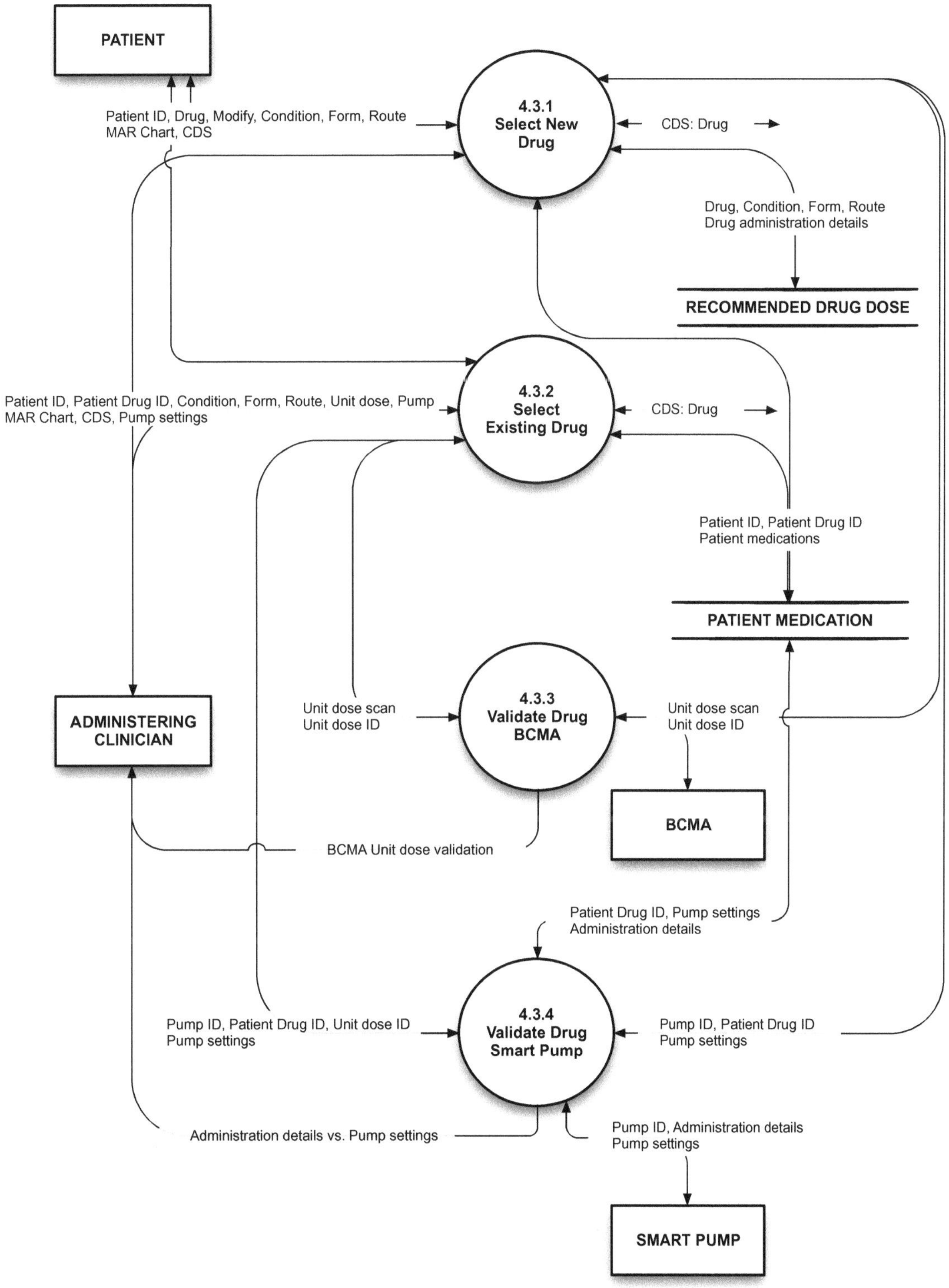

Figure 6.9 DFD 4.3: Select drug.

When challenged with a *Unit Dose* scan, *Validate Drug BCMA* may thus respond as follows:

- ✓ Drugs match: The drug in the *BCMA* system is the same as the drug in the *EHR/Medications* system.
- × Drugs do not match (Alarm!): The drug in the *BCMA* is different from the drug identified in *EHR/Medications.*
- ✓ Routes match: The route in the *BCMA* system is the same as the route in the *EHR/Medications* system.
- × Routes do not match (Alarm!): The route in the *BCMA* is different from the route identified in *EHR/Medications.*

Process 4.3.4 *Validate Drug Smart Pump* is responsible for matching the drug and route in the *EHR/Medications* system with the drug and route as they are represented in the *SMART PUMP* system. Process 4.3.4 needs to support the bidirectional communication between *EHR* and the *SMART PUMP* as two scenarios:

1. Transmit drug administration details from *EHR* to *SMART PUMP*:
   - → Receives a *Patient Drug ID* uniquely identifying a drug administration instance
   - → Receives a *Pump ID* uniquely identifying a pump
   - ← Sends *Patient Drug ID* to *PATIENT MEDICATION*
   - → Receives the drug administration details from *PATIENT MEDICATION*
   - ← Sends the drug administration details to *SMART PUMP*
   - → Receives *Pump settings*
   - ← Validates *Administration details* in the *PATIENT MEDICATION* table versus *Pump settings* in *SMART PUMP* system and sends results to user

2. Send pump settings to *MAR* chart:
   - ← Once a pump is active, 4.3.4 sends pump settings (e.g., every minute) to *PATIENT MEDICATION*

Process 4.3.4 reconciles two systems: *PATIENT MEDICATION/Administration details* versus *SMART PUMP/Pump settings.* When challenged with a new or an existing drug, process *Validate Drug Smart Pump* may respond as follows:

- ✓ Drugs match: The drug in the *SMART PUMP* system is the same as the drug in the *EHR/Medications* system.
- × Drugs do not match (Alarm!): The drug in *SMART PUMP* is different from the drug identified in *EHR/Medications.*
- ✓ Routes match: The route in the *SMART PUMP* system is the same as the route in the *EHR/Medications* system.

× Routes do not match (Alarm!): The route in the *SMART PUMP* is different from the route identified in *EHR/Medications*.

Process 4.3.4 is an additional layer of safety during the *Administer* workflow.

DFD 4.4 Select Dose/Prepare

Following a drug selection, the next step is to set the dose (Figure 6.10). Process 4.4 *Select Dose/Prepare* allows both *PATIENT* and *ADMINISTERING CLINICIAN* to review the *MAR* chart, and after a drug is selected in 4.3, select a dose, prepare the drug, and consider relevant *CDS* guidelines regarding dose and preparation. While in process 4.3 the user selects a drug, in process 4.4 the drug, form, and route are given, and the focus is on setting the dose parameters.

The information on dosing regimens resides in two tables: *RECOMMENDED DRUG DOSE* and *PATIENT MEDICATION*. Process 4.4 has the responsibility for selecting the *Right Dose* at the *Right Time*. *BCMA* and *SMART PUMP* participation in 4.4 *Select Dose* (Figure 6.11) is based on a similar validating role as in 4.3 *Select Drug*.

Process 4.4.1 *Review Drug Dose* provides the *PATIENT* and the *ADMINISTERING CLINICIAN* with the *MAR* chart, dosing regimens related to the patient's medications, other recommended options, and *CDS* considerations.

Process 4.4.2 *Prepare Drug Dose* receives a set of discrete data elements (including *Time/Duration*), and using several simple equations detailed in

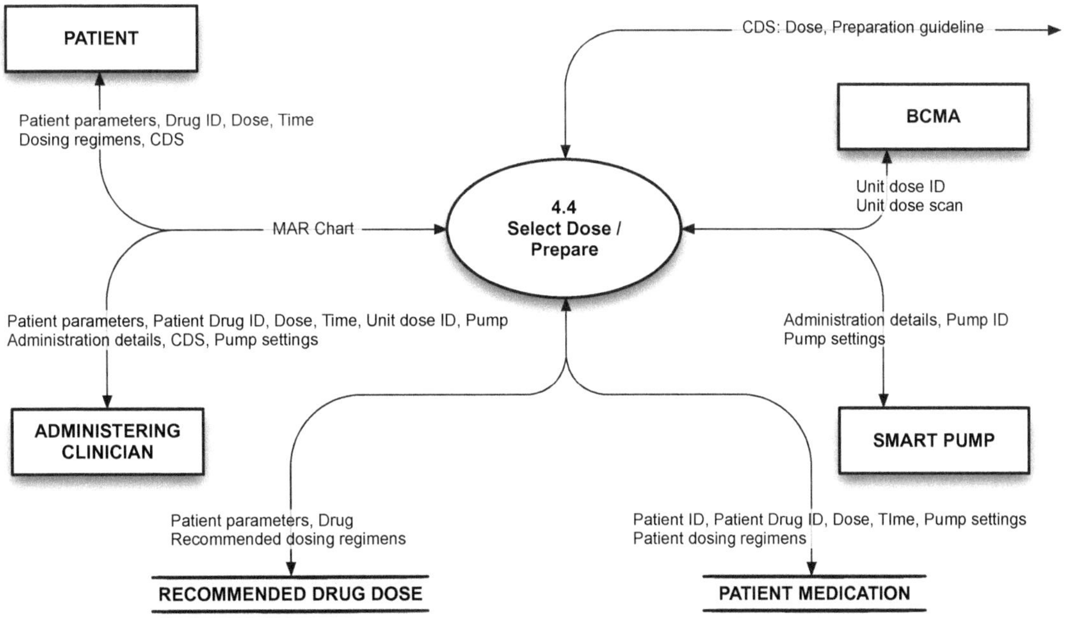

Figure 6.10 Select dose/prepare process: Isolated from DFD 4.

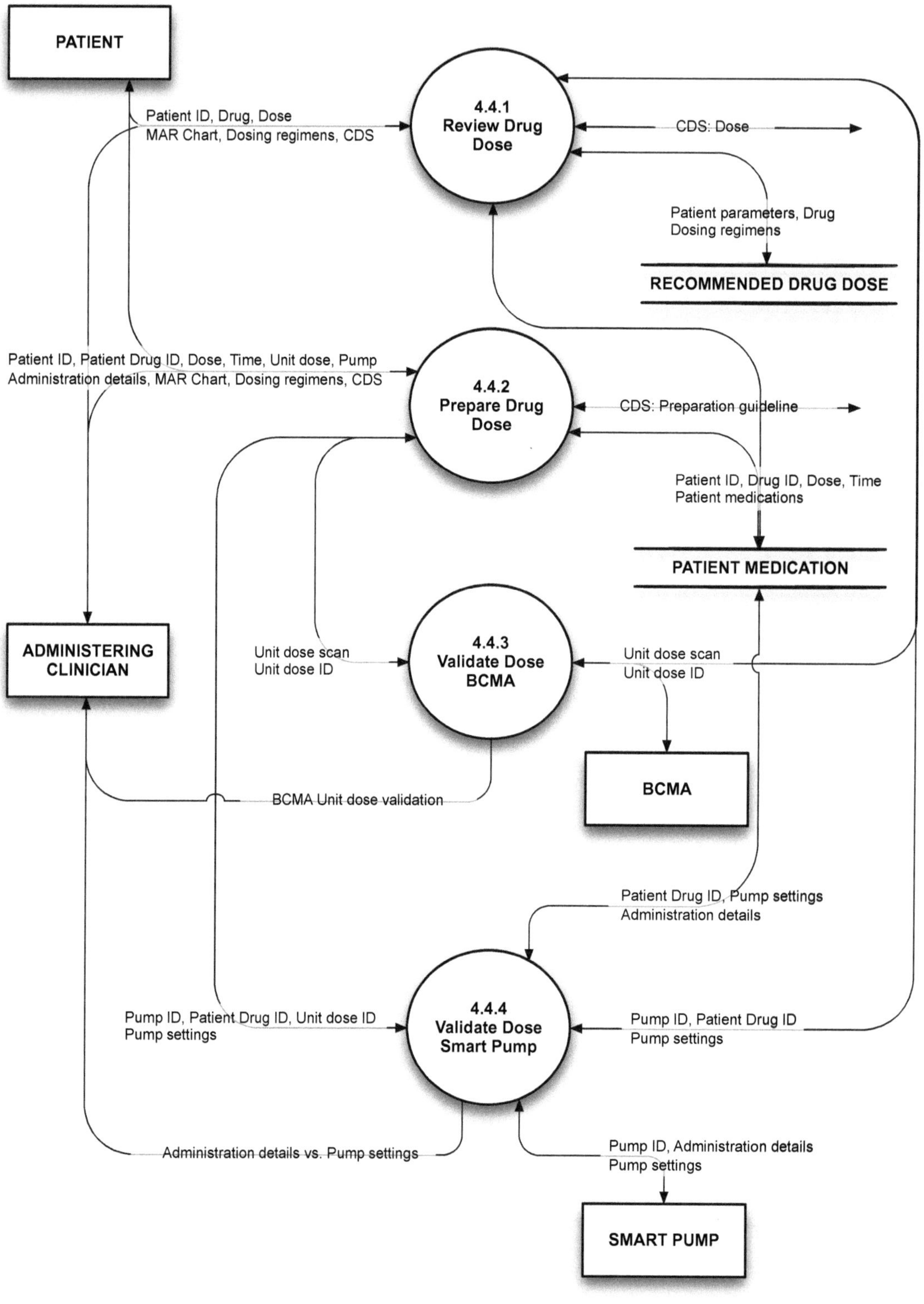

Figure 6.11 DFD 4.4: Select dose/prepare.

the following, it calculates the dosing parameters while considering *CDS* and *Preparation guidelines.*

With a *Smart pump* attached, the administration details can be downloaded at this stage from the *EHR* into the *SMART PUMP*, bypassing the need to program the pump manually.

The Process 4.4.3 *Validate Dose with BCMA*

→ Receives a *Unit Dose* scan
← Sends *Unit Dose* scan to *BCMA*
→ Receives *Unit Dose ID* from *BCMA*
← Validates *Unit Dose ID* in *BCMA* with *Patient Drug ID* in *PATIENT MEDICATION*

Thus, process 4.4.3 is reinforcing the *Right Dose* concept by deciding whether

✓ Dosing parameters match: The dosing parameters in the *BCMA* system are the same as those in the *EHR/Medications* system.
× Dosing parameters do not match (Alarm!): The dosing parameters in the *BCMA* are different from those in the EHR/*Medications* system.

Process 4.4.4 *Validate Dose Smart Pump* receives administration details from *EHR* into the pump and charts the *MAR* with the pump settings:

→ Receives the *Pump ID, Patient Drug ID, Unit Dose ID*
← Sends parameters to *PATIENT MEDICATION*
→ Receives administration details—*Dose* and *Time* parameters—and calculates *Pump settings*
← Validates *Pump settings* versus *Administration details* and sends *Pump settings* to *SMART PUMP, PATIENT MEDICATION,* and user

Process 4.4.4 is another validating actor in the *Administer* workflow, helping with the *Right Dose* and *Right Time*:

✓ Dosing parameters match: The dosing parameters in the *SMART PUMP* system are the same parameters in the *EHR/Medications* system.
× Dosing parameters do not match (Alarm!): The dosing parameters in the *SMART PUMP* are different from the parameters in the *EHR/Medications.*
✓ Time parameters match: The Time parameters in the *SMART PUMP* system are the same parameters in the *EHR/Medications* system.
× Time parameters do not match (Alarm!): The Time parameters in the *SMART PUMP* are different from the parameters in the *EHR/Medications.*

Using the processes detailed, the nurse may validate all the *6 Rights* during the *Administer* workflow with just four scans (Table 6.1).

Table 6.1 Administer Workflow Validating the Six Rights

Nurse Scans Bar Code On	*System Validates the Right*
1. Own badge	Clinician
2. Patient wristband	Patient
3. Unit dose	Drug, Dose, Route
4. Pump	Drug, Dose, Time

DFD 4.5 Interact with Storage

Drugs need to be pulled from or returned to a *DRUG STORAGE* area. These drug distribution systems, such as *ADC,* have been mentioned in this chapter and in the previous one. Medications are sometimes destroyed; when narcotics are destroyed, the process may require a cosignature. Isolated from DFD 4 (Figure 6.12), the DFD 4.5 process *Interact with Storage* is described next.

Given a *Patient ID, Patient Drug ID* or *Unit Dose ID,* the *ADMINISTERING CLINICIAN* may select to

4.5.1 *Pull Drug from Storage*
4.5.2 *Return Drug to Storage*
4.5.3 *Destroy Drug*

Both *DRUG STORAGE* and *PATIENT MEDICATIONS* are updated on these interactions (Figure 6.13).

DFD 4.6 Administer/Sign

While the drug and dose are being calculated and considered, nothing is committed. *PATIENT* or *ADMINISTERING CLINICIAN* needs to finalize the *Administer* workflow by committing the changes to the *PATIENT MEDICATION* and *DRUG COMMUNICATION* tables (Figure 6.14). The differences between the *PATIENT*

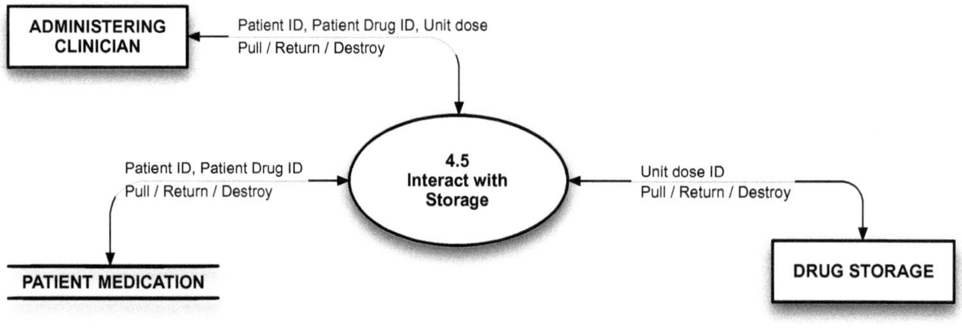

Figure 6.12 Interact with storage process: Isolated from DFD 4.

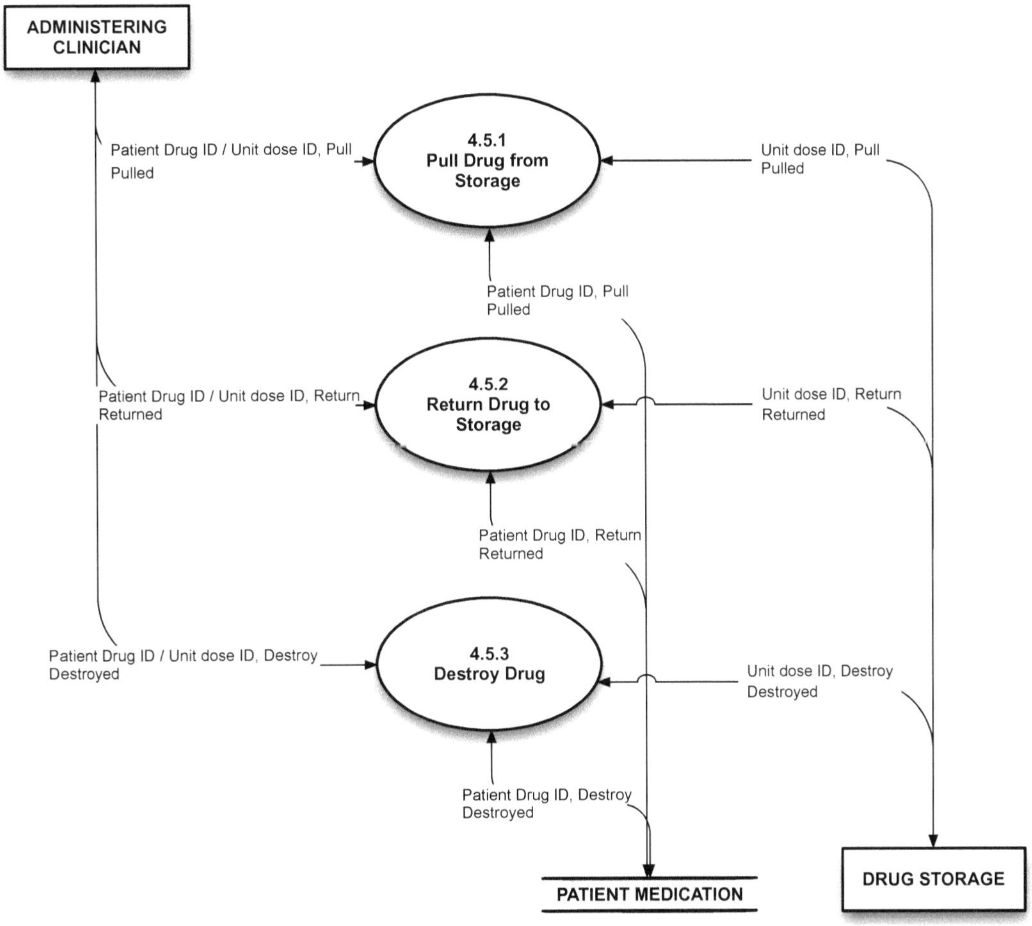

Figure 6.13 DFD 4.5: Interact with storage.

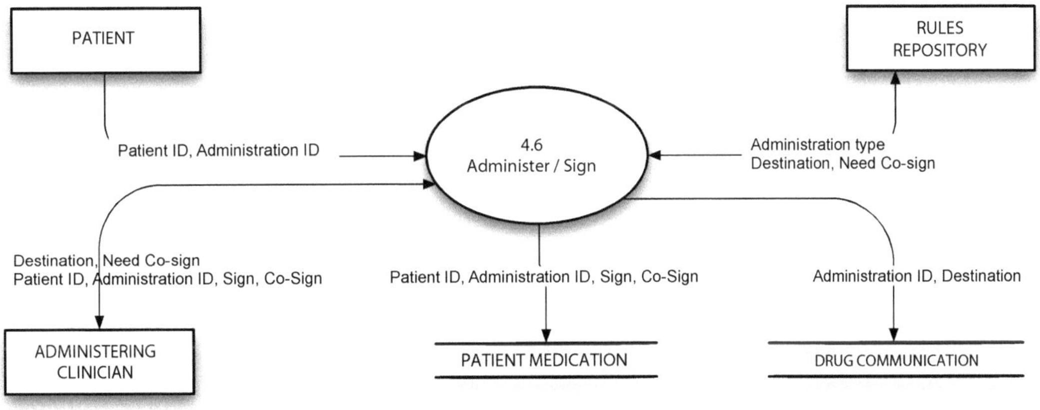

Figure 6.14 Administer/Sign process: Isolated from DFD 4.

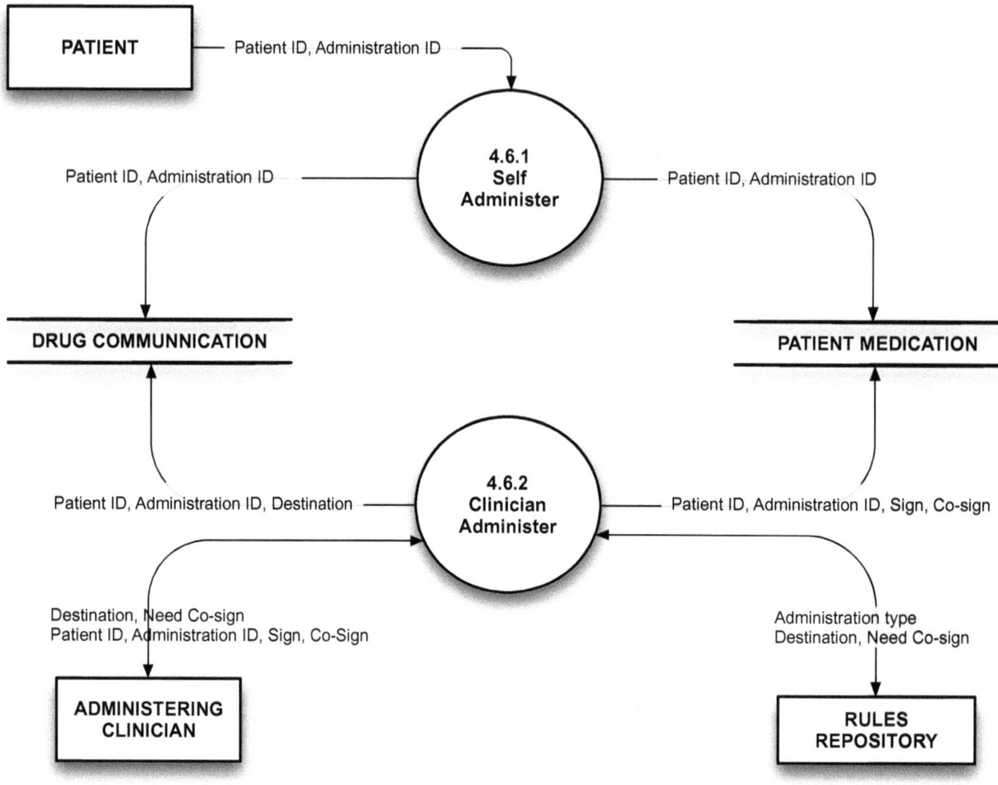

Figure 6.15 DFD 4.6: Administer/Sign.

and the *ADMINISTERING CLINICIAN* is that the latter may sometimes need to cosign an administration in a similar fashion as an ordering physician may need to cosign an order (Figure 6.15). Of course, the *PATIENT* does not need to *"sign"* while documenting self-administration, just finalize and commit the changes to the system.

The *RULES REPOSITORY* supplies the information on the need to cosign a drug administration (see Chapter 4).

Data Elements

Analyzing the dataflows in DFD 4.x, we extract the discrete data elements for the database of the *Medications* system, which is evolving from previous chapters as the data model (Figure 6.16). We should not model into tables the dataflows with external systems since by definition external actors are beyond the system boundaries and not within our control (Table 6.2).

The last two data elements *Unit dose ID* and *Pump ID* in Table 6.2 need to be added to the logical model as follows: Since an administration instance (one row) in the *PATIENT MEDICATION* table may have only one *Unit Dose ID* at a time and since one *Unit Dose ID* may be related to only one instance in *PATIENT MEDICATION*, the relationship is one to one. This means we can add *Unit Dose ID* as a field, an attribute (one column) in the *PATIENT MEDICATION* table.

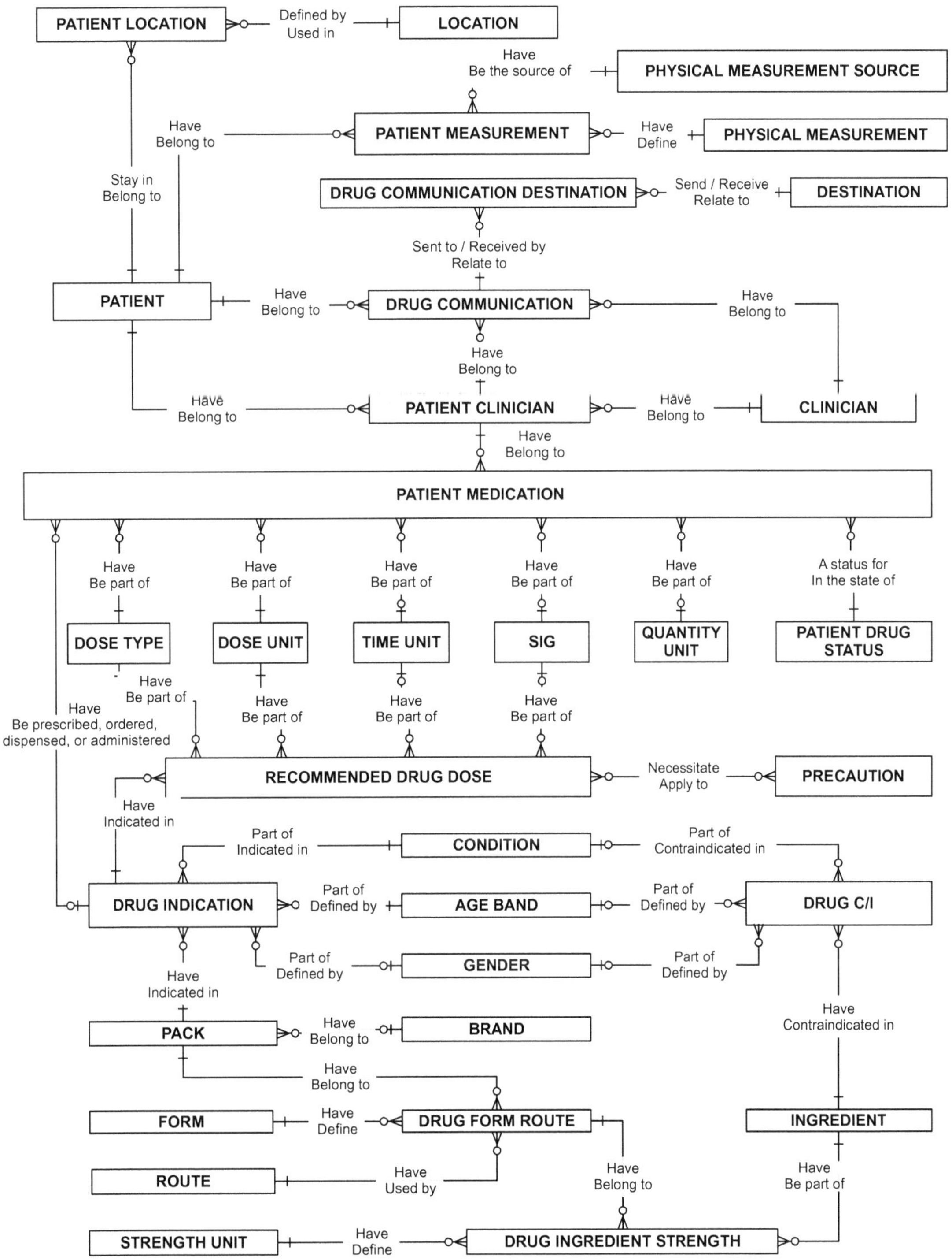

Figure 6.16 ERD for the medications domain.

Table 6.2 Data Elements Administer Workflow

DFD 4.x Data Elements	*ERD Table*	*External Actor*
Patient ID	*PATIENT*	*BCMA*
Clinician ID	*CLINICIAN*	*BCMA*
Administration ID	*PATIENT CLINICIAN*	
Communication	*DESTINATION* *DRUG COMMUNICATION* *DRUG COMMUNICATION DESTINATION*	*PATIENT* *CLINICIAN*
Measurement value	*PATIENT MEASUREMENT* *PHYSICAL MEASUREMENT* *PHYSICAL MEASUREMENT SOURCE*	*NON-DRUG REPOSITORY*
Patient parameters	*PATIENT MEDICATION* *PATIENT MEASUREMENT*	*NON-DRUG REPOSITORY*
Patient medications	*PATIENT MEDICATION* *PATIENT DRUG STATUS* *DRUG INDICATION* *PATIENT CLINICIAN* *DOSE TYPE* *DOSE UNIT* *TIME UNIT* *QUANTITY UNIT* *SIG*	*CDS* *BCMA* *SMART PUMP*
Drug	*BRAND* *PACK* *INGREDIENT* *DRUG INGREDIENT STRENGTH* *FORM* *ROUTE* *DRUG FORM ROUTE* *DRUG INDICATION* *PATIENT MEDICATION*	*CDS* *BCMA* *SMART PUMP*
Recommended drugs	*RECOMMENDED DRUG DOSE*	*CDS*
Patient Drug ID	*PATIENT MEDICATION*	

(*continued*)

Table 6.2 Data Elements Administer Workflow (Continued)

DFD 4.x Data Elements	*ERD Table*	*External Actor*
Dose	*PATIENT MEDICATION* *PATIENT DRUG STATUS* *RECOMMENDED DRUG DOSE* *DOSE TYPE* *DOSE UNIT* *TIME UNIT* *QUANTITY UNIT* *SIG*	*CDS* *BCMA* *SMART PUMP*
Unit dose ID	*PATIENT MEDICATION*	*BCMA* *SMART PUMP*
Pump ID	*PATIENT MEDICATION*	*SMART PUMP*

Same one-to-one relationship exists between an administration instance in *PATIENT MEDICATION* and *Pump ID,* allowing us to add *Pump ID* as a field (column) to the *PATIENT MEDICATION* table without breaking any normalization rules introduced in Chapter 1.

PATIENT MEASUREMENT is a versatile data store that accepts vital signs, lab values, notes, scales, assessments, and so on while monitoring the drug effects. The data model to support this versatility is discussed in Chapter 4 (Figure 4.17).

An important concept in the data model is the *PATIENT DRUG STATUS.* It allows each instance in *PATIENT MEDICATION* to have a different *Status*, which in turn enables tracking the whole life cycle of a medication, *Order* → *Dispense* → *Administer*. The entity of *PATIENT DRUG STATUS* monitors the drug (Table 6.3):

Status during its life cycle
Location
Next step in the workflow

Location and *Next Step* are fields that are missing and should be added to the table *PATIENT DRUG STATUS* in the logical model (see Figure 2.11).

The parameters involved in dose settings are numerous; actually, for setting the *5 Rights*, there is a need for most of the following discrete data elements:

- *Patient ID & Patient Weight*
- *Drug Names & Ingredients*
- *Ingredient Strength & Strength Unit*
- *Form*
- *Route*
- *Dose & Dose Unit*
- *Dose Type*

- *Frequency & Frequency Unit*
- *Time, Duration & Duration Unit*
- *Quantity & Quantity Unit*
- *Unit Dose ID*
- *Pump ID*
- *SIG*

Calculation of Dosing Parameters for a Continuous Drip

The drugs may be administered intermittently as tablets or in a continuous mode, such as fluid solutions. When process 4.4 *Select Dose/Prepare* is applied to continuously administered drugs, such as intravenous drips, units may need conversion, and the *Administration details* and *Pump settings* need to be calculated (Table 6.4).

Let us follow a medication during its life cycle—*Order* → *Dispense* → *Administer*—and see how units, dose parameters, and the mentioned formulae interact (Table 6.5). The drug ordered (*Intropin)* is a brand name, while the drug dispensed and eventually administered has the same ingredient, albeit under the generic name *Dopamine.*

The strength, volume, quantity, and other parameters are initially deferred; these data elements are not defined in the *Order* since the ordering clinician may

Table 6.3 Patient Drug Status

Patient Drug Status	*Next Step*
Ordered/Signed	*Pharmacy Processing*
Pharmacy Processing	*Pharmacy Dispensing*
...	
Pharmacy Dispensed	*En route: Pharmacy to Floor Data Storage*
...	
In Floor Data Storage	*Nurse pulls drug from Data Storage*
Nurse pulled drug from Data Storage	*Nurse validates the 5 Rights*
...	
Unit dose—BCMA validated	*Nurse validates the 5 Rights*
SMART PUMP settings validated	*Nurse validates the 5 Rights*
...	
The 5 Rights validated	*Administer start*
Administer in process	*Administer end*
Dose changed in EHR system	*Administer end*
Dose changed in SMART PUMP system	*Administer end*
...	

Table 6.4 Formulae for Calculating Infusion Dose Parameters

Formula	*Example*
1. *Quantity = Volume × Concentration* Or *Quantity = Volume × Strength*	*100 mg = 50 ml × 2 mg/ml*
2. *Drug Rate = Volume Rate × Concentration* Or *Drug Rate = Volume Rate × Strength*	*1 mg/min = 0.5 ml/min × 2 mg/ml*
3. *Drug Rate = Patient Weight × Dose*	*500 mcg/min = 100 kg × 5 mcg/kg/min*
4. *Drops Rate = Volume Rate (ml/min) × 60*	*12 drops/min = 0.2 ml/min × 60 drops/ml*
5. *Duration = Date & Time End – Date & Time Start*	*30 minutes = 2012.01.31.00:21 2012.01.30.23:51*
6. *Volume Total = Volume Rate × Duration*	*6 ml = 0.2 ml/min × 30 minutes*

not know these parameters before the drug is dispensed. Using formula 3 from Table 6.4, the pharmacist calculates the *Drug Rate:*

Drug Rate = Patient Weight × Dose = 100 kg × 5 microgram/kg/min = 500 microgram/min = 0.5 mg/min

Table 6.5 Data Elements during a Drug Life Cycle

	Order	*Dispense*	*Administer*
Patient weight	*220 pounds*	*100 kg*	*100 kg*
Drug	*Intropin*	*Dopamine*	*Dopamine*
Form	*Solution*	*Solution*	*Solution*
Route	*IV*	*IV*	*IV*
Strength		*1.6 mg/ml*	*1.6 mg/ml*
Dose	*5 mcg/kg/min*	*5 mcg/kg/min*	*5 mcg/kg/min*
Drug rate		*0.5 mg/min*	*0.5 mg/min*
Volume rate			*0.3167 ml/min*
Frequency/drops rate			*19 drops/min*
Duration			*60 minutes*
Volume total		*250 ml*	*19 ml*
Quantity		*1 Bag of 400 mg Dopamine in D5W*	*30.4 mg*
Unit dose ID		*1357-9233-4690-0081*	*1357-9233-4690-0081*
Pump ID			*776-3221*
Clinician	*Ordered—Dr. S*	*Dispensed—Pharmacist M*	*Administered—Nurse A*

Aiming at a *Duration* of 12–14 hours (*720–840 minutes*) at the dose ordered, the pharmacist estimates the total quantity of *Dopamine* needed as

Quantity dispensed = *0.5 mg/min* × *800 minutes* = *400 mg*

Using two ampoules of *200 mg Dopamine* each, the pharmacist prepares a solution of *400 mg Dopamine* in a total volume of *250 ml*. Formula 1 from Table 6.4 rearranged defines the *Strength* of the dispensed *Dopamine* solution as

Strength = *Quantity/Volume* = *400 mg/250 ml* = *1.6 mg/ml*

Pharmacy uniquely identifies this bag as *Unit dose ID 1357-9233-4690-0081* and dispenses it to the ward. At this stage, blanks still remain in Table 6.5 since the pharmacist does not know the duration, volume, or quantity of the *Dopamine* that would be actually administered to the patient; at this moment in time, the pharmacist does not know if the nurse is planning on diluting the *1.6 mg/ml* any further.

The nurse receives from the pharmacy a *Unit dose* of one bag of *250 ml Dopamine 1.6 mg/ml in D5W* (name of the solution diluent, *Dextrose 5% in Water*). After pulling the drug from storage, the nurse uses a pump for the continuous administration of this drug. The nurse may also dilute a drug received from the pharmacy or mix it with other drugs, but not in this case.

Note that while the nurse may change the *Strength* of a solution dispensed by the *Pharmacy*, the change can be only in one direction: Dilution = Lower *Strength*. Nurse may dilute a solution to decrease its *Strength*, but cannot increase its *Strength* by concentrating a given solution.

Another restriction that may limit the nurse's options when administering fluids and solutions is the total volume: One should not administer a high volume of fluids to patients who cannot tolerate *hypervolemia* (literally "*too much volume*") due to age, weight, or disease (e.g., neonates, preemies, children, or patients suffering from chronic kidney disease, on dialysis, with congestive heart failure, etc.).

Rearranging formula 2 from Table 6.4, the nurse calculates the *Volume Rate:*

Volume rate = *Drug rate/Strength* = *0.5 mg/min/1.6 mg/ml* = *0.3125 ml/min*

Pumps sometimes employ the unit of *Drop*:

1 drop = *1/60 ml* = *0.0167 ml*

Since the concept of "partial drop" does not apply, the minimal volume one can infuse with drop-based pumps is thus

1 drop/minute = *0.0167 ml/minute* = *1 ml/hour.*

Using formula 4 from Table 6.4, the nurse calculates the actual *Drop Rate* of the pump:

$$Drop\ Rate = Volume\ rate \times 60 = 0.3125\ ml/min \times 60\ drops/ml = 18.75\ drops/min.$$

Since there are no partial drops, the nurse sets the pump to *19 drops/min* (which equals *0.3167 ml/min).*

At the end of the *Dopamine* infusion, using formula 5 from Table 6.4, the nurse calculates the *Duration* of the *Dopamine* administration as 60 minutes. With the parameter of *Duration* and using formulae 6 and 1, the nurse calculates the *Volume Total* and the *Quantity Total*:

$$Volume\ Total = Volume\ Rate \times Duration = 0.316667\ ml/min \times 60\ min = 19\ ml$$

$$Quantity\ Total = Volume\ Total \times Strength = 19\ ml \times 1.6\ mg/ml = 30.4\ mg$$

The *Administer* workflow exposes numerous opportunities crying for automation. The calculations mentioned should be performed automatically. After all, not long ago a computer was a "*calculator*."

Let us sketch a generic *Calculation* process that may serve during *Order, Dispense,* and *Administer* (Figure 6.17). An algorithm to calculate dosing

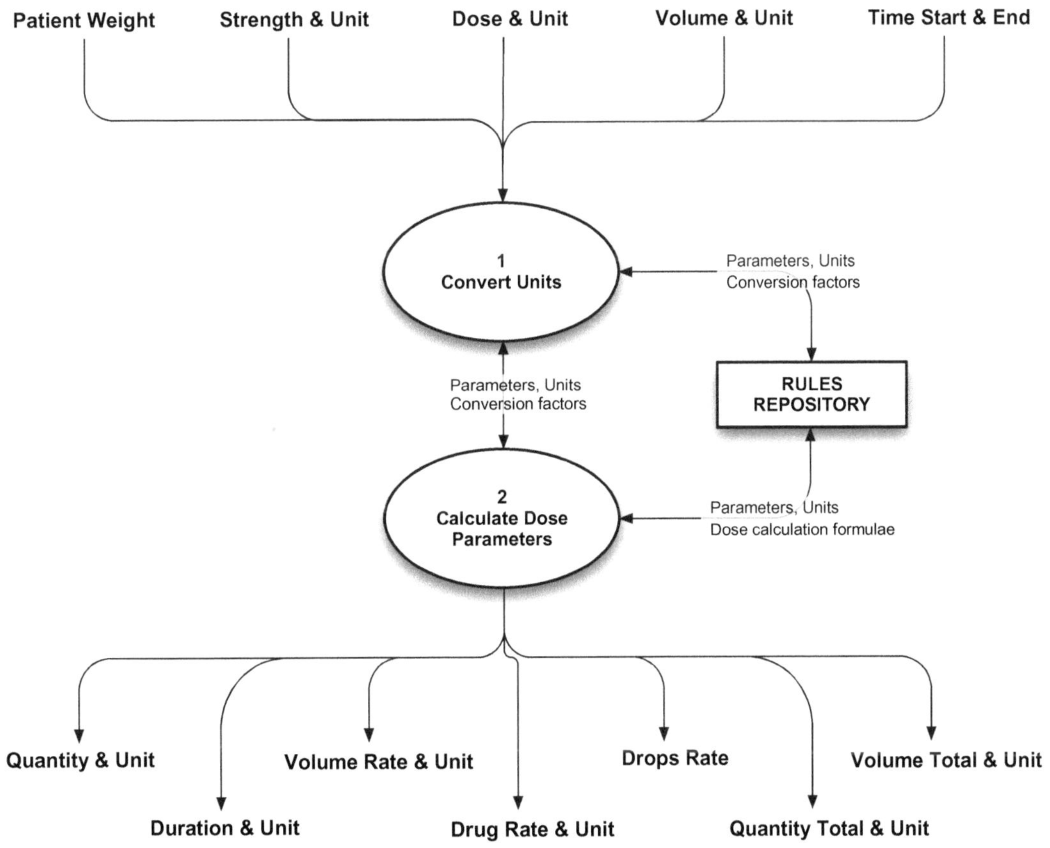

Figure 6.17 Calculation of dosing parameters for a continuous drug.

parameters needs first to convert and match units, such as *500 mcg = 0.5 mg,* and then using formulae similar to those six mentioned, it can calculate all the data elements necessary for the *Administration details* and *Pump settings.* The *RULES REPOSITORY* provides the conversion factors as well as the calculation formulae.

A data model that enables calculations such as those mentioned is an adaptive model: It adapts to many dosing regimens with different and changing units. It also supports the collaboration between the workflows of multiple clinicians—physician, pharmacist, and nurse—during the whole drug life cycle: *Order → Dispense → Administer.*

Summary

This chapter introduced the safety concept of *5 Rights* and the processes involved in validating the right patient, drug, route, dose, and time parameters.

Two technologies may help with automating this validation process: *BCMA* and *Smart Pump.* Their interactions with the *Medications* system were briefly introduced.

The patient is an active actor who may populate the *MAR* chart, like a *PHR,* while self-administering at home. In the hospital, the *Smart pump* should automatically chart a part of the *MAR.*

Analyzing the dataflows of the *Administer* workflow, we have identified some new discrete data elements and the consequent minor changes to the logical model that are required.

The calculation of the dose parameters for an intravenous infusion were detailed using six simple formulae and based on a full life cycle of a drug: *Order → Dispense → Administer.*

The chapter ended with a sketch of an automated calculation process for dosing parameters of continuous drugs.

References

1. Administration of medication. *Encyclopedia of nursing and allied health.* http://www.enotes.com/nursing-encyclopedia/administration-medication (accessed July 29, 2011).
2. Grigg, S.J., S.K. Garrett, and J.B. Craig. 2011. A process centered analysis of medication administration: Identifying current methods and potential for improvement. *International Journal of Industrial Ergonomics* 2011: 1–9. doi:10.1016/j.ergon.2011.01.014.
3. Vanderveen, T., S.R. Lewis, and S. Almeida. 2007. Reducing complexity: A strategic approach to optimizing the medication use process for all medications. *Patient Safety & Quality Healthcare.* http://www.psqh.com/sepoct07/complexity.html (accessed July 28, 2011).

4. Bates, D.W., D.J. Cullen, N. Laird, et al. 1995. Incidence of adverse drug events and potential adverse drug events: Implications for prevention. *Journal of the American Medical Association* 274: 29–34.
5. Paoletti, R.D., T.M. Suess, M.G. Lesko, et al. 2007. Using bar-code technology and medication observation methodology for safer medication administration. *American Journal of Health-System Pharmacy* 64: 536–43.
6. Pedersen, C.A., P.I. Schneider, and D.J. Scheckelhoff. 2009. ASHP national survey of pharmacy practice in hospital settings: Dispensing and administration—2008. *American Journal of Health-System Pharmacy* 66: 926–946.
7. Dalton, J.A. 2010. Bedside bar coding technology and implementation. In *Pharmacy Informatics,* ed. P.O. Anderson, S.M. McGuinness, and P.E. Bourne, 107–122. Boca Raton, FL: CRC Press.
8. Proceedings from the ISMP summit on the use of smart infusion pumps: Guidelines for safe implementation and use. Institute for Safe Medication Practices. http://www.ismp.org/tools/guidelines/smartpumps/default.asp (accessed July 2, 2011).
9. Eckel, S.F. 2004. "Smart" pump technology. *Pharmacy Times.* http://www.pharmacytimes.com/publications/issue/2004/2004–11/2004–11–4722 (accessed July 2, 2011).

Review Questions

1. In a fully automated environment and assuming the nurse is already logged in, what is the minimal number of steps to validate the 5 Rights and activate an infusion?
2. Which aspects of DFD 4 pertain to a *Personal Health Record (PHR)*?
3. How is *Pharmacy* communicating with the *Administering Clinician*?
4. While loading a pump with drug A, a nurse erroneously documents drug B in the *SMART PUMP* system. How can the *Medications* system prevent this error?
5. If a patient weights 110 pounds and the dose is 1.2 mcg/kg/min, what is the drug rate in milligrams/hour?
6. A neonate weighting 3 kg has an order for an *Adrenaline (Epinephrine)* infusion at 0.2 mcg/kg/min. The pharmacy has dispensed 1.8 mg *Adrenaline* in a 100-ml bag. What would be the pump drop rate?
7. What problem should be expected with a drug administered at 0.8 ml/hour with a drop-based pump?
8. Which drugs may a patient document as self-administered?
9. A drug at a strength of 10 mg/ml is administered at 6 drops/min to a patient weighing 100 kg. What is the dose in micrograms/kilogram/minute?
10. A nurse receives a solution of a drug with a strength of 0.001 mcg/ml. Concerned that the diluted solution would cause an overload of fluids, the nurse concentrates the solution so the same quantity can be administered in a smaller volume. Explain the calculations involved.

Chapter 7

User Interface

Form ever follows function.

Louis Sullivan

This chapter presents user interface (UI) concepts and usability principles and then illustrates the UI needs of an electronic health record (EHR) with emphasis on the medications domain.

The main goal is for the UI is to tell a clinical story at various levels of details; for ambulatory settings, the time frame is measured in years, while in an intensive care unit (ICU) environment, it is literally a matter of minutes. The UI also needs to satisfy different clinicians: physician, pharmacist, nurse. All have similar but not exactly the same requirements in terms of what should be displayed.

Most of the currently existing UI solutions tend to separate various parts of the EHR into different folders or tabs. As we discuss in this chapter, this practice does not support, and it may even break, the clinician's cognitive processes; the medications cannot be visually separated from the other clinical parts of an EHR: patient's history, diagnosis, problems list, physical exam, labs, imaging, and so on.

Thus, a UI solution needs to summarize vast amounts of underlying data and display rich clinical information using a high-density, graphic-intense UI. Using principles of information visualization, the chapter gradually develops several screen layout designs, UI for optimal cognitive support, while reviewing patient clinical information, prescribing, ordering, administering, and reconciling medications.

Usability

For many users, and rightly so, the UI is The System. A clinical UI solution, badly designed, may doom the best EHR to failure. The quality of a specific UI design is measured as usability, which is defined by the *International Organization for Standardization* as *"the effectiveness, efficiency and satisfaction with which the intended users can achieve their tasks in the intended context of product use"* (1). According to the *National Institute of Standards and Technology* (2):

- *Effectiveness* is measured as the percentage of tasks completed successfully during a time frame. An error is considered if the user intention was, for example, to prescribe drug X and instead drug Y was erroneously selected.
- *Efficiency* is measured in time units or number of steps: how long or how many clicks it takes to successfully complete a task. For example, product A allows 90% of users to complete an order for drug X in less than 55 seconds. If using product B and performing same task, it takes on average 145 seconds, then product A is more efficient than B. The number of steps (clicks, screens) is another measure of efficiency: If there are two alternatives to perform a task, but one takes 5 clicks and the other necessitates 14 clicks, then the 5-click path is obviously more efficient.
- *Satisfaction*, a subjective measure, employs standardized rating scales such as the *Software Usability Measurement Inventory (SUMI)* (3) or *System Usability Scale (SUS)* (4). These scales quantify the term *satisfaction* and allow stratification of the user experience.

Cognitive Load

Clinicians usually perform in environments saturated with sensory distractions, which compete fiercely for their attention. In addition, vast amounts of information need to be processed under time pressure while a mental picture of the patient's clinical situation has to be created and modeled in the clinician's mind (5). All these activities create a cognitive load, which may reach a point of negative impact on the quality of care and patient's safety.

The UI should support the clinician's cognitive processes (6); several ways the UI can help in this regard are detailed in this chapter.

Principles of Graphic Excellence

Lund published in 1997 a still-relevant list of 34 rules of thumb for designing a high-quality UI that does not pose significant cognitive risks (7). My favorites are:

Don't overload the user's buffers.
Minimize the need for a mighty memory.
User should always know what's happening.
User should be in a good mood when done.

In his influential book *The Visual Display of Quantitative Information*, Edward Tufte defined the principles of graphical excellence (8):

Show the data. Complex ideas communicated with clarity, precision and efficiency.
Induce the viewer to think about the substance rather than about the methodology, design or technology.
Avoid distorting what the data have to say.
Present many numbers in a small space.
Graphical excellence is nearly always multivariate.
Make large data sets coherent.
Encourage the eye to compare different pieces of data.
Reveal the data at several levels of detail, from a broad overview to the fine structure.
Graphical excellence is that which gives to the viewer the greatest number of ideas in the shortest time with the least ink in the smallest space.

Characteristics of excellent graphic design, such as data density; data/ink ratio; time series and small multiples; "spark lines" (intense, simple, word-size graphics); missing data; ornaments; vibrating grids and how to calm them; golden ratio; axis compression; icons and navigation apparatus; typography; and the use of color are thoroughly discussed in Tufte's other books (9–11), in his patent on a graphical summary of a patient status (12), Wong's *Wall Street Journal Guide Guide to Information Graphics* (13), Lidwell et al.'s *Universal Principles of Design* (14), and others (15,16).

An excellent discussion about graphic problems posed by time series (which are used extensively in clinical UIs) can be found in Bertin's *Semiology of Graphics* (17). Some of his elegant solutions are shown in Figures 7.1A and 7.1B.

Cognitively loaded with usability measures, rules of thumb for high-quality human–computer interaction, principles of graphic excellence, and the subtleties of time series, we can proceed now with the analysis of clinical UIs.

Characteristics of a Clinical Story

The main goal for a clinical UI is to tell a medical story correctly and succinctly, the characteristics of which are illustrated in Figure 7.1C.

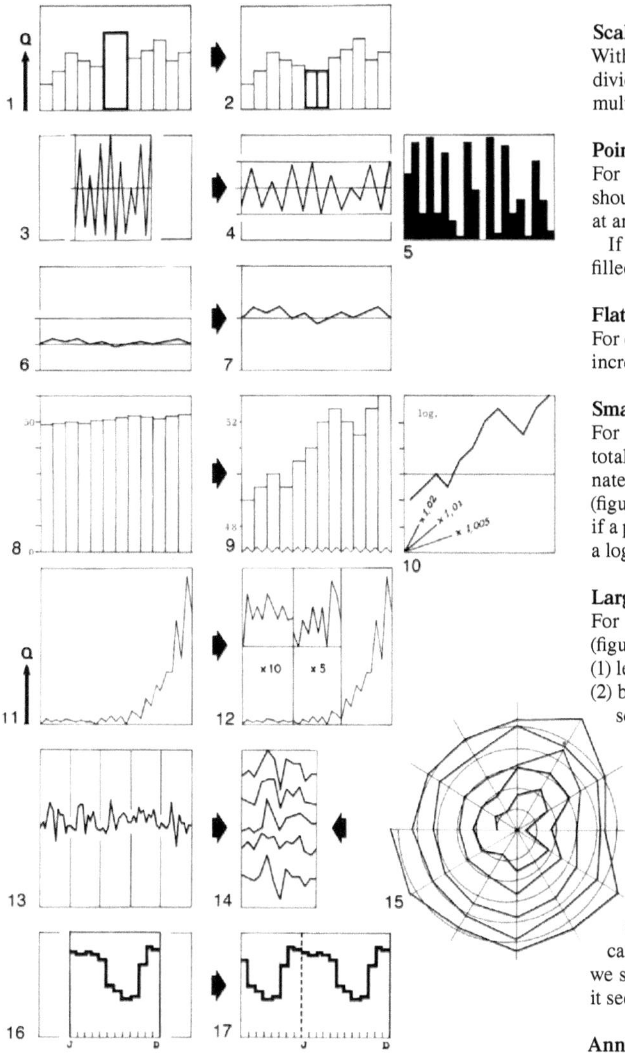

GRAPHIC PROBLEMS POSED BY TIME SERIES

Scale in years
With a scale in years, a two-year total (figure 1) should be divided by 2 (figure 2). A total for six months should be multiplied by 2.

Pointed curves
For overly pointed curves (figure 3), the scale of the Q should be reduced; optimum angular perceptibility occurs at around 70 degrees (figure 4).
If the curve is not reducible (large and small variations), filled columns can be used (figure 5).

Flat curves
For overly flat curves (figure 6), the scale of the Q should be increased (figure 7).

Small variations
For small variations in relation to the total (figure 8), the total loses its importance, and the zero point can be eliminated, provided the reader is made aware of this elimination (figure 9). The graphic can be interpreted as an acceleration if a precise study of the variations is necessary; here, we use a logarithmic scale (figure 10). (See also page 240.)

Large range
For a very large range between the extreme numbers (figure 11), we must either:
(1) leave out the smallest variations;
(2) be concerned only with relative differences (logarithmic scale), without knowing the absolute quantities;
(3) select different parts (periods) within the ordered component and treat them on different scales above the common scale (figure 12).

Obvious periodicity
If there is obvious periodicity (figure 13), and the study involves a comparison of the phases of each cycle, it is preferable to break up the cycles in order to superimpose them (figure 14). A polar construction can be used, preferably in a spiral shape (figure 15), but we should not begin with too small a circle. As striking as it seems, it is less efficient than an orthogonal construction.

Annual curves
For annual curves of rainfall or temperature, if a cycle has two phases (figure 17), why depict only one (figure 16)?

Figure 7.1A Graphic problems posed by time series. (From Bertin, Jacques. *Semiology of Graphics*. © 1983 by the Board of Regents of the University of Wisconsin System. 2010 by ESRI Press. Reprinted by permission of the University of Wisconsin Press.)

The main protagonist in this story—the patient—presents, complains, suffers from, or may show symptoms and signs. The physician suspects, considers, examines, orders lab tests and imaging studies, and prescribes or orders medications or procedures. Medications are dispensed and either administered by a clinician or self-administered by the patient. Lab results and imaging reports return, and this in turn may change previous decisions. The patient may react to treatment, drugs may have side effects, and the clinical story can easily go through several similar cycles of iteration for each and every problem the patient has. The story is created by multiple clinical events happening during time, usually in the format of "patient does" versus "done to the patient."

A contrast
Unlike what we see in figure 18, the pertinent or "new" information must be separated from the background or "reference" information. The background involves: (a) the invariant, highlighted by a heading (Port St. Michel); (b) the highly visible identification of each component (tonnage and dates). The new information (the curve) must stand out from the background (figure 19).

Reference points
It is impossible to utilize a graphic such as figure 20, except in a general manner. There is confusion concerning the position of the points, and no potential comparison is possible, as it is in figure 21.

Precision reading
A precision reading (utilization on the elementary level, as in figure 24) is difficult in figure 22, which results in a poor reading of the order of the points, and in figure 23, where there is ambiguity concerning the position of the points. On the other hand, figure 22 does favor overall vision (correlation).

Null boxes
Curves accommodate null boxes poorly (figure 25). Columns (figure 26) are preferable.

Unknown boxes
The drawing must indicate the unknowns of the information in an unambiguous way (figures 28 and 30).The reader might interpret figure 27 as a change in the structure of the curve and figure 29 as involving null values.

Very small quantities
Except in seeking a correlation (quite improbable here) the number of ships entering into a port is represented better by figure 33 than by figures 31 or 32. The reader can perceive the numerical values at first glance.

Positive-negative variation
This is in fact a problem involving three components O, Q, ≠ (+ –), and it must be visually treated as such. Figure 34 can be improved by utilizing a retinal variable (in figure 35 a value difference: black–white) to differentiate the ≠ component and thus highlight positive-negative variation.

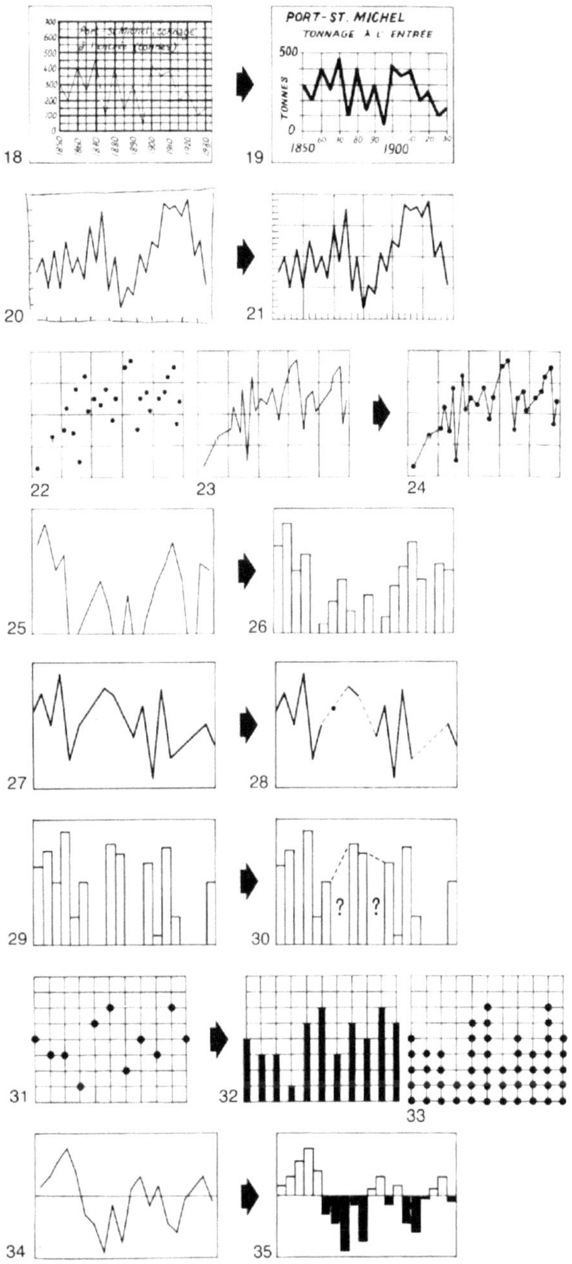

Figure 7.1B Graphic problems posed by time series, continued. (From Bertin, Jacques. *Semiology of Graphics.* © 1983 by the Board of Regents of the University of Wisconsin System. 2010 by ESRI Press. Reprinted by permission of the University of Wisconsin Press.)

The time axis at the bottom of Figure 7.1C is the common denominator for all the events in the clinical story, each event having a date and time stamp. The UI also shows, on the right side, tentative events in the future: next appointment, medications to be taken by/administered to the patient, pending orders, imaging or lab tests, and so on.

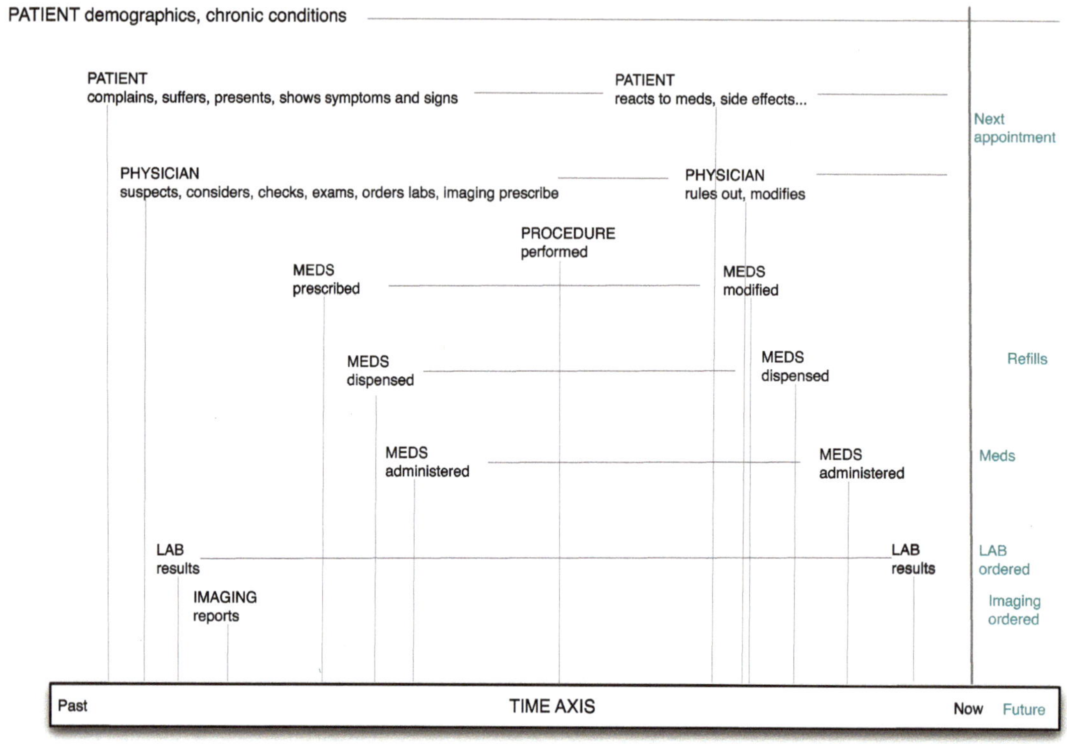

Figure 7.1C High-level clinical story.

Since the clinical story is usually quite complex, the UI needs to support clinicians in grasping the temporal relationships between the clinical events. One possible arrangement would be to anchor the clinical events to the time axis vertically and group them in their own "bands" horizontally.

Cause and Effect

In addition to allowing the user to browse clinical events during time, the UI needs to display cause-and-effect relationships. Imagine a screen with many numeric lab results and a different screen with the medications a patient is taking; the clinician needs to invest mental resources to model the story from these two disparate sources, lab and meds.

The displayed clinical story in Figure 7.2 tells us that diet alone was not enough to lower this patient's cholesterol. There is a temporal relationship between the institution of *Lipitor* (a cholesterol-reducing drug) and the lowering of the patient's cholesterol. This temporal relationship becomes clearer if we graph the cholesterol levels in relation to the diet and *Lipitor* (Figure 7.3).

We can use the same framework and increase the number of displayed parameters as well as the time resolution, showing minutes instead of months,

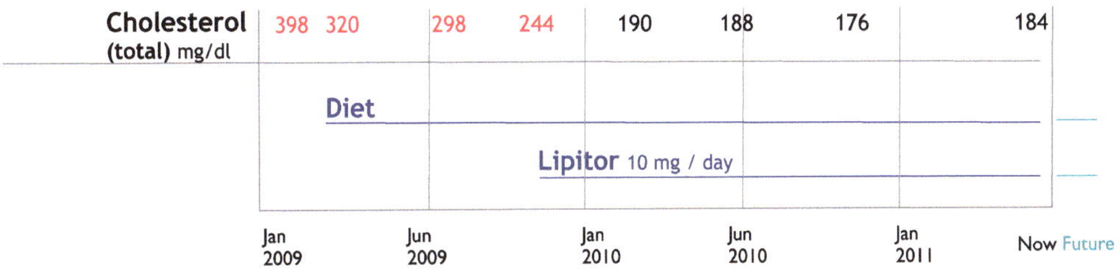

Figure 7.2 Cause and effect: Months scale.

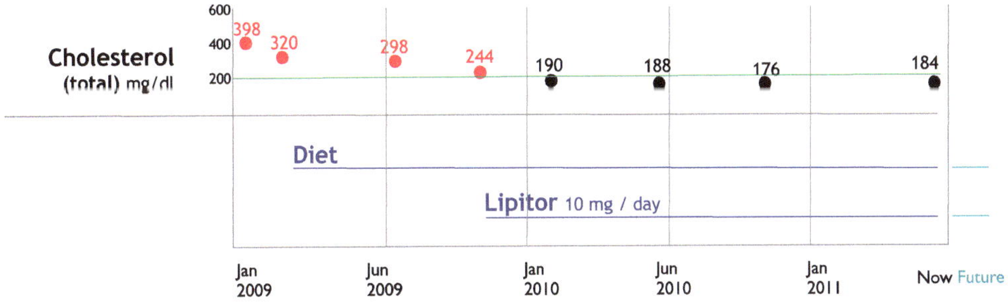

Figure 7.3 Cause and effect: Graph.

and still be able to clearly tell a clinical story (Figure 7.4). There are five parameters in Figure 7.4: patient heart rate, two drugs (*Atropine* and *Isoproterenol*), one electrocardiogram (ECG), and the insertion of a temporary pacemaker—a procedure. The time axis scale is 1 hour.

And, what about the clinical story? The patient's heart rate is spiraling down; the downward trend of the heart rate graph coupled with the use of color make this part of the story evident. Minimum and maximum values as well as last measurement are tagged with values. At one point in time (02:23) *Atropine* is injected intravenously (IV), with a larger dose repeated at 02:28 to

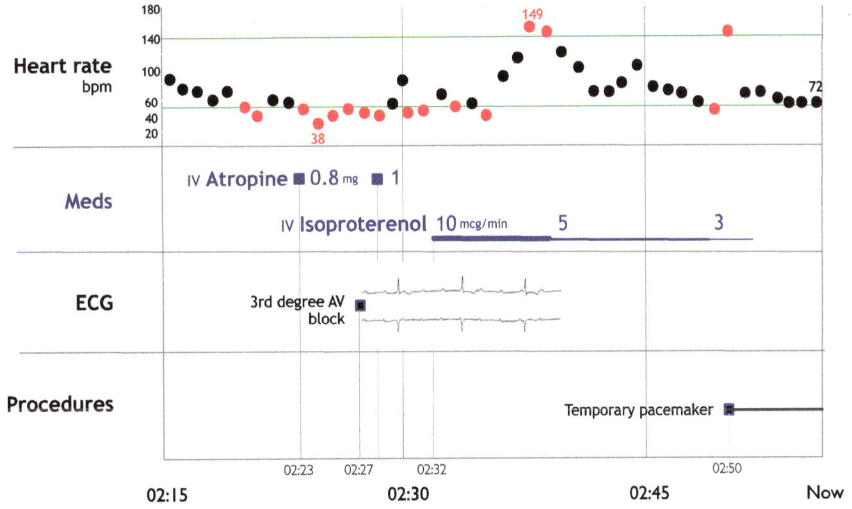

Figure 7.4 Cause and effect: Minutes scale.

treat the low heart rate. At 02:27, the ECG revealed a *third-degree AV (atrio-ventricular) block*—most probably the reason for the low heart rate. The relationships between the ECG and the drugs used, the patient's reaction to these medications, and the need for a temporary pacemaker are illustrated as a story: a series of clinical events, some discrete and some continuous but all relating to the same time axis.

Note the ECG is still decipherable even after significant shrinkage in size; graphical representations can be drastically downsized and the human eye can still interpret them.

Titrate to Effect

The infusion of *IV Isoproterenol*, in Figure 7.4, is initiated at 02:32 at 10 mcg/kg/min. It is then titrated to effect: The drug administration is changed in response to the patient's heart rate. At first, the administration is decreased to 5 mcg/kg/min since the heart rate was too high, the effect was too strong; then, the administration was later adjusted to 3 mcg/kg/min. Eventually, after the pacemaker insertion, the infusion of this drug was stopped since it was no longer needed.

Titrating to effect happens on the scale of years as well: Treating a patient's hypertension or diabetes is usually an exercise in titration to effect; the components of the medication cocktail and the medications doses are changed iteratively to achieve a specific effect gradually, with goals such as blood pressure lower than 140/90 mmHg or *HbA1C* (indicates amount of glucose in the blood over 3 months) lower than 7%.

In Figure 7.4, the UI invites visual comparison along the time axis between multiple parameters: heart rate, drugs, ECG, pacemaker, and the temporal relationships between them.

To support the clinician's cognitive processes, the UI needs to display multiple variables in parallel and make the task of visually comparing between them easy so cause-and-effect and titration-to-effect phenomena become evident.

Consider the alternative: Vital signs are in one folder, medications are in a different folder, the ECG is in its own application, and procedures are in a separate computer/system. This arrangement will necessitate the user investing energy in hunting down the pieces of the clinical story and then using brain power just to create a mental picture similar to Figure 7.4.

The EHR should take responsibility over the low-level processes of data collection, analysis, summary, and (using the UI) display of the information. Computers excel at these tasks. The clinician's mind should be free to search for patterns, compare events and trends, and grasp relationships in the clinical story—high-level cognitive activities that humans (still) perform better than machines.

Parallel Channels of Information

The need for a graphical representation of multiple parallel channels of information is not new or specific to medicine. One does not need to actually understand the opening notes of Beethoven's Fifth Symphony to see the analogy with a clinical UI (Figure 7.5). Each musical instrument has its own "information channel" depicted horizontally with musical notes. The name of the instrument is on the left. Channels run in parallel for 12 instruments sharing a single time axis; imagine how the music would sound if the different instruments were not fully in sync along the time axis.

Let us apply this analogy to the UI of an EHR (Figure 7.6). The various components of an EHR—patient history and problem list, lab, medications, and so on—replace the orchestra instruments, while the clinical events (small numbered circles) replace the musical notes. The time axis becomes the glue that keeps events in the correct temporal perspective and the clinical story elements synchronized. With parallel representation of clinical components, the UI facilitates visual comparison between multiple events and multiple time series.

Figure 7.5 Orchestra: Multiple parallel channels of information. (From Ludwig Van Beethoven *Symphony #5* © Copyright by Motoshi Katoh, 2007. All rights reserved. http://www.basso-continuo.com/, 1/1808.)

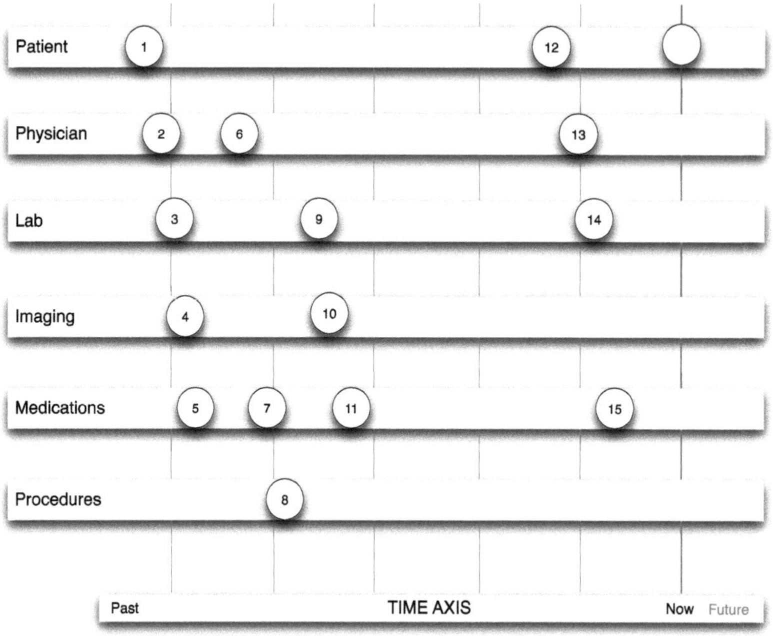

Figure 7.6 EHR: Multiple parallel channels of information.

UI Main Elements

The UI needs to summarize and display the patient's past as well as allow interactions related to the present clinical situation (Figure 7.7). Assuming parameters such as name, date of birth, address, and so on do not change often, the UI displays them in the topmost banner. The banner should always be visible to all users, so it is a good place to display other slow-changing dimensions, such as allergies and alerts.

The patient's clinical past is summarized and represented in multiple horizontal channels of information corresponding to various EHR domains. The time axis ends on the right side with the present date and time; the details are on the right side. The clinical parameters and events are displayed in a fashion that facilitates visual comparison between them as well as between the present and the past.

Time Axis

The user can navigate the time dimension (*X*-axis) with simple gestures, such as swiping left or right. These hand gestures are used extensively as touch screen UIs in mobile phones and computer tablets. Alternatively, arrows to click on with a mouse would achieve the same effect (Figure 7.8).

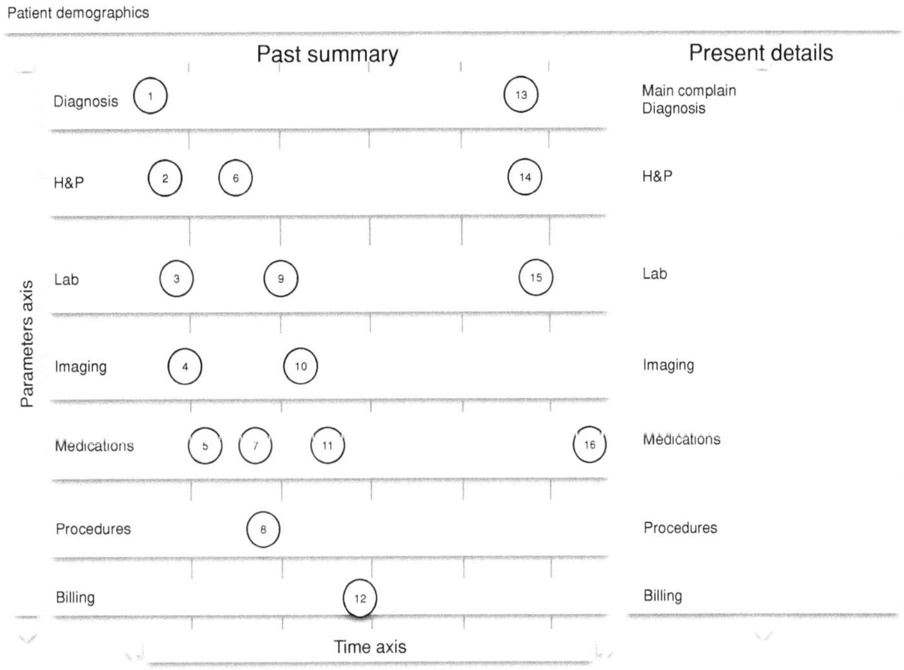

Figure 7.7 UI main elements.

The summary display acts like a carousel: It can be rolled/swiped/panned toward recent events or to the distant past (Figure 7.9). While the carousel-like display is moved left or right, the time axis maintains the correct temporal relationships between the clinical events (the small elliptical shapes).

We have identified the need for the time axis on a clinical UI to be scalable, to accommodate scales of minutes as well as years. Another hand gesture can

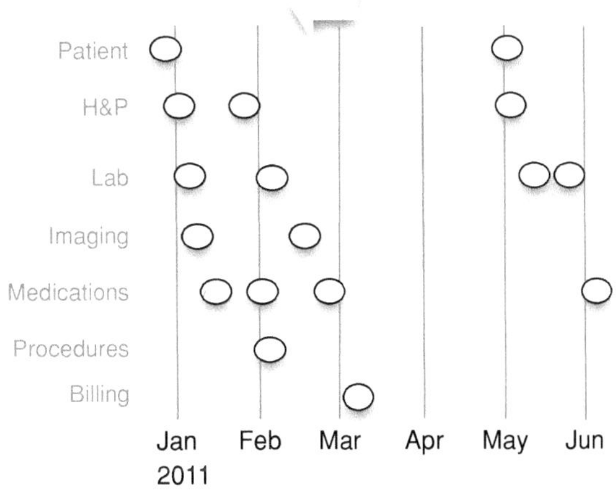

Figure 7.8 Swipe time axis to the present.

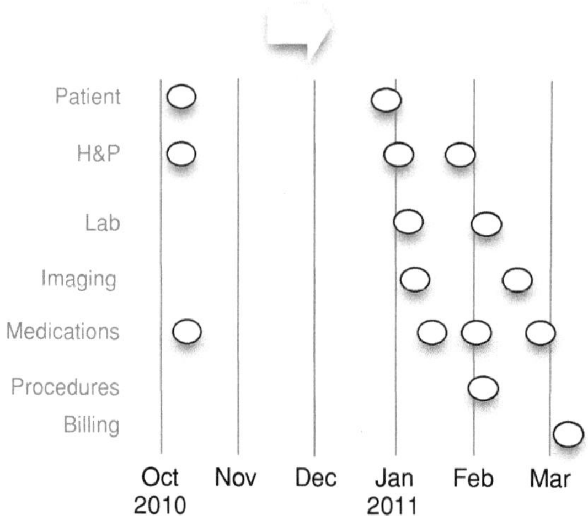

Figure 7.9 Swipe time axis to the past.

be used to expand the time scale (Figure 7.10). On the left side of Figure 7.10, a triangular shape shows the zoom level; in this case, it is the months scale.

Repeating the expansion hand gesture zooms in on more detailed views. The same zoom-in effect can be achieved by moving the bar on the zoom level triangular shape from months to weeks, days, or hours. This would change the time resolution accordingly.

The opposite hand gesture ("pinch") will contract the time scale and zoom out (Figure 7.11). The zoom scale shows now years instead of months on the time axis. The size of the clinical events should change accordingly: What seemed large on a scale of hours will contract to smaller symbols and even icons on a scale of years. Continuous parameters such as infusions or heart rate may appear as multiple graphs on a scale of days and as single discrete dots on a scale of years. The UI zoom in/out should change the time axis scale as well as the size of the clinical events displayed.

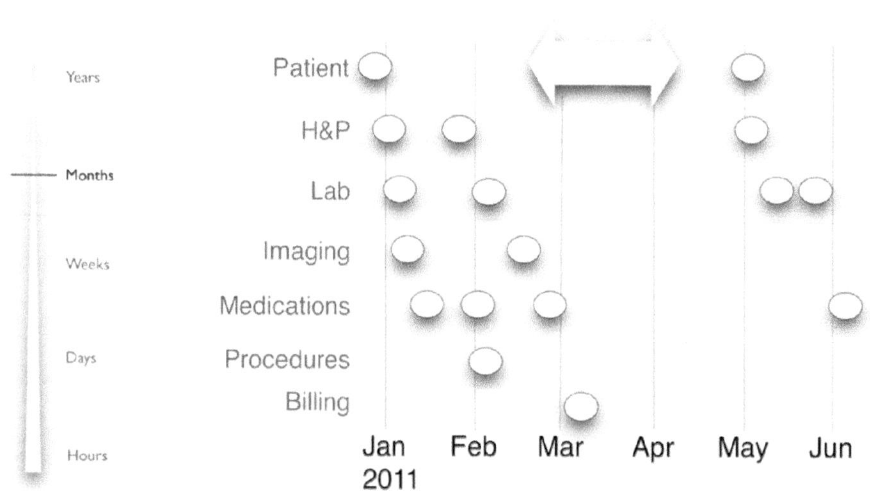

Figure 7.10 Expand time scale: Zoom in.

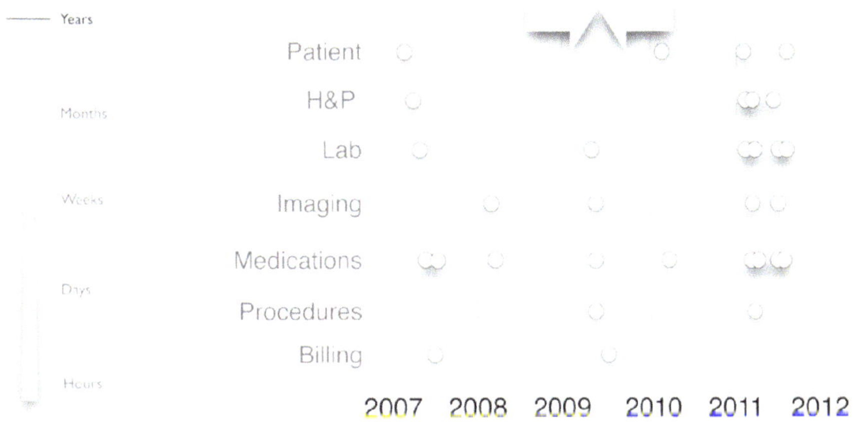

Figure 7.11 Compress time scale: Zoom out.

Caution: The Direction of Time Axis

Sometimes, the software configuration allows too much flexibility, such as allowing a reverse chronological time axis. The reason for this dubious practice is to display the parameter name as closely as possible to the last, most recent value. First, we discuss the danger inherent in a reverse chronological time axis, and then we search for better UI alternatives for showing the most recent value of a parameter.

Consider the clinical story in Figure 7.12. There are two parameters in the clinical story in Figure 7.12: blood glucose and regular insulin infusion (a drug that lowers blood glucose). A tired, distracted clinician eyeballing this UI may not notice the numbers on the time axis and be under the impression that blood glucose is on the rise. The clinician may further reason that the insulin infusion has been incorrectly titrated to effect: the higher the glucose, the less

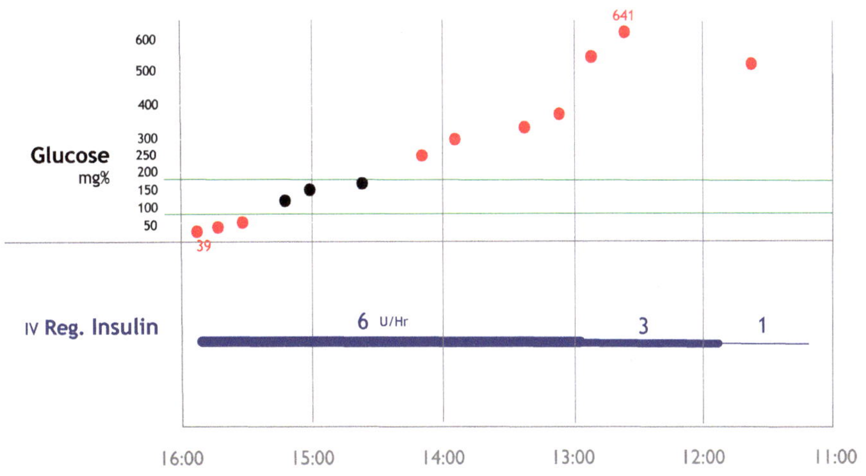

Figure 7.12 Reverse chronological order.

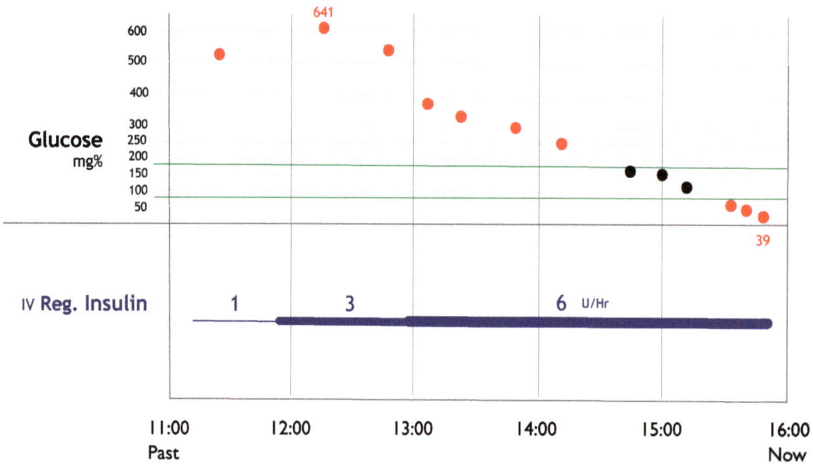

Figure 7.13 Correct chronological order.

insulin is administered. Thus, the clinician's erroneous conclusion is that more insulin is needed.

The actual clinical story is exactly the opposite (Figure 7.13). The patient is receiving too much insulin, and the blood glucose level is actually dangerously low. In such a situation, the administration of more insulin would lower the glucose further and could prove fatal for the patient.

The direction of the time axis on a graph is reflexively assumed to be from left/past to right/present. This reflex, ingrained in most of us coming from scientific backgrounds, is the result of viewing thousands of graphs during our education, professional training, and career. Anything to the contrary necessitates the clinician to invest cognitive resources to fit the UI (with a reverse time axis) to the familiar mental picture of the expected chronological direction.

A reverse chronological time axis presented to a tired or distracted clinician or even one who is apparently alert but has worked for years with systems that have the UI with a left-to-right time direction may cause serious UI-induced clinical errors.

Parameters Axis

In Figure 7.7, the various components of the EHR, diagnosis, physical exam, labs, medications, and so on are arranged in horizontal bands as channels of information. There are many components anchored to the *Y*-axis, each with its own long list of parameters, so one cannot expect to show all in one screen. There is a limit to how much even a great UI may compress graphs before the meaning of the underlying data changes. The UI should allow a clinician to view the data with minimal design distortion. For example, a certain aspect ratio between the *X*- and *Y*-axis should be maintained (Figure 7.1A).

Let us use the same carousel metaphor we used with the horizontal time axis, this time on the vertical *Y*-axis (Figure 7.14). Using a mouse or vertical swipe

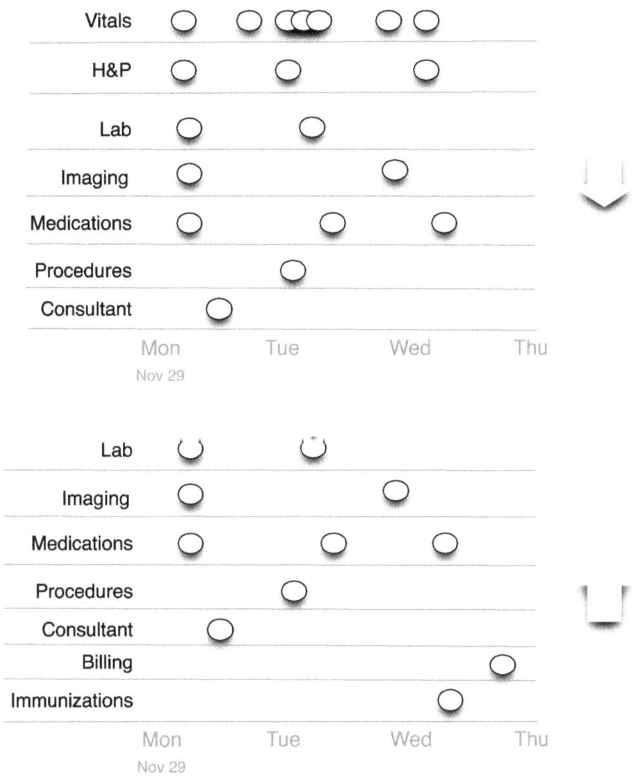

Figure 7.14 Panning vertically through EHR UI components.

gestures on a touch screen, a clinician can browse in Figure 7.14 through vitals, labs, and procedures and then scroll down to billing and immunizations.

The order in which the various EHR channels of information are arranged should be easily configured to best fit the clinician and patient type. If one prefers the H&P (*History and Physical exam*) at the top and the vitals at the bottom of the UI, this should be a simple "drag-and-drop" rearrangement of the UI (Figure 7.15).

Different specialties have different needs: Some like to see the patient's temperature, white blood cells count and their differential, the culture/sensitivity together with the antibiotic treatment—all on the same screen. Others prefer the procedure report with labs and imaging to be tightly connected and so on. According to Lund (7), "*Let people shape the system to themselves, and paint it with their own personality.*" A UI that allows easy rearrangement and configuration of the screen would be useful to a large number of medical specialties and various personalities, without the need of complex, time-consuming, and expensive configuration sessions.

Caution: Hidden Information

The UI proposed, in which the clinician can browse the time dimension horizontally and EHR domains vertically, may still hide abnormal results. Consider a case where the surgeon is viewing the patient's EHR and misses a low hemoglobin lab result,

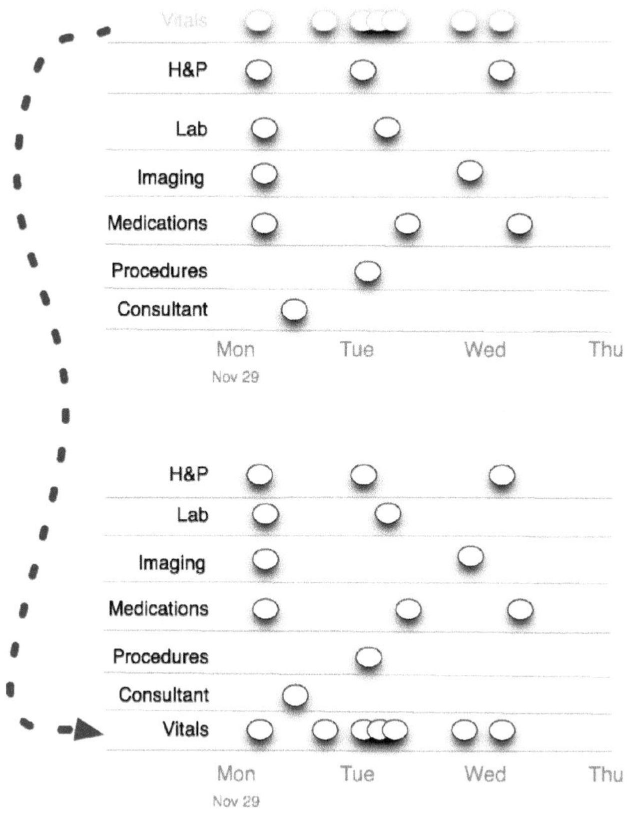

Figure 7.15 Rearranging parameters with drag and drop.

indicative of potential internal bleeding, because the result is hidden in a screen part that is not visually displayed. Just because the user did not scroll to the abnormal result is no excuse for the UI to potentially induce an omission error and cause a patient harm. Thus, a UI mechanism should be in place to prominently display recently received abnormal results, at least until the user acknowledges reviewing them. In other words, if never reviewed by the clinician, abnormal results should not be hidden by the UI.

Tabular Versus Graphical Display of Data

The top rectangle in Figure 7.16 displays three parameters: heart rate and systolic and diastolic blood pressure during 14 minutes. Even with the abnormal values in red, it is difficult see the clinical story; numbers tend to visually blend into the surrounding numbers, and the user needs to invest brainpower resources to create the clinical story told by the values.

If a clinician would like to see the parameters for one day (1,440 minutes), the user needs to scroll this rectangle more than 100 times. It is only human nature to dislike a UI option that necessitates a lot of navigation, endless, repetitive mouse clicks and multiple screens and still not be able to see the proverbial forest for the trees.

Tabular: 14 minutes

Heart rate	99	120	112	108	102	99	82	84	81	75	49	52	58	57
Sys BP	76	89	100	88	72	112	108	107	118	105	100	103	84	82
Dia BP	51	46	60	52	43	82	79	76	83	68	59	57	54	48
	12:00					12:05					12:10			

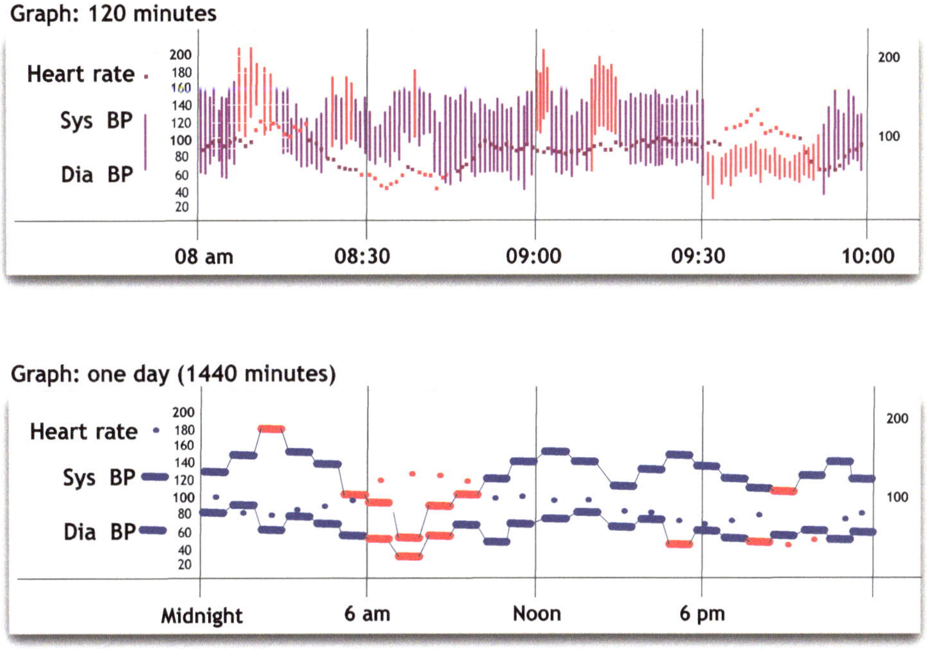

Figure 7.16 Tabular versus graphical display of data.

The middle rectangle in Figure 7.16 shows one-per-minute measurements of the same three parameters, this time during 2 hours. The data density of the (same area) middle rectangle is 120 min/14 min = 8.6 times higher than the tabular, numeric arrangement. If a user needs to view these parameters for the previous 24 hours, "only" 12 clicks are needed to see the whole picture.

The lower rectangle in Figure 7.16 uses the same area to display information for 1 day. The UI displays the hourly average for each parameter instead of the discrete once-per-minute values. The same area rectangle displays 24 averaged values of the three parameters, which represents information during 1,440 minutes. This is a data density representation, which is more than 100 times higher than the tabular format. Obviously, the user does not need to click even once to view the whole day.

The UI needs to display large amounts of information correctly, completely, and unambiguously. To do that, the UI needs to summarize (or average in the example) the data displayed. The more work the UI does in this regard, the less is the mental effort and cognitive load for the user.

Number of Clicks and Data Density

From the example, a simple relationship between data density and number of clicks becomes evident: A higher data density in a UI translates into a lower number of mouse clicks or screens to view and vice versa.

Let us assume I is the total amount of information a UI needs to display, and it is measured in discrete data elements. D is the data density measured as the average number of discrete data elements represented per UI area unit (square inch or square centimeter). Let S be the number of screens. The relationship between these variables is

$$I = D \times S$$

This formula shows that the number of screens a user needs to click through is inversely related to the data density. It means the UI with a low data density display translates into a UI with a high number of clicks to view many screens. This simple equation does not illustrate another important fact: Too many screens or clicks mean a clinician may experience a broken cognitive process.

The UI should summarize underlying data, thus decreasing I, as well as present data in a high-density, graphic-intense format, so D is maximized. Only then can the vendor of the EHR boast and celebrate having the lowest number of clicks or screens on the market.

Trends Are Nice, but Where Are the Numbers?

Clinicians like viewing trends of data and other graphical displays of information since it helps them create a mental picture of the clinical story, but they also need to know the numbers behind the graphs. A UI mechanism should enable clinicians to review a graphical display and still be able to see the actual values making the graphs (Figure 7.17).

A possible solution, used by several financial Web sites to monitor stock changes, may be a vertical bar on the UI, which can be moved right or left over the graphical display, and while it rests or hovers over a point in time, it shows the values of the graphs corresponding to that time. Such a UI mechanism coupled with the capability to expand and contract the time axis will cover all the resolution spectrum: from years to minutes. It is a graphically intense, high-density, dynamic UI with a wide visualization scalability range.

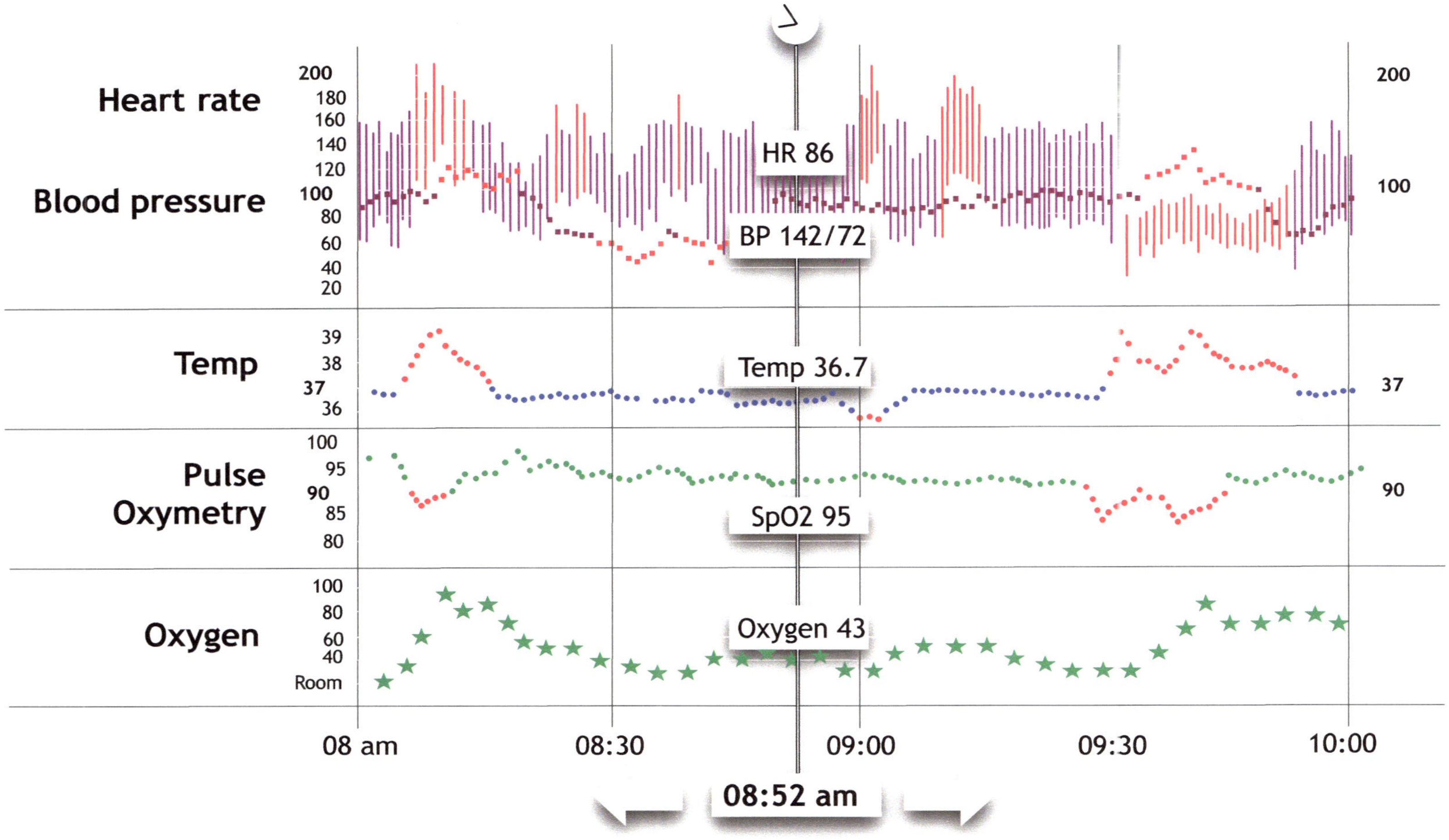

Figure 7.17 UI mechanism to view graph values.

Layers of Information

Not all the discrete data elements have the same clinical importance. The route, name, strength, and dose of a medication are clinically more important than the name of the clinician who ordered, dispensed, or administered the drug. This second layer in turn is more important than the brand names, price of the drug, or formulary alternatives and so on. Clinical information comes in layers of importance. This fact should be represented in the UI with different sizes and types of font, careful use of color, hiding clinically unimportant information, and a mechanism to drill down the data if necessary.

Clinically, the most important part of the UI in Figure 7.18 is to convey the first layer of information: The patient is receiving an *IV infusion of Isoproterenol* at specific rates during time. Hovering over the name of the drug reveals the second layer of information: who has ordered this drug and when, who has administered and when, and so on. Clicking on the drug name allows drilling into even more details. To decrease the cognitive load, the UI should not be cluttered with unnecessary information, so the second and third layers of information are not visually present, at least initially. Still, the UI needs to facilitate drilling down the layers of information if a user needs to see more details.

Review Patient Data: Ambulatory UI

Let us check how a UI with an ambulatory time frame of years may look for patient *Lastname Firstname*, a 69-year-old Caucasian male (Figure 7.19). The uppermost banner contains the patient demographics and other dimensions that change infrequently, such as weight, height, calculated body mass index (BMI). Since the patient banner is expected to be visible at all times and by all clinicians, it has alerts such as allergy information clearly shown in a central location.

The time axis displays almost 6 years of information in horizontally arranged channels: history, medications, labs, imaging, ECG, and immunizations. More EHR components are available but not visually displayed unless a user scrolls down the patient summary part of the screen. If there is any abnormal value in a horizontal band, the UI should make it visible until a user acknowledges reviewing the information.

The patient's history shows he suffers from *asthma, hypertension,* and *diabetes*, each chronic condition with the corresponding time frame. It also shows when the patient quit smoking as well as how and when he was treated for benign prostatic hypertrophy by *TUR/P* (*TransUrethral Resection of Prostate*). Uppermost there are two episodes of *pneumonia*: *July 2006* and *October 2009*.

The drugs the patient is taking for his chronic conditions and any changes in the dosing regimens are detailed in the next horizontal channel of medications. The antibiotic used for treating the pneumonia in the past is illustrated as *Azithromycin* in 2006 and *Amoxicillin/Clavulanate and Clarithromycin* in 2009.

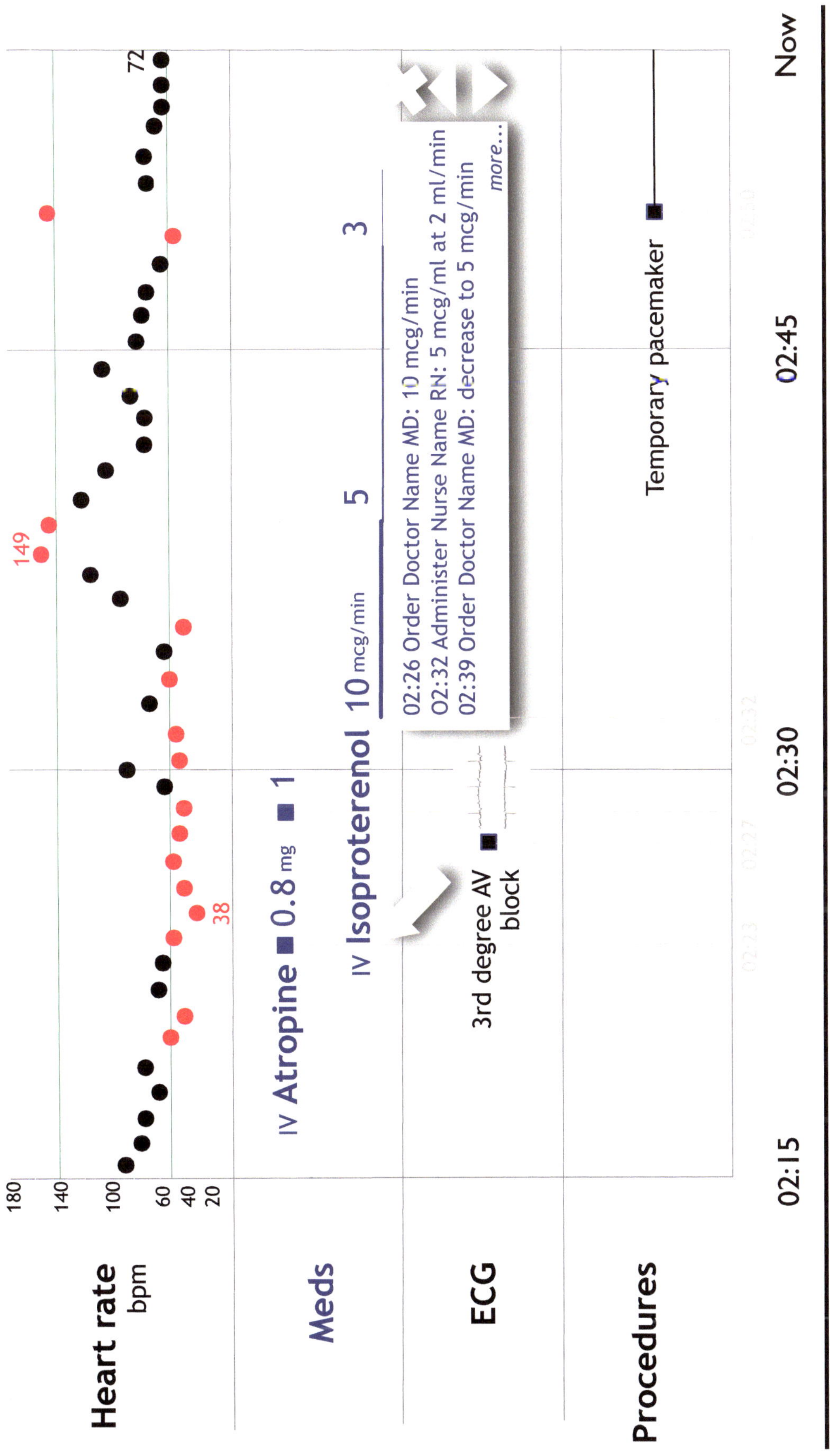

Figure 7.18 Second layer of information.

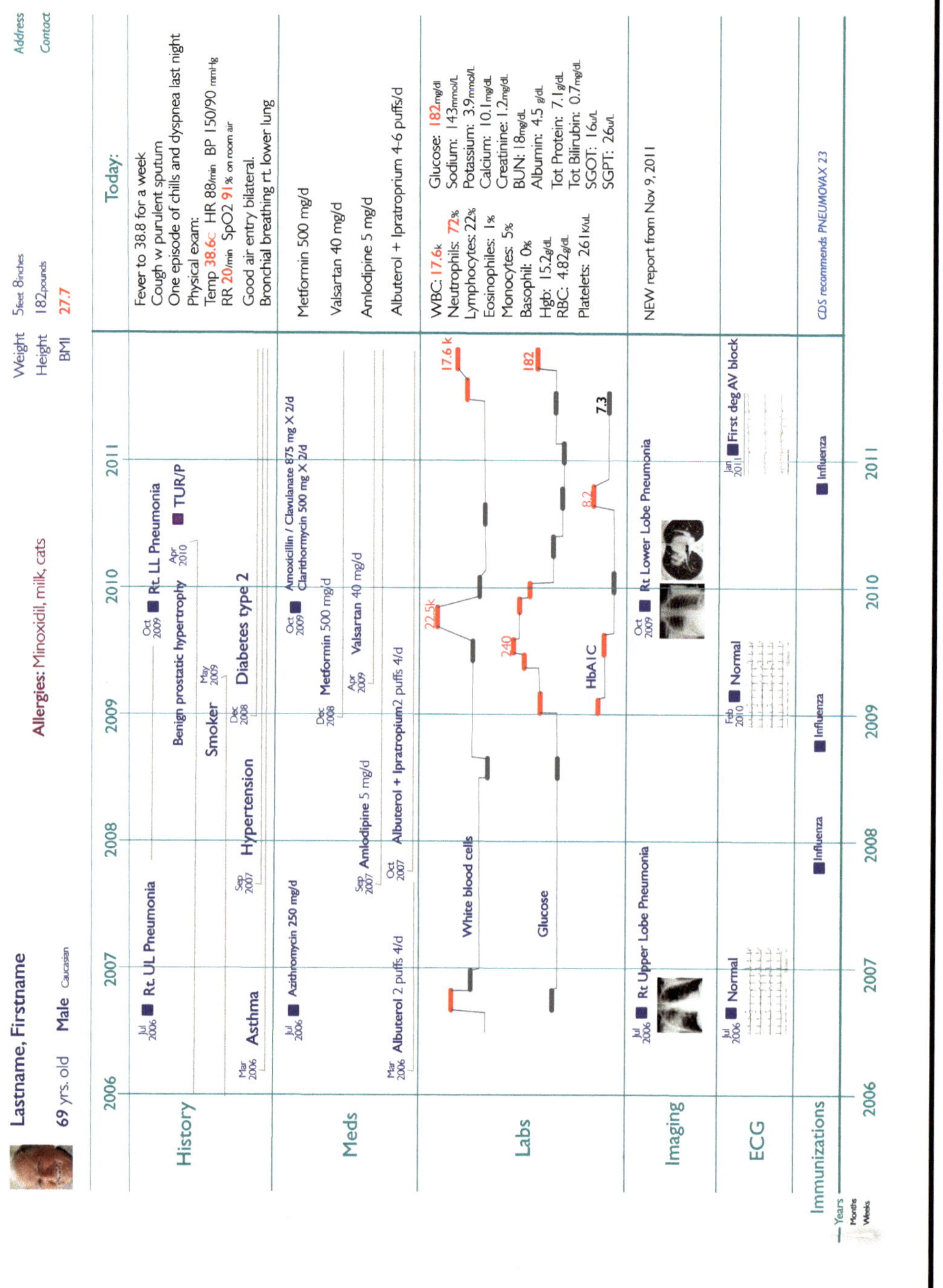

Figure 7.19 Ambulatory UI for reviewing years of patient data.

The temporal correlation between the history (pneumonia), medications (antibiotics), labs (white blood cells), immunizations (lack of influenza vaccine in 2009), and imaging done at that time (chest X-rays and computed tomographic [CT] scan) are visually displayed on the same time axis, facilitating the cognitive processes of building a mental model of the patient's clinical story.

Each of these discrete data elements can be drilled down; for example, clicking on the July 2006 icon of the chest X-Ray will bring the imaging study performed at that time and the relevant radiology report.

The rightmost part of the screen is dedicated to the present details. The history, medications, and the like follow the same arrangement as the past summary to facilitate visual comparison.

The red color is used for abnormal or dangerous values, black for the present and blue for past clinical details and navigation buttons, while teal is used for system-related UI elements such as titles, names of elements and categories, clinical decision support (CDS) recommendations, formulary alternatives, hyperlinks, and so on.

Refill a Medication

Clicking anywhere within the *Meds* horizontal band brings up an electronic version of a prescription (Rx) pad (Figure 7.20). The clinical story of the patient is visually present while the clinician prescribes a medication, so there is no need to visit several folders to review pertinent clinical information. To refill drugs, a clinician has to go through a short-and-simple process: Select the drug(s) from a list of medications the patient is taking, click on the "Refill" button, and then review and sign the prescription.

Modify a Medication

Let us examine the prescription pad a bit closer (Figure 7.21). For each drug, the route, form, brand and generic names, strength, and dose as well as the current status are presented. The topmost horizontal menu of the prescription pad displays options for a user to apply to the items selected by the check marks: refill, renew, stop, undo last changes, save modifications, and an "info button," which will drill down into the pharmacological details of drugs selected. These processes and the database structure to support them have been analyzed in previous chapters.

The clinician may modify the strength or dose of a medication using drop-down menus, which present only relevant, constrained options. For example, *Tab Diovan* comes as 40, 80, 160, or 320 mg/tab, so these values only will populate the drop-down list of strengths for this drug. CDS (the subject of the next chapter) alerts in real time on the high dose considered. Clinician can drill

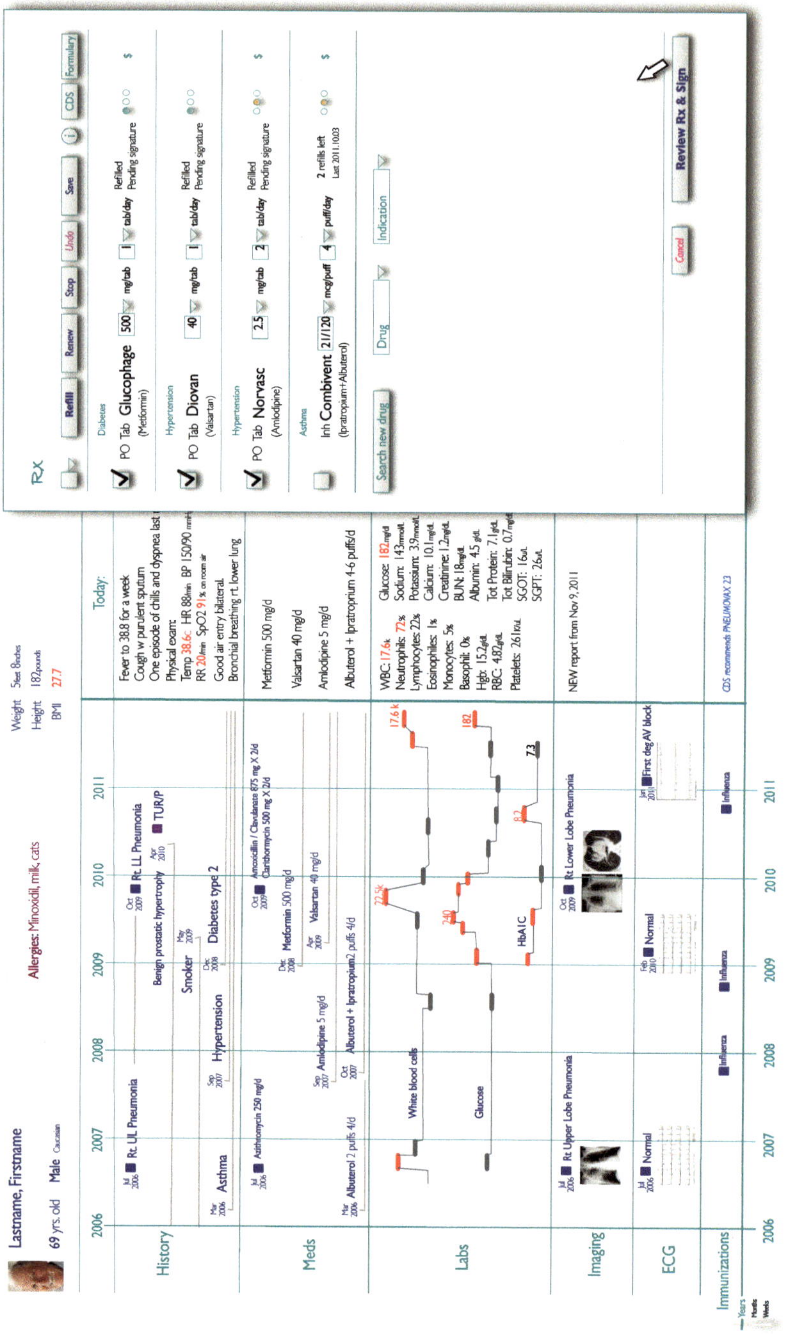

Figure 7.20 Refill medication.

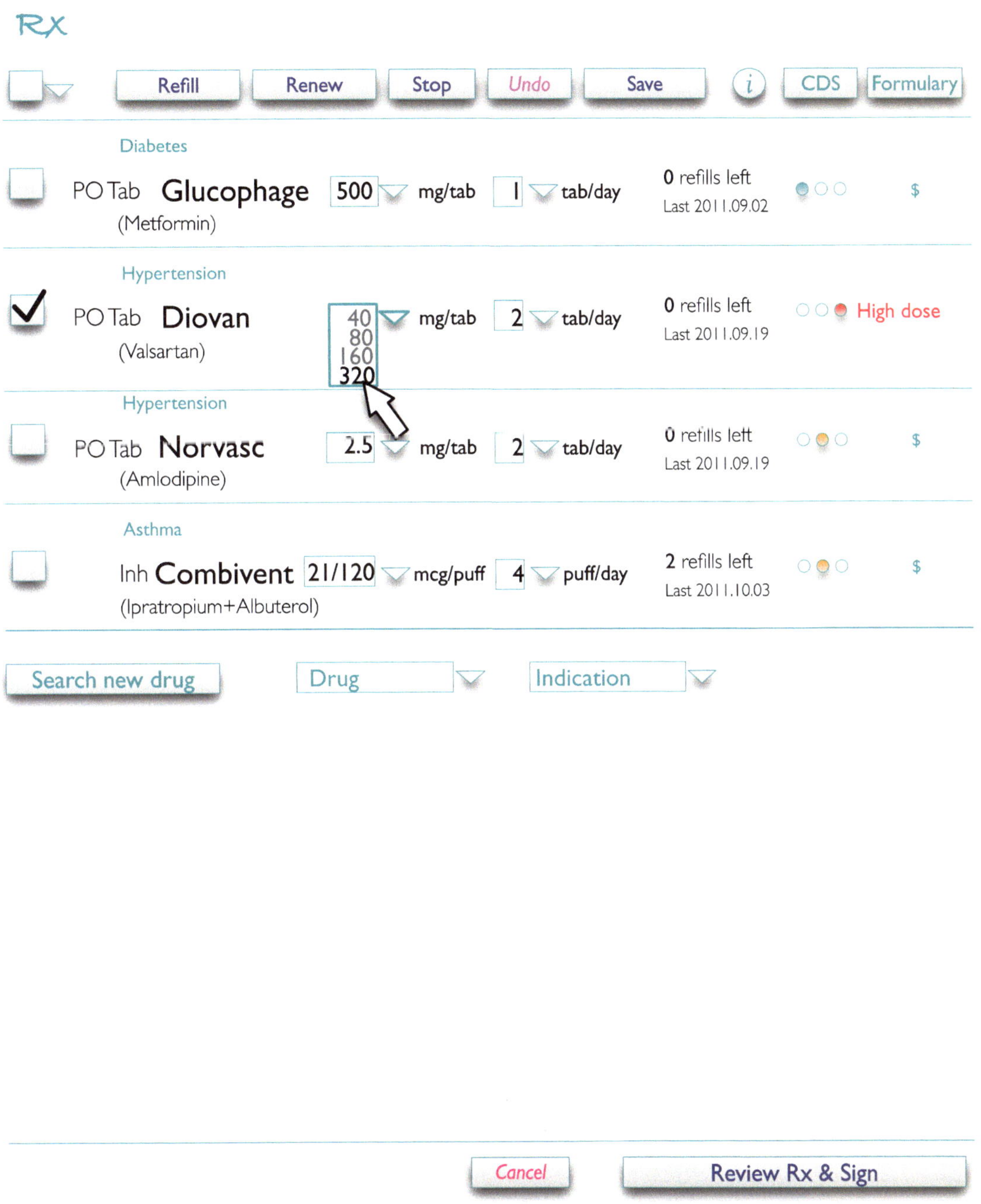

Figure 7.21 Modify medication.

down by clicking on the CDS icon and review drug interactions, age, gender, weight-related recommendations, and so on.

For *Glucophage, Norvasc,* and *Combivent* medications in Figure 7.21, there is a $ sign, indicating that a less-expensive therapeutic alternative may exist under the patient's health plan formulary. The user may review this information by hovering over or clicking on the $ sign.

In this UI example, the CDS and formulary information is not forced on the user, but it is still readily available for a more detailed view if necessary.

Prescribe a New Medication

As mentioned in previous chapters, the clinician may prescribe a new drug, which can be searched using the drug name (generic or brand), ingredient, or classification. Alternatively, a user may search for a medication with the patient's condition or indication, such as *Hypertension* in Figure 7.22.

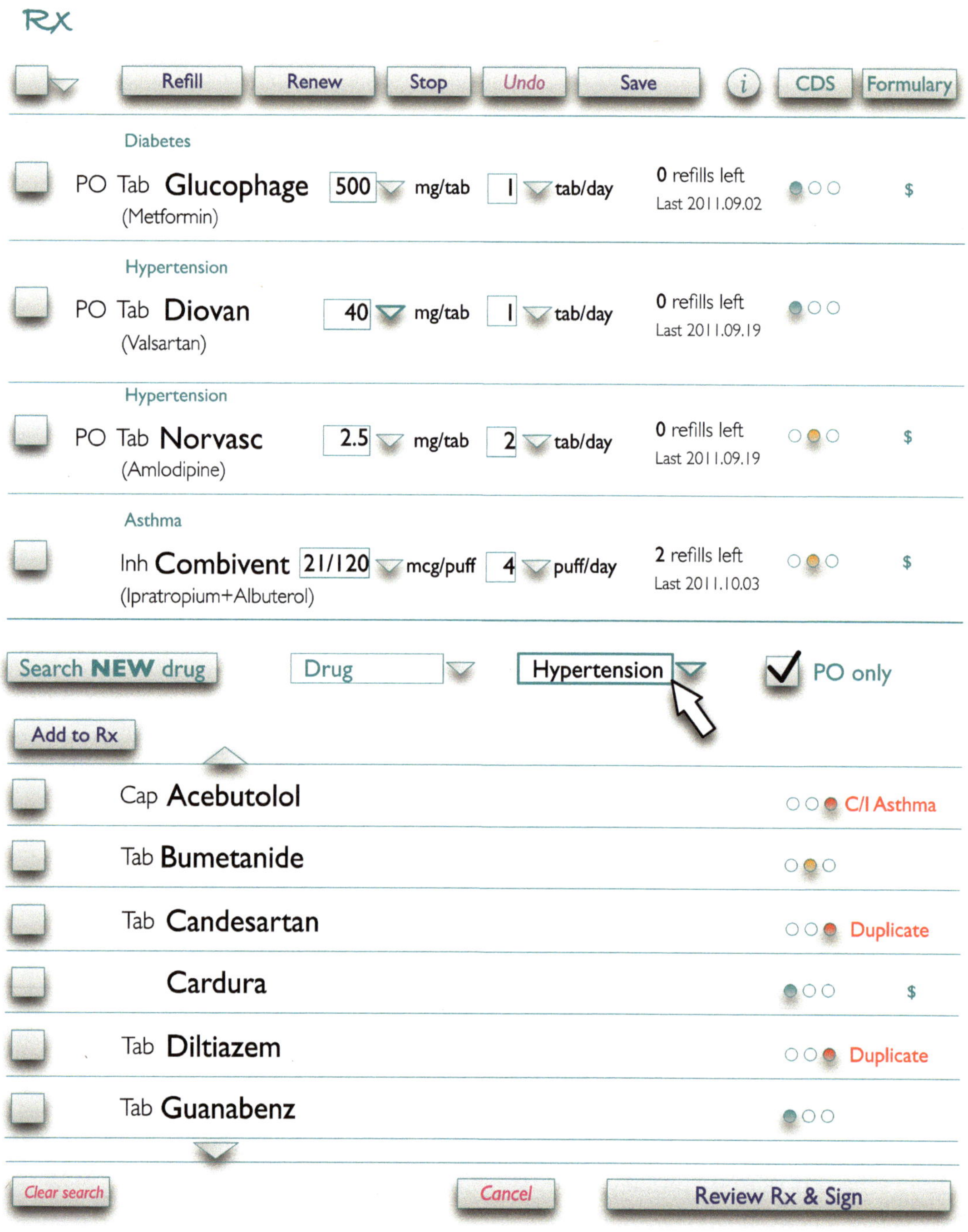

Figure 7.22 New medication prescription.

The UI responds to the query on oral (PO) drugs indicated for *Hypertension* with a list of medications, which in Figure 7.22 is arranged alphabetically. At this stage, the medication strength and dosing regimens are not shown. Several CDS alerts are clearly displayed, such as *Acebutolol C/I [contraindicated] in asthma* or *Candesartan* and *Diltiazem* being considered *duplicate therapy* with the patient's existing medications. The user can drill down and review the CDS reasoning in depth by clicking on the CDS icon. Alerting the clinician is not totally color dependent; it can be visualized also with the three-circle icon. The clinician may also rearrange this list so the CDS-filtered, safest drugs appear topmost.

Once selected, a medication moves to the upper list and joins the other drugs the patient is taking. The clinician can then define the drug strength, dose, SIG (signature), and number of refills.

Order Set

Let us follow our patient into the hospital as he is being admitted for *community-acquired pneumonia,* and his clinician uses an order set (Figure 7.23). An order set may include non-medication-related information, such as level of activity, monitors to be employed, and so on (see Chapter 4). The medications section of the *Admission: Pneumonia order set* in Figure 7.23 displays a list of *IV (intravenous) antibiotics* considered as "good practice" by evidence-based medicine (EBM) for treating *community-acquired pneumonia. SC (subcutaneous) Heparin,* an agent used to prevent *deep vein thrombosis* (*DVT*) is being shown as selected by the user. The clinician checks the medications deemed appropriate for the patient, sets the strength and dose, and then reviews and signs the order set.

Usually, ordering physicians are not involved in the selection of the type and size of the fluid bag used to prepare a solution, the intravenous access type, or injection location; these are covered by the pharmacy dispensing or the nurse administering the drug (see the next section).

Formulary information is not displayed, assuming the standing order set was considered in this regard by the hospital pharmacy and formulary committee, before its activation.

Medication Administration

The nurse administering a drug has information needs similar to other clinicians: review the patient's clinical story, select medications, medication strengths and administration regimens, and so on, which can be satisfied with a UI modeled as in Figure 7.20. Instead of using a prescription pad, the nurse documents the administration of a drug in the *Medication Administration Record* (MAR) (Figure 7.24).

There are specific requirements for medication administration, which need to be represented in the MAR UI as detailed in Chapter 6. The UI the nurse is using

Figure 7.23 Order set.

has other parameters exposed, which are pertinent to the *Administer* workflow. Drugs that can be administered as a bolus (single time, discrete) as well as continuously (drip) have a selection button, like *Heparin* in Figure 7.24.

For medications administered only as a continuous drip, such as *Dopamine*, the following calculator is exposed in the UI:

- The nurse may enter the medication quantity, and the system calculates volume or vice versa.
- A weight-based dose may be entered, and the system translates this into pump rate or vice versa.

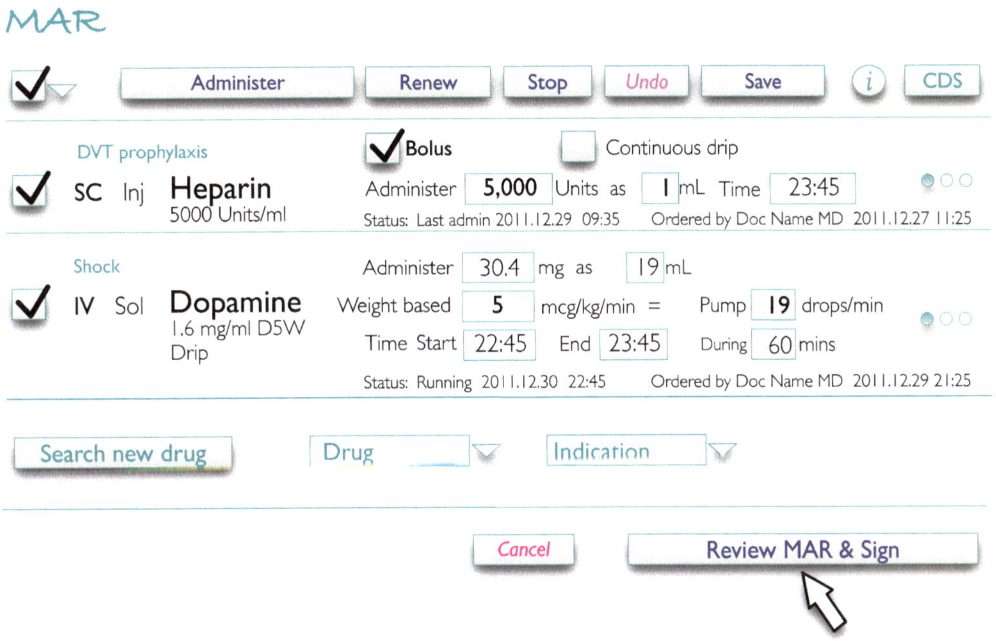

Figure 7.24 Medications administration.

- A start/end time frame is provided, and the system calculates the duration of administration, or
- A duration is given, and the system calculates a time frame (using current time as the start point).

Combinations in which only part of the parameters are being input, such as weight-based dose and start time only, are allowed, and the missing parameters are calculated using the formulas detailed in the previous chapter.

Medication Reconciliation

Either because of the high-quality healthcare experience or in spite of it, our patient is much better and ready to go home (the "D" in *ADT—Admission Discharge Transfer*). As previously discussed in Chapter 4, usually around *ADT* events, there is a need for medications to be reconciled. The process of *Meds Rec* involves the comparison of drugs and dosing regimens across venues of care: home, emergency department (ED), long-term care, ICU, ward, and so on. The UI should thus support the cognitive evaluation of time frames, locations, and their respective drugs and doses (Figure 7.25).

The clinician performing the *Meds Rec* can review on the left side of the display in Figure 7.25 the drugs the patient was taking before admission and the medications administered during the time of hospitalization. Medications selected on this side can be moved to the right-sided discharge prescription pad using the

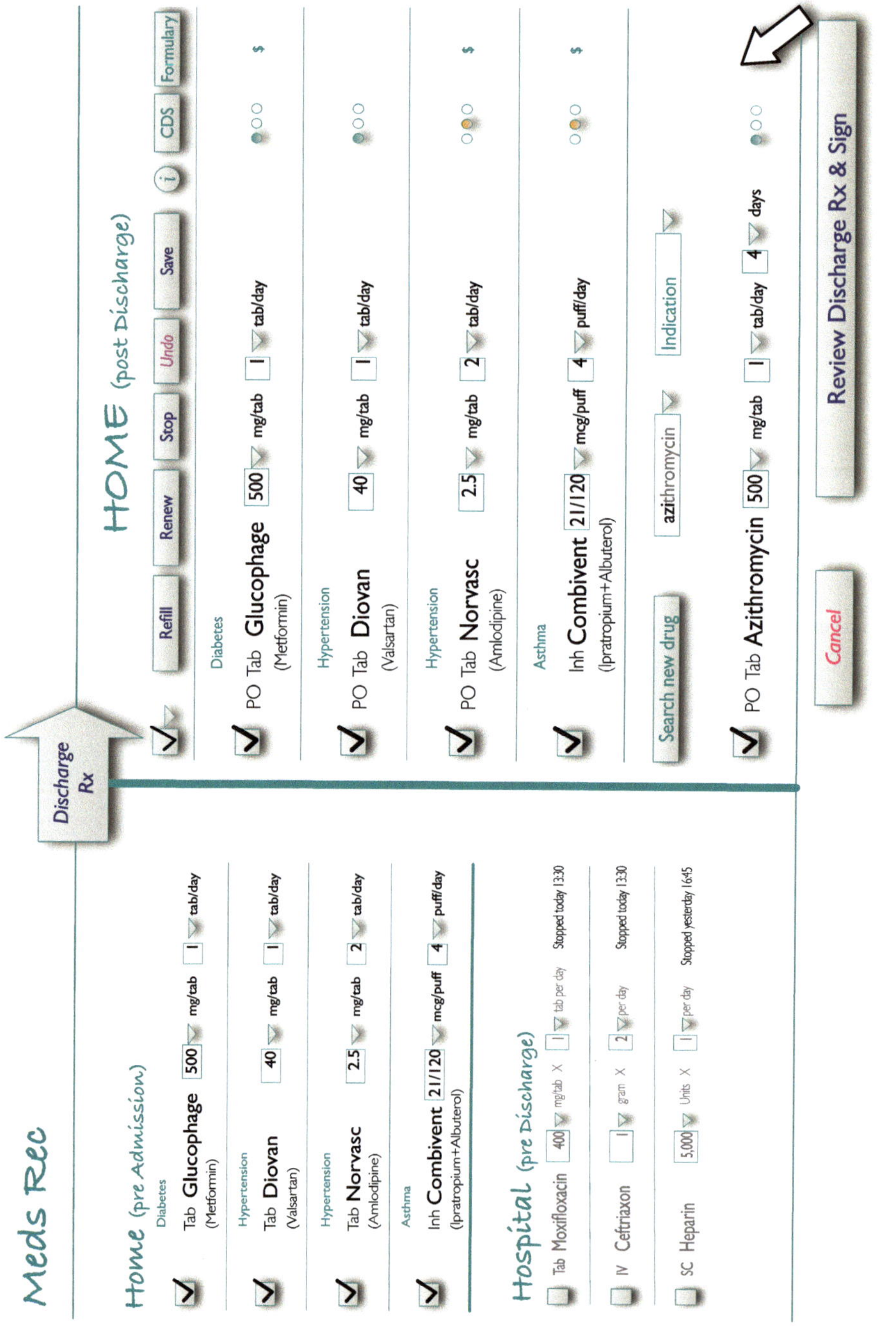

Figure 7.25 **Medication reconciliation.**

large arrow button at the top. Once in the prescription pad, strength and dosing regimens may be adjusted, medications may be refilled or renewed, and so on. New drugs may be prescribed using a mechanism described previously. CDS and formulary information is one click away. Finally, the user reviews and signs the discharge document and prescription and wishes well to our patient.

Summary

This chapter defined usability, presented several graphic principles, and then analyzed a clinical UI.

The chapter explained the need to support a clinician's cognitive processes, hence the requirement for the system to summarize and for the UI to display large amounts of information within one eye span. The utilization of the time axis as the glue to keep all EHR components in sync and the zooming—years to minutes—were illustrated.

From general clinical UI requirements, we focused then on the visualization needs specific to the medications domain: prescribe, order, administer, and reconcile—workflows and data structures that have been analyzed in previous chapters.

References

1. Ergonomic requirements for office work with visual display terminals (VDTs)—Part 11: Guidance on usability. ISO 9241-11:1998.
2. Schumacher, R.M., and S.Z. Lowry. 2010. NIST guide to the processes approach for improving the usability of electronic health records. National Institute of Standards and Technology NISTIR 7741 (November 2010). http://nist.gov/itl/hit/upload/Guide_Final_Publication_Version.pdf (accessed April 18, 2011).
3. Software Usability Measurement Inventory (SUMI). http://sumi.ucc.ie/ (accessed April 17, 2011).
4. System Usability Scale (SUS). http://usabilitynet.org/trump/documents/Suschapt.doc (accessed April 10, 2011).
5. Stead, W.W., and H.S. Lin (editors). 2009. *Computational technology for effective health care—Immediate steps and strategic directions*. Washington, DC: National Academies Press.
6. Belden, J.L., R. Grayson, and J. Barnes. 2009. *Defining and testing EMR usability: Principles and proposed methods of EMR usability evaluation and rating*. HIMSS EHR Usability Task Force. Chicago: HIMSS.
7. Lund, A.M. 1997. Expert ratings of usability maxims. A study of the heuristics design experts consider important for good design. *Ergonomics in Design* 5(3): 15–20.
8. Tufte, E.R. 2001. *The visual display of quantitative information*. Cheshire, CT: Graphics Press.
9. Tufte, E.R. 1997. *Visual explanations: Images and quantities, evidence and narrative*. Cheshire, CT: Graphics Press.
10. Tufte, E.R. 2001. *Envisioning information*. Cheshire, CT: Graphics Press.

11. Tufte, E.R. 2006. *Beautiful evidence*. Cheshire, CT: Graphics Press.
12. Powsner, S.M., and E.R. Tufte. 1994. Graphical summary of patient status. *Lancet* 344: 386–389.
13. Wong, D.M. 2010. *The Wall Street Journal guide to information graphics: The dos and don'ts of presenting data, facts and figures*. New York: Norton.
14. Lidwell, W., K. Holden, and J. Buttler. 2003. *Universal principles of design. 125 ways to enhance usability, influence perception, increase appeal, make better design decisions and teach Through design*. Minneapolis, MN: Rockport.
15. Tidwell, J. 2006. *Designing interfaces*. Sebastopol, CA: O'Reilly.
16. Scott, B., and T. Neil. 2009. *Designing web interfaces*. Sebastopol, CA: O'Reilly.
17. Bertin, J. 1983. *Semiology of graphics*, 214–215. Redlands, CA: Esri Press.

Review Questions

1. Product A: 88% of trained users select the correct drug and dose in less than a minute using 12 clicks. Product B: 86% of trained users select the correct drug and dose in less than a minute using 4 clicks. Which EHR is more effective, and which one is more efficient?
2. What are the potential UI issues in the following figure?

History Physical exam Diagnosis Problems list **Medications** Prescriptions Labs Imaging Procedures Consultations Reporting Billing CDS Formulary	**Medications** Tab Glucophage 500 mg/tab 1 tab/day Tab Diovan 40 mg/tab 1 tab/day Tab Norvasc 2.5 mg/tab 2 tab/day Inh Combivent 21/120 mcg/puff 4 puffs/day

3. What difficulties should one expect with the UI in the following figure?

History
Physical exam
Diagnosis
Problems list
Medications
Prescriptions
Labs
Imaging
Procedures
Consultations
Reporting
Billing

Labs 01/12/2011

	Result	Unit	Range
Sodium	140	mmoL/L	133-143
Potassium	4.3	mmoL/L	3.6-5.2
Chloride	103	mmoL/L	98-112
*Glucose	139	mg/dL	70-100
BUN	14	mg/dL	7-18
*Creatinine	1.4	mg/dL	0.6-1.3
Calcium	10.1	mg/dL	8.8-10.5
*Troponin-I	1.8	ng/mL	0.0-0.06
Total bilirubin	0.5	mg/dL	0.0-1.0

4. What may interrupt the clinician's cognitive processes in the following figure?

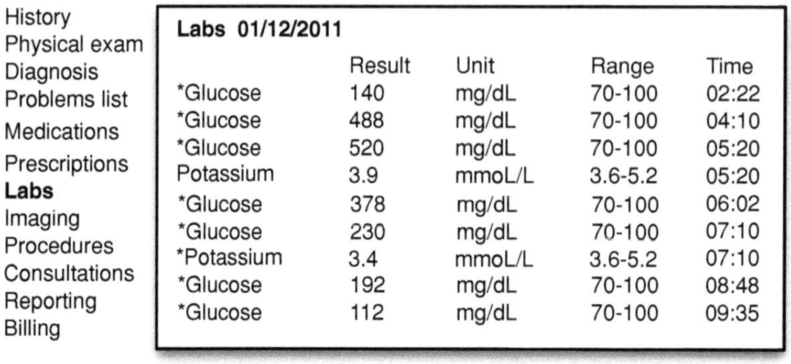

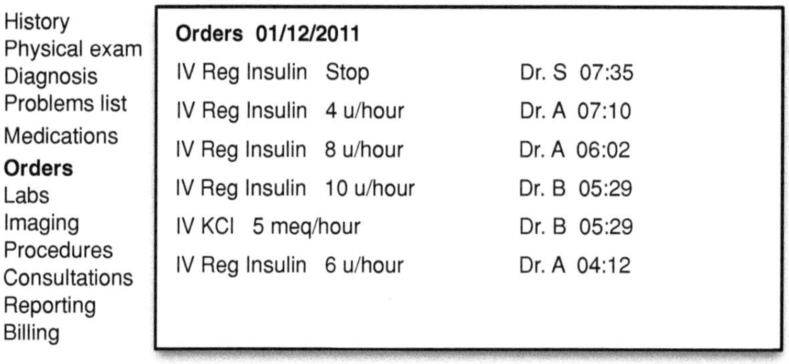

5. Assuming there are 3,600 discrete data elements to be displayed for a specific time frame, how many screens need to be reviewed if the average data density is (a) 120 elements/screen? (b) 1,800 elements/screen?
6. How would you improve the UI in the following figure?

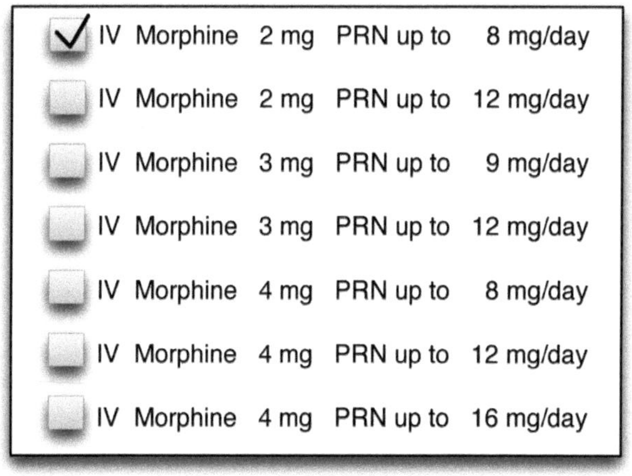

7. Why is the system assuming a rectal suppository should be used in the right eye according to the figure in Question 6? What went wrong?

8. Tab Diovan comes in strengths of 40, 80, 160, and 320 mg/tab. Why is the Diovan strength drop-down menu showing different values in the figure in question 6?

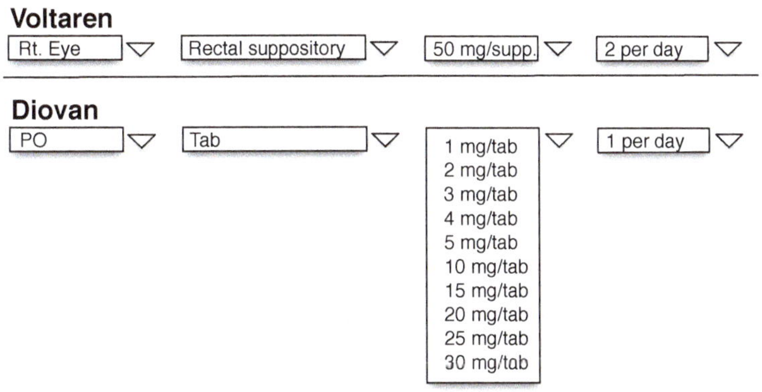

What would you suspect is not working correctly?

9. How would you improve the UI of the patient banner in the following figure?

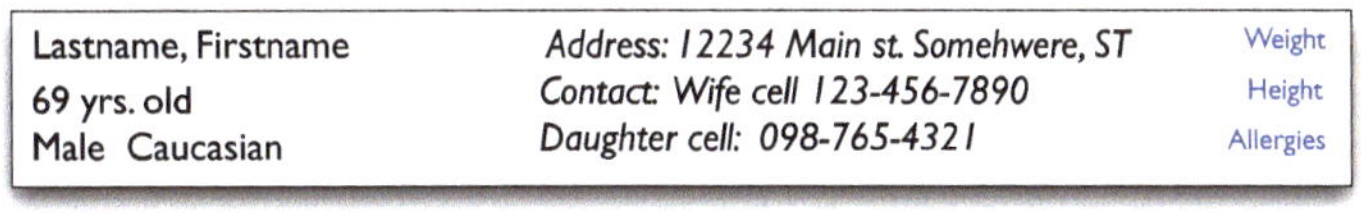

10. "User should be in a good mood when done." Why wouldn't a clinician be in a good mood viewing the display in the following figure?

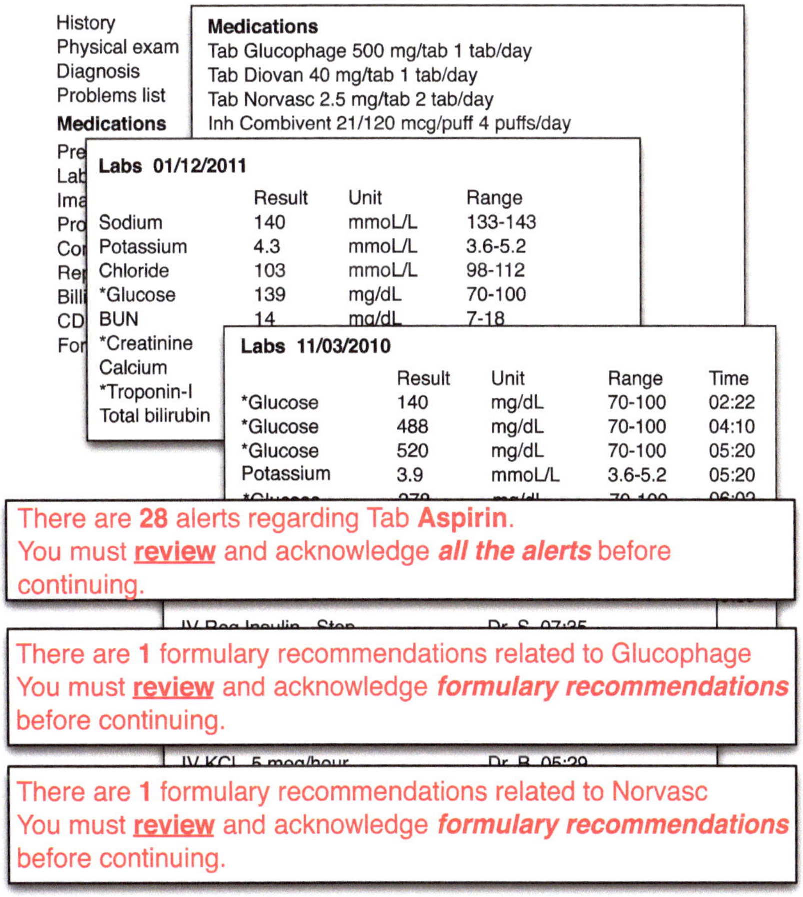

Chapter 8

Clinical Decision Support

We made too many wrong mistakes.

Yogi Berra

In this chapter, we define *Clinical Decision Support* (*CDS*), explain why it is needed, and examine best practices to develop and implement CDS. General CDS characteristics are discussed, and then we focus on the medications domain.

The issues of false positive versus false negative, "alert fatigue," and the high percentage of alert override by clinicians are presented, followed by several tentative solutions to these problems, such as tiering the alerts by severity, frequency, and quality of supporting documentation, allowing CDS to be configured at the institution as well as the clinician levels and automating the CDS workflow instead of following an unnerving and time-consuming dialog paradigm between clinician and computer.

CDS workflow, as shown in previous chapters, has three main parts: support the selection of the safest drug for a given patient, adjust the optimal dose, and provide educational material. During the optimization of drug and dose, CDS has to process rules, recommendations, and considerations related to patient demographics (age, weight, gender, etc.); conditions (such as therapeutic intent, evidence-based medicine, allergies, concurrent diseases, pregnancy); other medications a patient is taking (interactions, duplicate therapy, intravenous admixtures); and laboratory results.

These CDS processes and the relevant data structures are analyzed with tools that should be familiar by now—dataflow diagrams (DFDs) and entity-relationship diagrams (ERDs)—while emphasizing the automated aspects of CDS.

Common barriers to a successful implementation of CDS are summarized with some practical recommendations on how to initiate a CDS program.

Finally, we catch a glimpse into the near future: personalized medicine based on pharmacogenomics CDS.

What Is CDS?

Clinical decision support (CDS) is blessed with many definitions. The concise one is as follows: "*A computer-based system that assists physicians in making decisions about patient care*" (1). A more detailed definition emphasizes the need to connect between the individual's clinical information and the relevant medical knowledge:

> *Software that is designed to be a direct aid to clinical decision-making, in which the characteristics of an individual patient are matched to a computerized clinical knowledge base and patient-specific assessments or recommendations are then presented to the clinician or the patient for a decision. (2)*

In a retrospective review of 70 studies, Kawamoto et al. (3) identified the essential factors leading to a successful utilization of CDS, as measured by improved patient outcomes: CDS should be fully integrated with the clinical workflow, and it must support the decision-making process with recommendations, not just assessments, at the time and location of care, rather than prior to or after the patient encounter.

Types of CDS

CDS comes in several flavors, depending on the clinical domain it is supporting:

Diagnostic: using symptoms and signs to enhance diagnostic accuracy
Medication: the subject of this chapter
Preventive: optimization of preventive care
Testing: optimal utilization of lab and imaging tests
Procedure: surgical and anesthetic decision support

Common to all these systems is the need to provide in real time, at the point of care, relevant high-quality medical information to support a clinician's decision-making process.

Why Is CDS Needed?

In the MEDLINE and Cochrane repositories, 1,500 medical articles and 55 new trials are published every *day* (4). The exponential increase of knowledge makes it impossible even for the best-intentioned and well-read clinician to keep abreast of the current best practices and EBM. This rapid growth of medical information is expected to accelerate with the dawning of the era of genomics and personalized medicine.

The *Institute of Medicine* 1999 report *To Err is Human* estimated that between 44,000 and 98,000 people die every year in the United States due to medical errors, and a significant proportion of these errors are related to medications (5).

The *"Roadmap for National Action on CDS"* calls for *"optimal, usable and effective clinical decision support ... widely available to providers, patients, and individuals where and when they need it to make health care decisions"* (6). CDS is expected to help reduce medical errors, especially in the medications domain (7).

Clinical Decision Characteristics

The significance of a clinical decision is usually based on several characteristics (8):

Severity of the situation: severe, moderate, minor
Frequency and predictability of occurrence: highly probable, common, rare
Quality of the documentation supporting the decision (see the discussion in the next section)

CDS alerts and recommendations should be tiered according to these parameters. Attaching numeric values, different weights to severity of a phenomenon, frequency of occurrence, and quality of documentation allows an algorithm to stratify decisions and present only the most relevant ones to a clinician. Tiered alerts are better accepted by clinicians and have a higher compliance rate and less override than a CDS with nontiered alerts (9).

Trustworthy Medical Information

Successful CDS is based on high-quality medical knowledge. How can one measure the quality of clinical knowledge?

A meta-analysis (a study that combines several other studies to research a hypothesis) of thousands of patients is "worth" more than a case report on an individual; double-blinded prospective studies are considered more important, having more academic weight and relevancy than retrospective studies.

How do we make sure the transformation from research articles into real-world practice, in the form of a CDS recommendation, was properly done? *Clinical Practice Guidelines We Can Trust,* published by the National Academies Press, is a concise compendium of current best practices and recommended standards related to the identification, development, and evaluation of clinical practice guidelines (10). A detailed method for rating the strength/value of medical information was recommended by the authors for evaluating the quality of clinical documentation supporting a CDS rule.

CDS Configuration

The hospital implementing (and paying for) as well as the clinicians using CDS should have input into the system configuration. Which alerts should be shown and to whom?

A pediatrician may be interested only in a specific age band and have no interest in alerts related to geriatric patients. A geriatrician may have no interest in pregnancy and lactation considerations.

What are the severity, frequency and quality of documentation thresholds beyond which an alert should fire? Should severe but rare drug interactions be displayed as "critical"? Which alerts are considered important enough and thus allowed to stop workflow and interrupt a clinician? Which CDS rules may individual clinicians modify, and which ones are set in stone by the institution?

Recommendations for an initial CDS configuration setting of an *eRx* system can be found in the work of Van Ornum (11).

Both the hospital and its practicing clinicians should have the capability to monitor, tweak, and modify the relative weights attached to the medical knowledge used by CDS and thus alter the algorithm performance profile. These changes should be made following a clear and transparent set of policies, defined by the governance structure in charge of the CDS.

CDS configuration cannot be the sole responsibility of the vendor. Clinicians and the hospital must actively participate in the process, even if it means bridging between sometimes-conflicting requirements. The deliverable of this configuration process is a repository of configuration parameters a database CDS can use and optimize its own performance at both the hospital and the individual clinician levels.

CDS Adaptability

Clinicians are familiar with a number of patients, a limited number of drugs, and the diseases pertinent to their specialty. Alerts related to a combination of familiar parameters the clinician is experienced with are considered superfluous. For example, an alert about the potential respiratory depression *IV Morphine* may cause is too obvious to be shown to an anesthesiologist, who is ordering this drug to treat severe pain many times each day. This is a CDS false alarm from the clinician's perspective, even if it is based on high severity, high frequency, and high quality of documentation.

Alerts unrelated to the characteristics of the patients a clinician is treating are another source of false alarms and clinician's irritation. For example, these alerts might include offering smoking cessation to a neonate or considering drug effects on pregnancy in an 82-year-old patient.

Waiting for the vendor to change the CDS configuration usually takes too long. The hospital information technology department may be swamped by other issues and not be able to respond fast enough to the clinician's mentioned needs. Thus, a mechanism should be in place to allow CDS to automatically adapt to the workflow of the individual clinician. The user should be able to suppress some of the alerts for a certain time frame (month, year) at the level of a specific CDS process (adverse reaction: *Morphine and respiratory depression*) or a combination of patient-drug characteristics (patient has taken this medication in the past without any adverse reaction).

For each CDS process defined in the following (age, gender, indication, contraindications [C/I], etc.), the configuration file of the individual clinician would state his or her alert preferences. A CDS considering a clinician's preferences, entered as real-time feedback, is a "learning system": It adapts and optimizes itself to best fit the user's needs.

A simple rating system may accompany the alerts; this makes possible identifying CDS rules considered "useful" versus those despised as a "nuisance." Users' evaluation of CDS may help tweak, adapt, and optimize alerts to the clinicians' workflows. As mentioned by Lund: *"Let people shape the system to themselves, and paint it with their own personality"* (12).

CDS performance needs continuous monitoring, tweaking, and maintenance, which should take into consideration the needs of individual clinicians as well as the hospital/healthcare organization. The process of CDS optimization can and should be automated with a closed-loop preferences feedback mechanism between the clinician and the system.

CDS—A Binary Classification System

A binary system classifies members of a given set into two groups, based on each member's property or quality. In medicine, for example, the decision whether a lab result means sickness or health is a binary classification. A binary classification has a "gold standard" considered the source of truth (such as the biopsy result when *prostate* cancer is suspected). A "test" result (such as the *PSA, [Prostatic-Specific Antigen]*, a blood laboratory value) is compared to the gold standard. The gold standard and the test may agree or disagree.

True positive and true negative results happen when the test and the gold standard are in full agreement (e.g., both the prostate biopsy and the PSA test agree whether a patient has cancer). When the test and the gold standard disagree, it creates false-positive and false-negative results. Statistical methods, beyond the scope of this chapter, such as sensitivity, specificity, positive and negative predictive values, receiver operating characteristics, and F-measure calculate different aspects, ratios, and characteristics of a binary classification system.

Since the decision whether a CDS rule should fire an alert is binary, we may regard CDS as a binary classification system and use statistical calculations when tweaking and configuring the system.

False-Positive versus False-Negative Alerts

A false positive is crying "Wolf!" when there is none around. If a CDS rule decides on alerting the user when there is no need to worry, it is a false positive alert. False positives are notorious in causing *alert fatigue*—the user eventually disregards most of the alerts. For example, in one study clinicians overrode 88% of the "critical" drug interaction alerts presented by a CDS (13). Other articles reported similar levels of alert override (14,15). The main reasons cited for these alert overrides were a high level of false positives, irrelevancy of alert to the patient condition, and unsafe, unclear, or inefficient alert handling by the software. When alerts are of the disruptive, workflow-stopping kind—pop-ups that interrupt work and cognitive processes—the clinician may consider the whole CDS a nuisance (16).

If *Canis lupus* approaches and nobody cries "Wolf," if the CDS did not fire an alert when it should have, this is a false negative. The price of this type of error is obviously different from a false positive: It induces a dangerous illusion that "everything is fine" while a patient may be harmed by a medication error (or a wolf).

An ideal binary classification system would have no false positives and no false negatives. Unfortunately, this can only rarely be accomplished since most of the time there is some overlap between the distributions of the populations of true "positives" and true "negatives."

In a binary classification system (sick/healthy, cry wolf/do nothing), a test cutoff point classifies a result into positive (right side of the vertical lines in Figure 8.1) or negative (the left side in the figure). A certain amount of false positives and false negatives is *always* present when overlap of the positive and negative populations exists. Moving the cutoff point of a test in Figure 8.1 to the left increases false-positive results while decreasing the false negative and vice versa: Moving the cutoff point to the right increases false-negative results while decreasing the false positive.

Figure 8.1 illustrates why the ideal situation of zero false positives and zero false negatives cannot be theoretically achieved if there is any degree of overlap between the two populations. This overlap in turn is proportional to the spread of each graph around its average value and inversely related to the distance between the graphs' means.

Decreasing one type of false results increases the other. Trying to achieve a balanced trade-off between false positives and false negatives is a more realistic goal than trying to eliminate both types of errors. False results in the decision-making process of a binary classification system should be expected, and a CDS is no exception in this regard. CDS recommendations and assessments should be stratified to best fit the gray areas of these false results using a tiering system based on the severity, frequency, and quality of supporting documentation related to the alert.

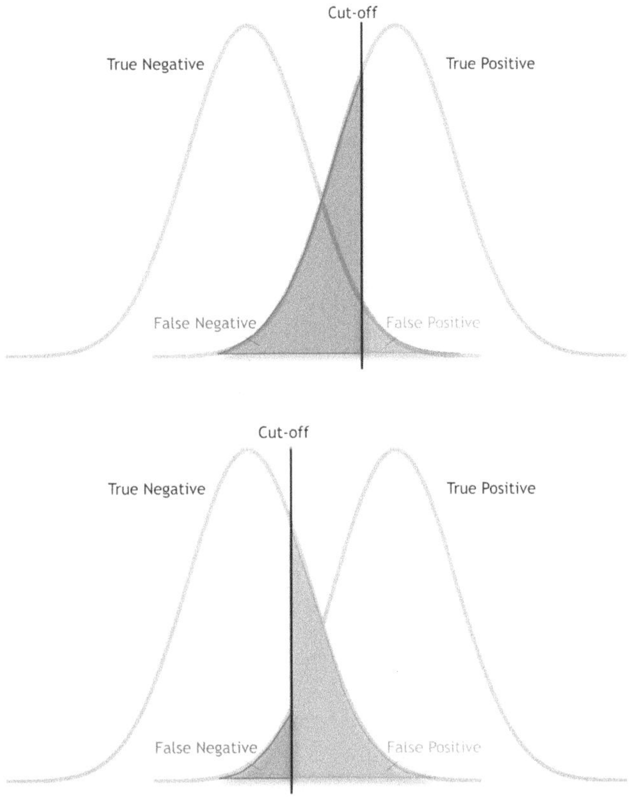

Figure 8.1 False positive and false negative: A trade-off.

Medication Errors

Defined as any mistake in prescribing, ordering, transcribing, dispensing, administering, or monitoring of a medication, medication errors are common: More than three-quarters of a million patients are injured or die in U.S. hospitals from medication-related issues (17). It is estimated that about a third of these errors are preventable mishaps (18,19).

Medication CDS

As previously mentioned, the CDS main processes related to the medications domain help a clinician

- Filter drugs and select the safest medication for a given patient
- Optimally adjust the dose for a patient's characteristics
- Educate the patient or clinician

Optimizing medications and doses is a multistep workflow; it has many subprocesses, which can be categorized as follows:

Demographics:
- Age
- Gender
- Weight
- Height
- Ethnicity

Patient condition:
- Indication/therapeutic intent
- Evidence-based medicine
- Allergies
- Contraindications and special precautions
- Adverse reactions and side effects
- Pregnancy and lactation
- Drug-food
- Drug-vitals
- Drug-procedure

Patient drugs:
- Specific medication use in the past
- Drug-drug interactions
- Duplicate therapy
- Drug-setting
- Intravenous admixtures
- Pharmacological alternatives

Lab:
- Results influencing drug or dose
- Lab tests to monitor while on a specific medication
- Drug interference with lab results

Both the filtering algorithm, which helps select the safest drug, and the adjusting dose algorithm, which optimizes a dosing regimen for patient characteristics, use (almost) the same processes. Age and weight are considered when selecting a drug and when adjusting for dose. Indication and concurrent diseases are used as a filter for selecting drugs and for adjusting the dose and so on. We represent the dual roles a process may have during the analysis and decomposition of the CDS workflow.

Dialog Paradigm

Clinicians usually interact with CDS using a dialog paradigm: A clinician asks a question, such as, "*Which drugs for hypertension are indicated to be taken by mouth (PO)?*" The CDS replies with a list of drugs indicated for treating the

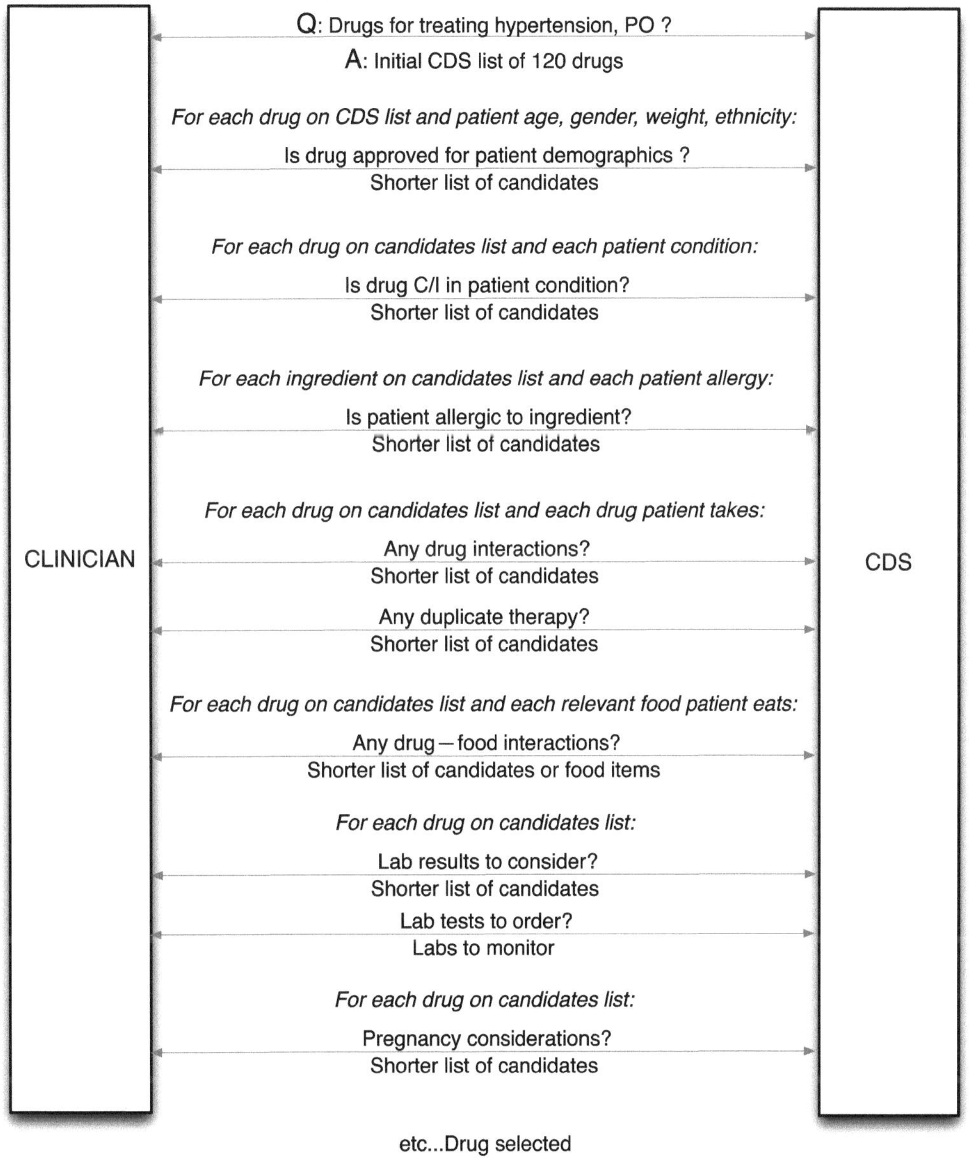

Figure 8.2 Dialog paradigm: Drug selection.

condition. A user then asks another question: "*Is medication X safe for a patient with asthma?*" The system replies with an assessment such as, "*No; drug X is not safe for patients with asthma.*" The user continues sifting through the initial candidate list of drugs, which may be quite long. Selecting a different drug, the clinician asks: "*Is the patient allergic to any of the medication's ingredients? Is the medication interacting with other drugs the patient is taking?*" This filtering dialog may continue for a while, as illustrated in Figure 8.2.

Consider the following scenario: A 69-year-old male with hypertension, diabetes mellitus type 2, and asthma is already taking *Amlodipine* and *Valsartan* for hypertension, *Metformin* for diabetes, and an *Albuterol/Ipratropium* inhaler for asthma. The patient is allergic to cats, milk, and *Minoxidil*. His blood pressure is still not under control, and the physician is looking to add a drug to the patient's medication regimen.

Initially, the CDS may recommend a list of more than 100 drug candidates: pharmaceutical agents in various product combinations indicated for the treatment of hypertension in this age group. The clinician needs to mentally eliminate medications contraindicated in asthma. Next, the clinician will have to subtract from the remaining list the medications contraindicated in patients with diabetes. From the remaining list of candidates, the clinician has to eliminate those classified as *calcium channel blockers* and *angiotensin receptor blockers* (*ARBs*) since these drugs will be considered duplicate therapy; the patient is already taking a *calcium channel blocker* (*Amlodipine*) and an *ARB* (*Valsartan*). The list of drug candidates becomes shorter with each filtering/elimination step. Next, the clinician needs to weight whether to prescribe an *angiotensin-converting enzyme* (*ACE*) *inhibitor*; this class of drugs may interact with the class of *ARBs*, which the patient is already taking (*Valsartan*). Depending on the severity of the interaction, frequency of occurrence, and quality of supporting documentation, the initial list of antihypertensive medications becomes shorter. Following additional filtration steps, the clinician is eventually presented with 10 potential medications that may be a good fit to the patient's characteristics.

There are several issues with this CDS dialog paradigm:

The clinician needs to go through an annoying, time-consuming process of questions and answers, while the clinician's role is to provide information about a patient. Why isn't CDS extracting this information automatically from readily available data stores maintained by the electronic health record (EHR)?

Each drug on the initial candidates list proposed by the CDS needs to be filtered by similar eliminating processes: age, allergy, gender, contraindications, side effects, and so on. Some of these filtering steps need to loop every drug on the candidates list with each and every drug the patient is taking (drug interactions); each and every condition/diagnosis of the patient (contraindications, adverse reactions); each and every allergy the patient has, and so on. It may prove a tiresome mental burden that not many clinicians will be able to support.

While the CDS identifies medications that are to be avoided, the remaining drugs considered "safe" and remaining on the list of candidates need to be memorized by the clinician for the next round of questions, the next filtration step. This necessitates the clinician employing a difficult cognitive task: subtracting a list of drugs to be avoided, from the larger initial drug candidates list.

After a lengthy dialog, the clinician eventually selects a medication filtered by the CDS dialog processes and considered safe for the given patient. Now, an unpleasant task awaits the clinician: To adjust the dose, almost the same dialog steps need to be revisited: age, weight, indication, other conditions patient may have, and so on. Another similar unnerving dialog is imposed now on the clinician for adjusting the optimal dosing regimen (Figure 8.3):

The clinician's time and cognitive resources should not be wasted in a lengthy dialog with CDS. The clinician should not be responsible for providing information to the CDS since the CDS can extract all the data it needs from the EHR without the clinician's help. The burden of data processing and temporary storing the filtered information should be transferred from the clinician to the CDS. The dialog paradigm mentioned for selecting a drug and then adjusting its dose can and should be automated (Figure 8.4).

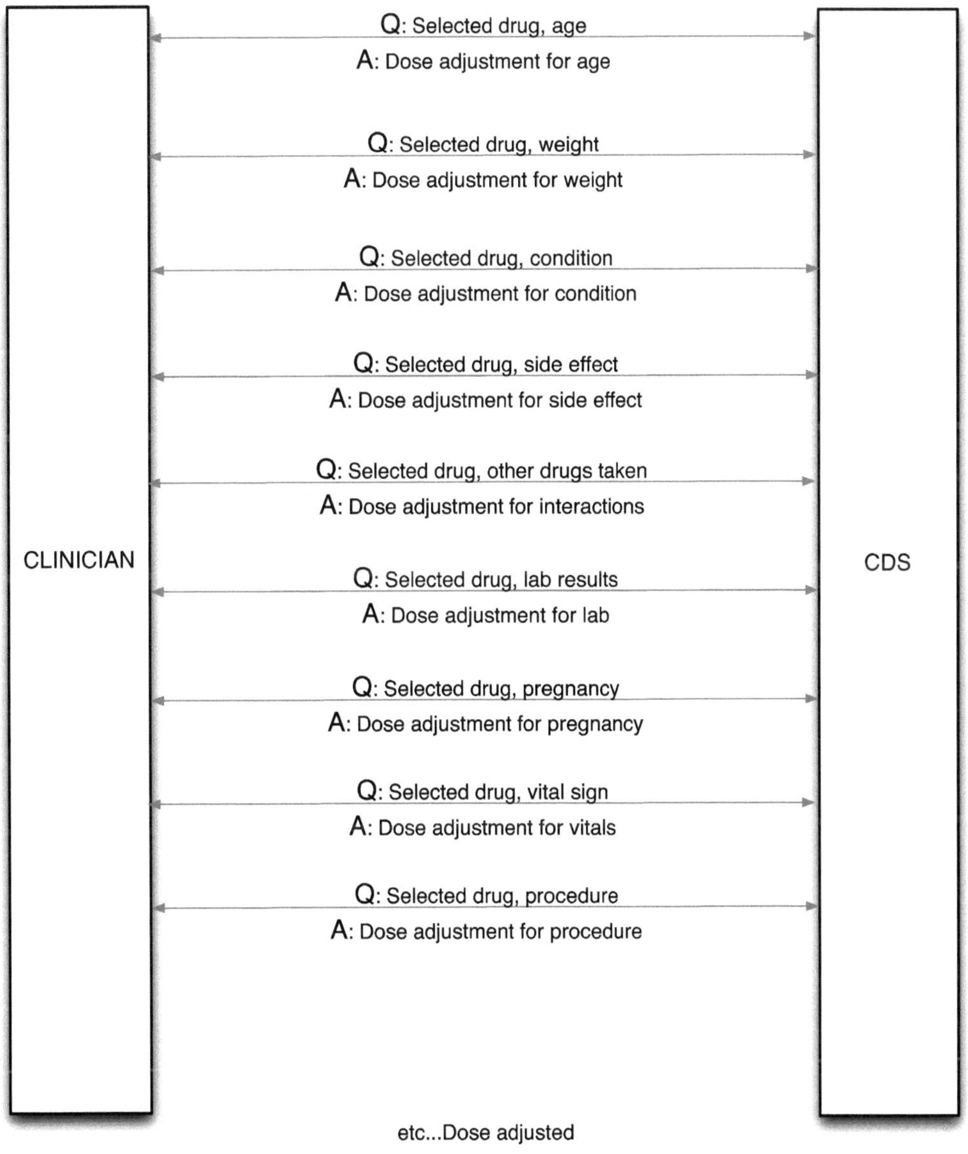

Figure 8.3 Dialog paradigm: Dose adjustment.

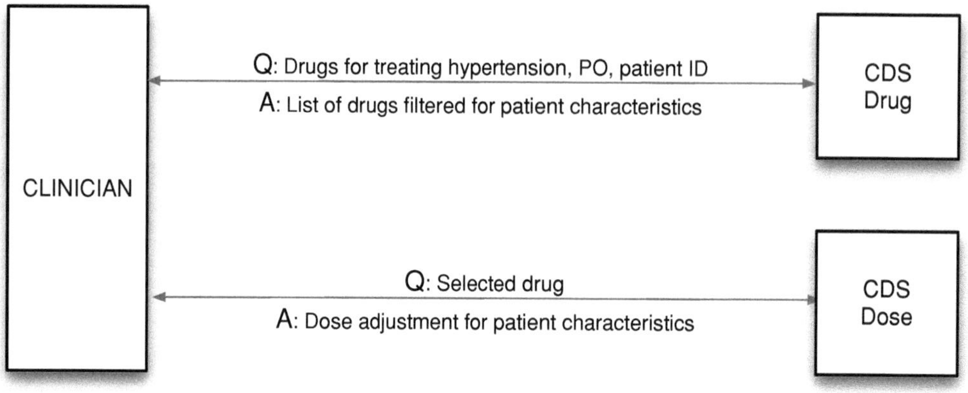

Figure 8.4 Automated CDS.

The collection of patient discrete data elements from various EHR sources, the computation, looping through logic cycles, filtering, subtraction, and elimination of drugs should be automatically performed by the CDS, with minimal clinician input. CDS recommendations should be available to the clinician together with the rationale behind the automated decision-making process. The clinician may or may not wish to view these recommendations and the logic behind the machine reasoning. Depending on the hospital and the individual preferences settings, an unobtrusive type of decision support, offering alternatives and not just assessments, should be displayed. Finally, only severe alerts, occurring frequently and supported by high-quality documentation, should be allowed to interrupt a clinician's workflow.

Automated CDS Algorithm Outline

Let us sketch first the algorithm outline, represented by the two small boxes in Figure 8.4. Basically, each drug CDS considers for a specific patient has to be filtered by multiple processes. Each process receives input from database tables and external actors and delivers at least one decision as an output. Some processes compare one drug with other drugs in different combinations (drug interactions); other processes compare one drug with multiple conditions (drug contraindications). These comparisons will result in multiple decisions for a single drug and for one algorithm step.

The name of the drug (e.g., *Atenolol*) and the CDS process (such as *Consider duplicate therapy*) as well as the details of the individual step considered (*Atenolol* and *Tenormin*) are all part of the expected algorithm decision for the specific step. These decisions, which may be numbered in the hundreds per patient session, are tiered (severe, moderate, etc.) and have details on the machine reasoning process, and each one may have information about frequency of occurrence and quality of documentation.

In a similar fashion, the CDS uses multiple processes for dose adjustment. Again, the algorithm passes through multiple steps, and at each step, one or more decision may be taken, such as *decrease dose by 30%* or *increase intervals between doses to 24 hours.* The machine reasoning and the documentation it is based on are attached to the decision.

So the clinician is not burdened with a dialog such as illustrated in Figure 8.2 or 8.3 and with hundreds of bits and pieces of information (many of them potentially false alarms), the CDS needs a temporary storage for the stepwise algorithm decisions. These temporary tables are discussed next.

Having CDS decisions in a tiered tabular format allows mathematical models to be applied to the results, such as attaching different weights to different steps or to the graded severity, frequency, and so on. The modular structure of the algorithm allows users to bypass some of the steps if they are considered irrelevant to the clinician's specialty or practice.

Let us take a closer look at the CDS processes.

Processes

DFD 5 CDS Workflow

We isolate first the CDS workflow number 5 from DFD 0, noting the other workflows that interact with CDS: *Prescribe, Order, Dispense,* and *Administer* (Figure 8.5).

Let us zoom in on the main CDS processes identified (Figure 8.6). The clinician is interacting with three processes: 5.1 *Filter Drug,* 5.2 *Adjust Dose,* and 5.7 *Educate.* Processes 5.1 and 5.2 are *control processes*; like air traffic controllers, they activate other processes (5.3–5.6) while keeping score of the CDS workflow results, such as the drug candidates list mentioned.

There are two temporary tables used for this purpose—*Temp CDS DRUG* and *Temp CDS DOSE*—in which CDS stores the results of the other processes to be eventually presented to the clinician in a summary format. These temporary tables are created for a single session, and then their contents are either deleted or stored for further analysis by the personnel responsible for monitoring and maintaining the CDS.

DFD 5.1 Filter Drug

The DFD 5.1 *Filter Drug* process collects discrete data elements from data stores, then activates four other processes (5.3–5.6), receives and stores their decisions in a temporary table, and displays the summary to the clinician. This process may use two paths, paralleling the ways a clinician may filter drugs: by indication or by drug (name, ingredient, or class) (Figure 8.7).

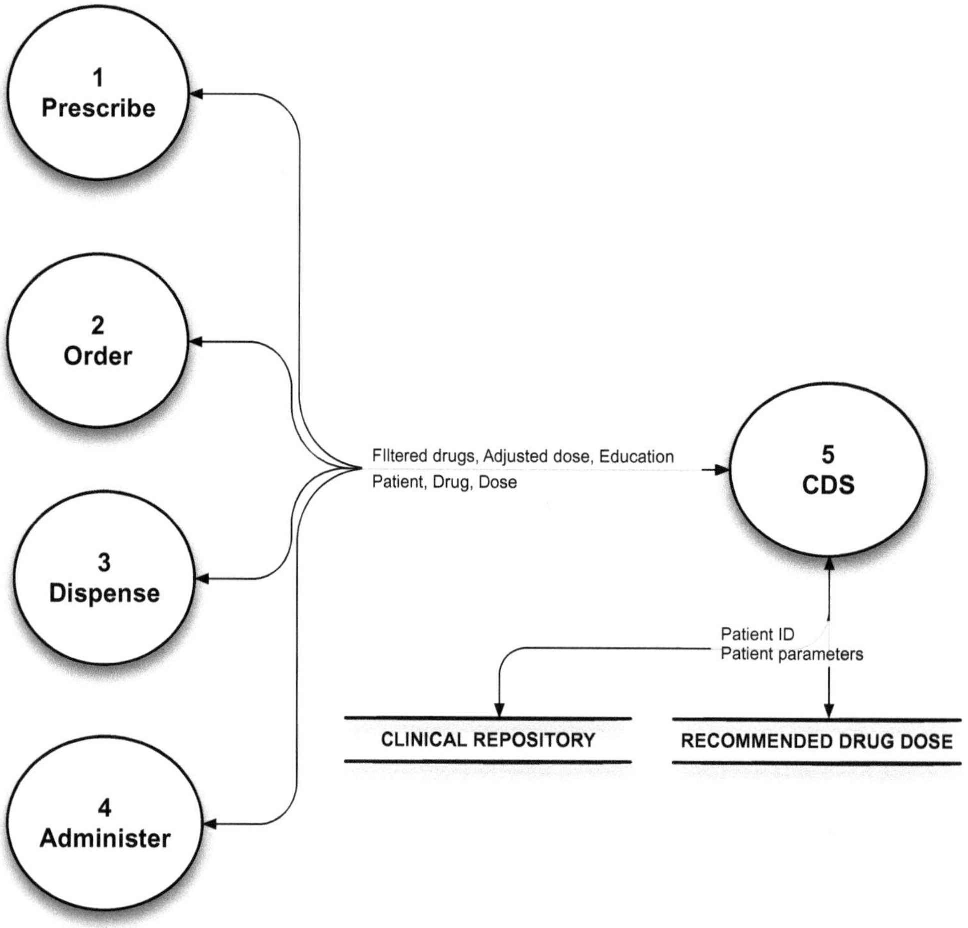

Figure 8.5 CDS workflow: Isolated from DFD 0.

Using the *Patient ID*, subprocess 5.1.1 collects discrete data elements from the medications database (e.g., whether a specific drug was administered to the patient in the past and the patient's reaction to it) as well as the *NON-DRUG REPOSITORY* (diagnosis, allergies, lab results, etc.). These data elements are then sent for further processing to 5.3–5.6.

In the first case, filtering drugs indicated for a condition, such as hypertension, 5.1.2 creates initially a list of drug candidates in the *Temp CDS DRUG* table and then filters the individual candidates by tagging each one with the decisions made by processes 5.3–5.6.

In the second case, evaluating a specific drug the clinician has selected, 5.1.3 filters the selected drug through the same 5.3–5.6 processes and stores the resulting information for display to the clinician.

The inputs 5.1.1 accepts are the patient identifier and either a drug or an indication. The output 5.1.4 eventually displays is a tiered list of decisions regarding medications, made by the processes 5.3–5.6.

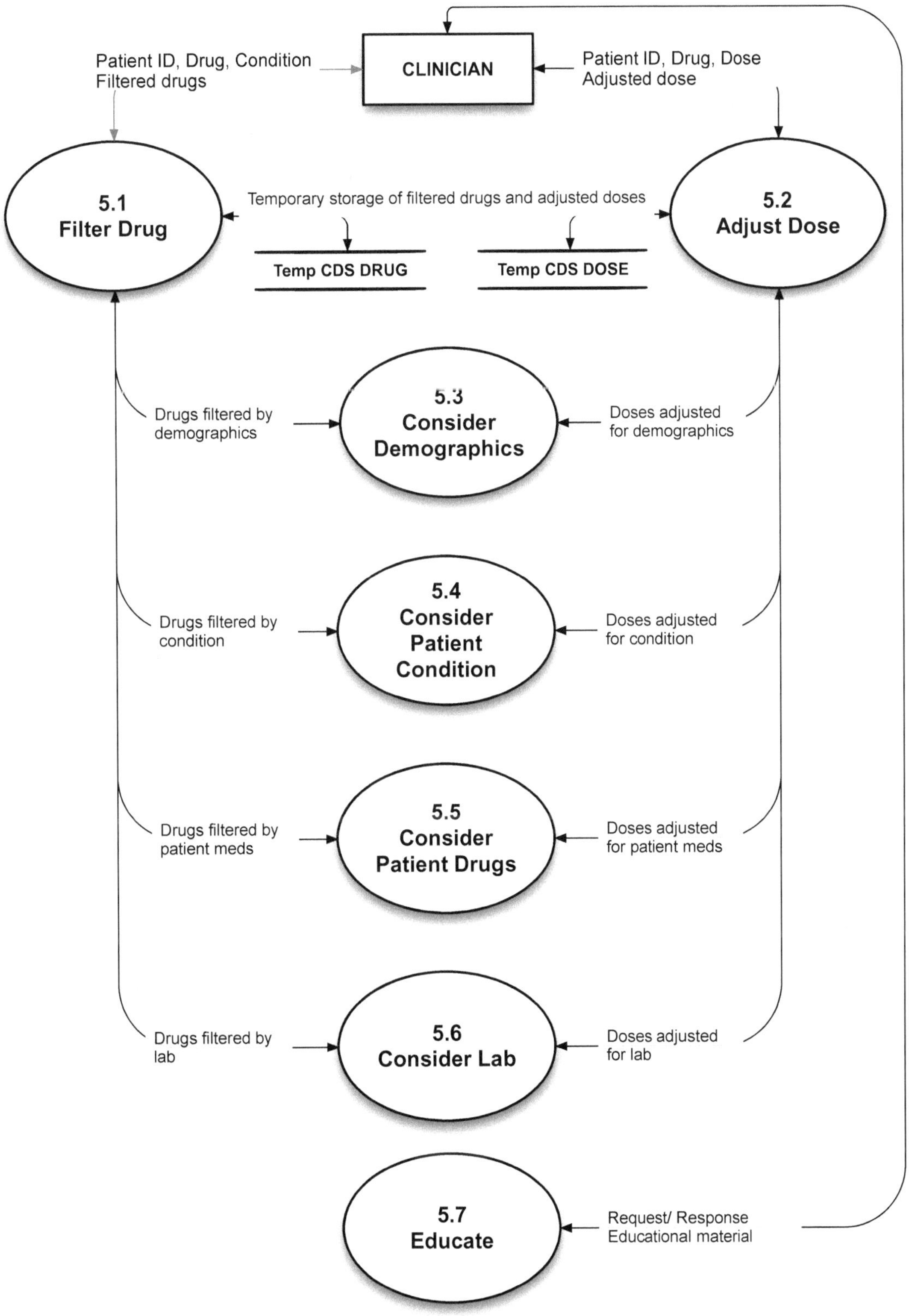

Figure 8.6 DFD 5: CDS workflow.

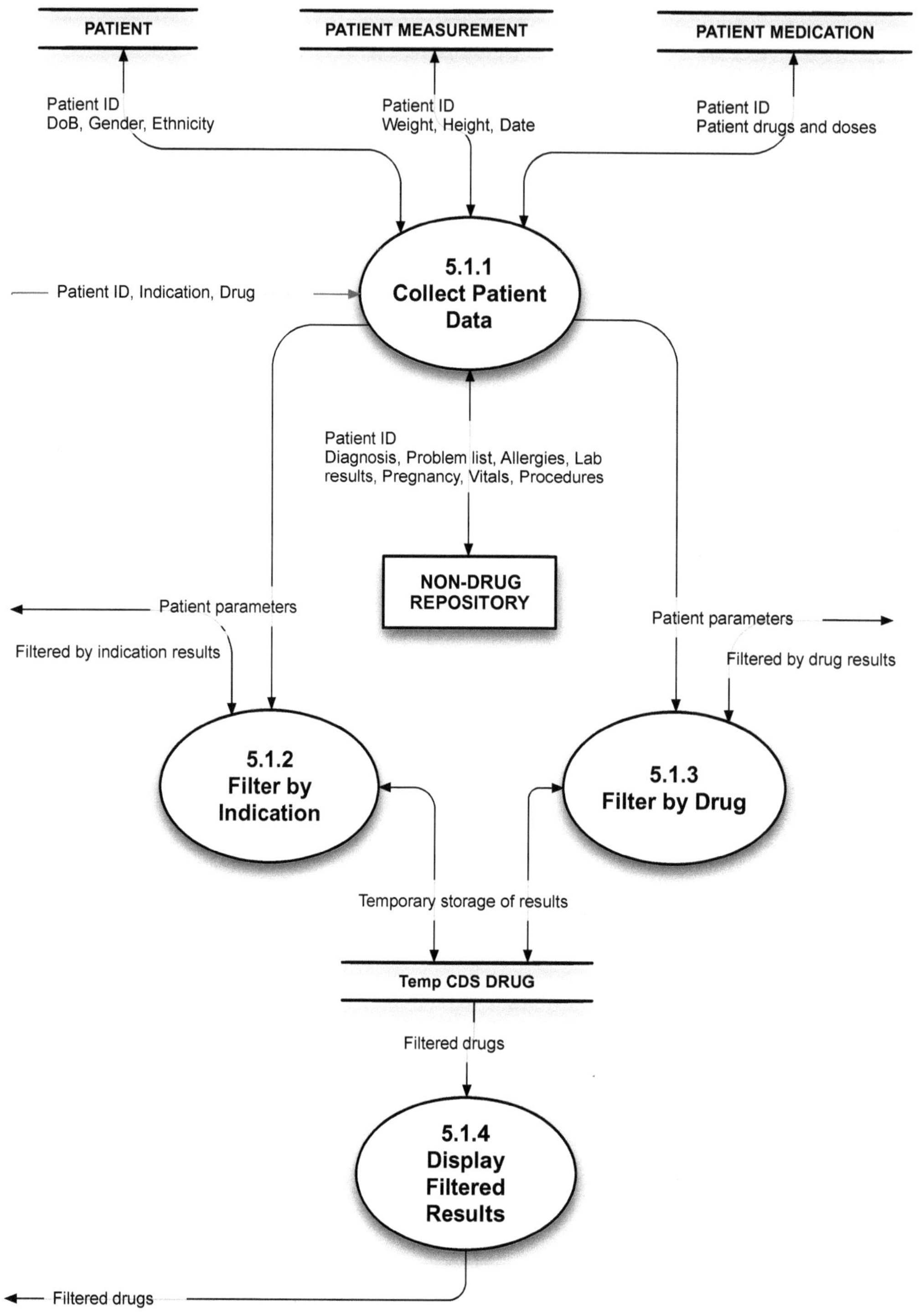

Figure 8.7 DFD 5.1: Filter Drug.

DFD 5.2 Adjust Dose

In a similar fashion, process 5.2 *Adjust Dose* is responsible for collecting patient information, activating processes 5.3–5.6, storing their calculations and decisions, and adjusting the dose of a single drug for the patient's characteristics (Figure 8.8).

Process 5.2 accepts as input the patient identifier, one drug, and potentially a dose the clinician has selected. The output of this process is a list of decisions related to the maximal/minimal dose, other dose adjustments needed for the specific patient (increase/decrease dose or frequency of administration). These decisions are stored temporarily in *Temp CDS DOSE* and presented as a summary to the user by 5.2.3.

DFD 5.3 Consider Demographics

Discrete data elements are passed to the 5.3.x processes from either 5.1 or 5.2 as detailed. Age, gender, weight/height, and ethnicity are considered by the DFD 5.3 *Consider Demographics* processes 5.3.1, 5.3.2, 5.3.3, and 5.3.4, respectively (Figure 8.9).

5.3.1 Consider Age

Date of birth is translated to a specific age band, which is used to decide whether a drug is indicated/contraindicated for the patient as well as setting the initial dose for the age band (together with the therapeutic intent; see further discussion in this chapter). Process 5.3.1 *Consider Age* is responsible for tagging each drug on the initial candidates list with a tiered decision, which may range from "Recommended as best practice" to "Do not use this drug".

For example, which chemotherapeutic agents are recommended for treating *acute lymphatic leukemia* in children 5 years of age? Conversely, process 5.3.1 may be asked whether *Adriamycin* is indicated for treatment of bladder cancer in a 72-year-old patient.

The same process may also calculate and consider the lifetime-accumulated dose for the chemotherapeutic agent *Doxorubicin*, for example. The tables providing information used by this process (*AGE BAND, DRUG INDICATION, DRUG C/I,* and *RECOMMENDED DRUG DOSE*) have been discussed in previous chapters.

5.3.2 Consider Gender

A similar algorithm is employed for CDS gender considerations in 5.3.2 *Consider Gender*. For example, find the optimal dose of *Bicalutamide* for treating *hirsutism* (abnormal hair distribution) in a 12-year-old female; or, is *Leuprolide* contraindicated in females?

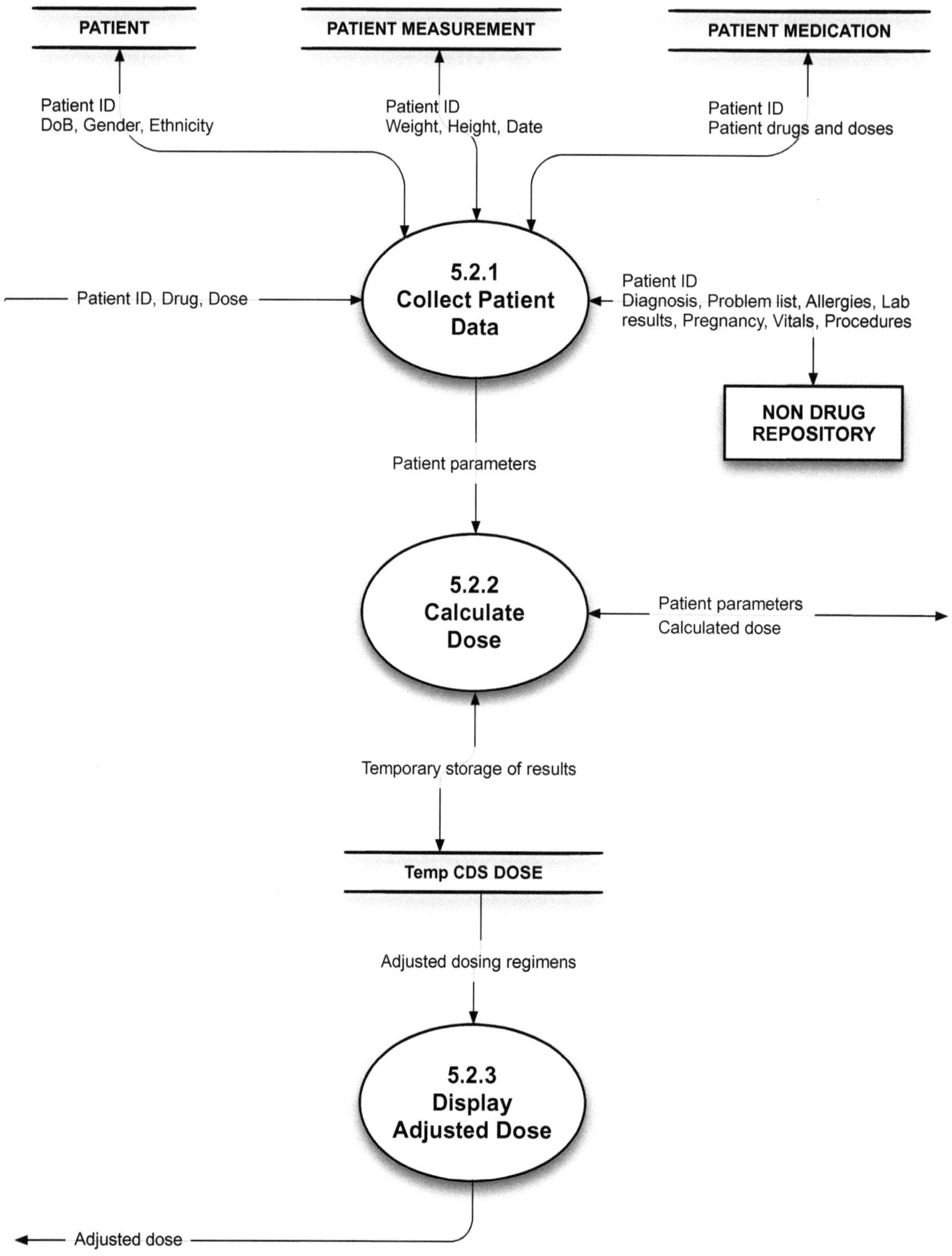

Figure 8.8 DFD 5.2: Adjust Dose.

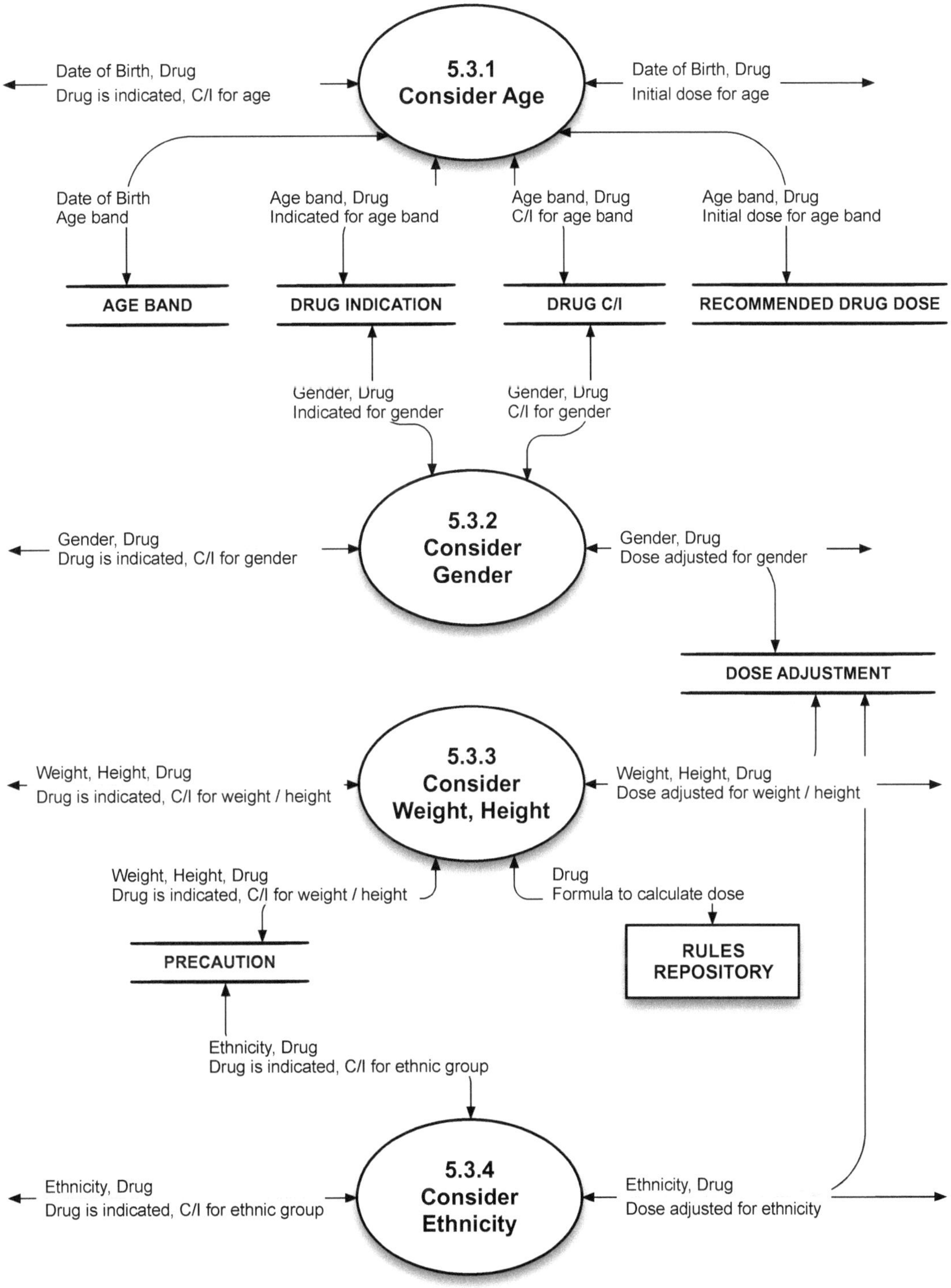

Figure 8.9 DFD 5.3: Consider Demographics.

5.3.3 Consider Weight, Height

The *RULES REPOSITORY* provides the formula to be used for dose calculations: weight, ideal body weight, body mass index (BMI), and body surface area (BSA). BMI and BSA consider both weight and height in their formulae. For example, many chemotherapeutic dosing regimens, such as for *Ifosfamide* or *Vincristine,* are based on the patient's BSA and not the patient's weight. There are other situations, such as for severely obese patients and neuromuscular paralyzing drugs, when dosing is optimally calculated for an ideal, "lean," and not the actual body weight. The initial dose is calculated by process 5.3.3 *Consider Weight, Height,* and the result is then adjusted by other processes and summarized by 5.2.

DOSE ADJUSTMENT (a new entity introduced in the data structures section in this chapter) and *PRECAUTION* are tables used by process 5.3.3. For example, dose adjustments for renal or hepatic diseases or immunocompromised patients are considered using information from these two tables.

5.3.4 Consider Ethnicity

Issues related to ethnicity, such as the benefit of using *Bidil* for black patients with *congestive heart failure,* are being considered by CDS 5.3.4 *Consider Ethnicity* and added to the other recommendations or contraindications to be summarized and presented to the clinician.

The results of these processes, personalized to specific patient characteristics, are temporarily stored in *Temp CDS DRUG* for the filtered medications and in *Temp CDS DOSE* for the adjusted dosing regimens. Each and every drug the CDS has considered is temporarily tagged with at least one tiered decision on whether to use or not for a specific patient and at least one other decision regarding the optimal dose.

DFD 5.4 Consider Patient Condition

The DFD 5.4 *Consider Patient Condition* process filters medications and adjusts dosing regimens considering patient diseases, allergies and other conditions, adverse reactions and side effects, and so on (Figure 8.10).

5.4.1 Consider Indication and EBM

Process 5.4.1 *Consider Indication and EBM* provides information related to indications, guidelines, best practices, and EBM protocols as well as the Food and Drug Administration (FDA) drug status (on label/off label) as explained in Chapter 3.

Given a therapeutic intent or indication, such as hypertension, and the patient's characteristics, process 5.4.1 answers questions such as the following: *Which drug should be added to existing antihypertensive medications to control*

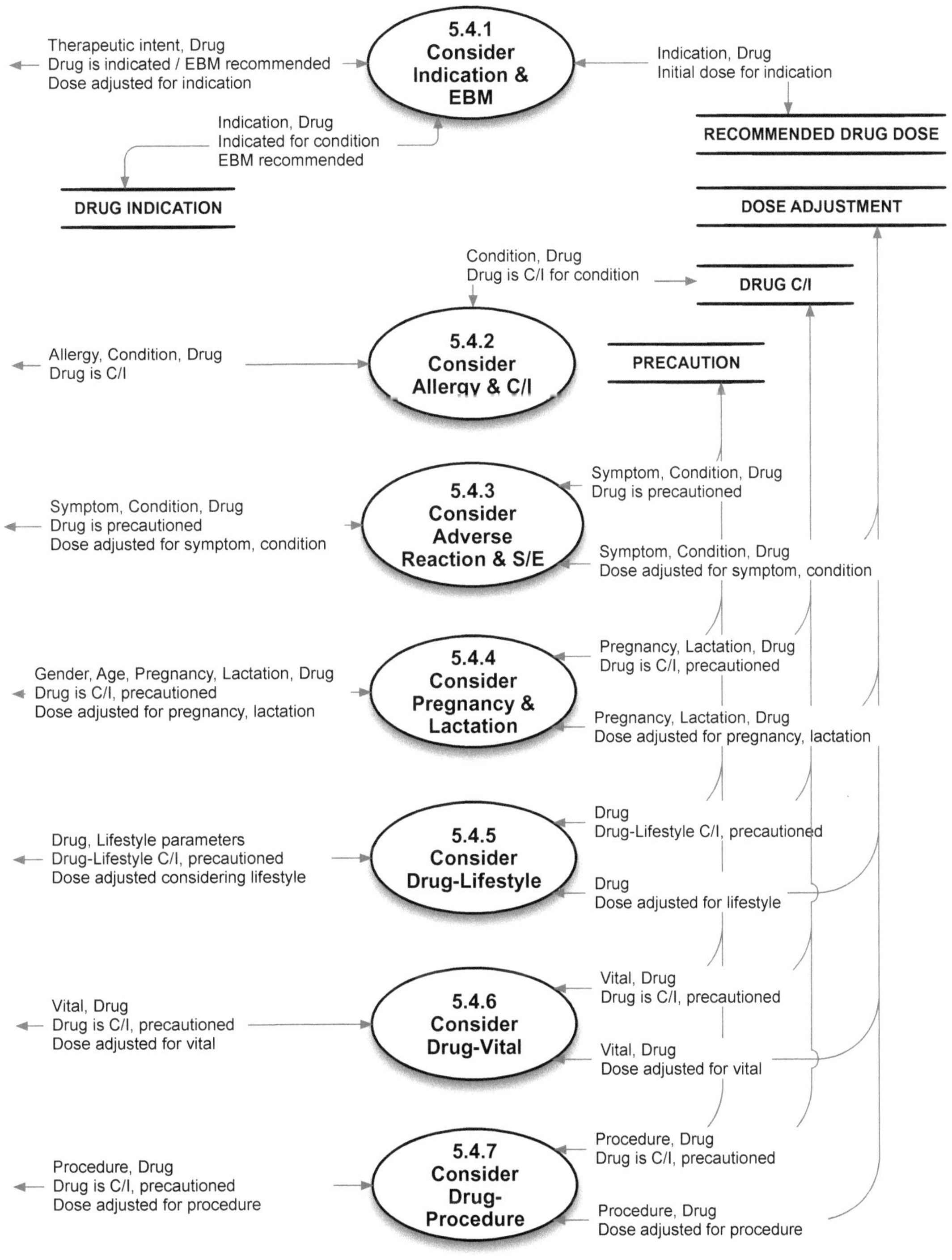

Figure 8.10 DFD 5.4: Consider Patient Condition.

the blood pressure? At what dose? And vice versa, *is a given drug indicated for a specific condition?*

For example, *what is the IV Sildenafil recommended dose for a 2-week-old preemie with pulmonary hypertension? What are the evidence-based recommendations for treating community-acquired pneumonia in a 69-year-old male?*

Is *Clopidogrel* part of an EBM protocol in preventing *deep vein thrombosis* in surgical patients?

5.4.2 Consider Allergy and C/I

All active and nonactive ingredients of the drugs considered by the CDS are checked against the patient's allergies in the 5.4.2 *Consider Allergy and C/I* process. Class sensitivity, such as "allergic to all *penicillins,*" and cross sensitivities, such as *penicillins/cephalosporins,* are considered during this process. Cross sensitivity may exist with food allergy as well; for example, patients with a peanut allergy may be intolerant to *Proventil,* a bronchodilator used to treat asthma.

Each condition, disease, or diagnosis of a patient is checked against the information residing in table *DRUG C/I.* For example, *Propranolol* is contraindicated in patients with *asthma.* Conversely, a clinician may ask the CDS about medications contraindicated in a specific condition, such as *Meniere's disease.* Due to the important hepatic and renal roles in the metabolism and excretion of most drugs, the CDS checks for the diagnosis of liver or kidney diseases in a given patient and whether relevant contraindications or dose adjustments apply to a specific drug. The decisions of these processes are tagged accordingly in the temporary tables mentioned and passed to 5.1 and 5.2.

5.4.3 Consider Adverse Reaction and Side Effect

As an example of the 5.4.3 *Consider Adverse Reaction and Side Effect,* for certain patients, such as professional singers, *Alosetron* is known to cause coughing and may not be the best choice. Or, similarly, if a patient already has diarrhea (symptom), there is really no point in prescribing a drug that is known to cause diarrhea as a side effect.

Adverse reactions to drugs and side effects should be considered by the CDS in the other direction as well: Is a symptom caused by one of the drugs the patient takes? For example, is it possible that the new *diuretic* prescribed is the cause for the large amounts of urine the patient complains of passing?

5.4.4 Consider Pregnancy and Lactation

It seems embarrassingly obvious to mention the fact that process 5.4.4 *Consider Pregnancy and Lactation* should be used for child-bearing age female patients *only.* However, too many CDS systems currently do not employ this simple filtering mechanism and display initially funny, then increasingly annoying, alerts to clinicians. Providing alerts on medications contraindicated in pregnant *males* or menopausal female patients may induce alert fatigue.

The FDA classifies drugs into five groups (A, B, C, D, X) regarding their potential benefits versus risks to the fetus in a pregnant patient. Multivitamins, considered safe for mother and fetus, are usually in class A, while *Thalidomide* is infamously in class X, being known to cause severe limb malformations in the newborn.

Medications may be secreted in a mother's milk and affect the breast-feeding child. There is no FDA drug classification related to safety in lactating mothers, so currently the agency is revising the whole pregnancy and lactation drug classification system.

The CDS should also consider drug safety as related to yet-unknown pregnancy as well as future pregnancy. For example, *Acitretin* in a young woman considering pregnancy in the near future should be flagged by the CDS as the wrong choice since it is a potent teratogenic agent that may harm the fetus.

5.4.5 Consider Drug-Lifestyle

A patient's food and other lifestyle aspects, such as alcohol and smoking, may influence medication metabolism and effect. CDS 5.4.5 *Consider Drug-Lifestyle* identifies and considers interactions such as *Diclofenac* and alcohol, *Warfarin* and spinach, or *Sildenafil* and grapefruit consumption.

In addition, the CDS considers herbal remedies, nutritional supplements, over-the-counter (OTC) drugs, and vitamins a patient is taking and their potential interactions with other medications. For example, *St. John's Wort*, a plant used for treating depression or anxiety, may decrease the efficacy of *oral contraceptives*.

5.4.6 Consider Drug-Vital

If a vital sign is abnormal, such as *bradycardia* (slow heart rate), CDS 5.4.6 *Consider Drug-Vital* considers this fact when a clinician orders a beta-blocker that is known to further decrease the heart rate, according to the adage: "*Don't let bradycardia slip into no cardia.*" Conversely, an observed abnormal vital sign should be checked against known effects and side effects of the drugs a patient is taking to see if it may be caused by one of the medications. For example, one of the drugs a patient is taking may be the cause for the observed slow heart rate. Process 5.4.6 checks both directions: a new drug and the patient's vitals as well as an abnormal vital sign and the drugs a patient is already taking.

5.4.7 Consider Drug-Procedure

Some procedures necessitate changes in the drugs and dosing regimens for a patient. The CDS advises a clinician on the best practices, having access to the information about a scheduled procedure and the patient characteristics. For example: *Metformin* should be stopped 1 day before elective surgery; specific *antibiotics* should be administered prior to colon operations; *anticoagulants* should be administered to patients undergoing major orthopedic surgery to prevent *deep vein thrombosis*. The timing as well as the specific details of the drugs to be stopped or initiated in relation to surgery are recommended by process 5.4.7 *Consider Drug-Procedure.*

In addition, some imaging studies employ contrast materials that need specific considerations related to the patient's renal function and hydration status as well as the medications a patient is taking; this is the responsibility of process 5.4.7 as well.

DFD 5.5 Consider Patient Drugs

Drugs may interact *in vivo* (inside the patient) or *in vitro* (outside the patient). The first type of interaction is process 5.5.1 *Consider Interaction* responsibility, while the second kind is for process 5.5.5 *Consider IV Admixture* to decide. In between, the CDS considers whether drugs are duplicate therapy (5.5.2), tentative pharmacological alternatives (5.5.3), and the patient setting (in the hospital vs. at home) (5.5.4) (Figure 8.11).

5.5.1 Consider Interaction

The potential interaction between two drugs is a complex pharmacological issue involved in the 5.5.1 *Consider Interaction* process. The *pharmacokinetics*, which involves absorption, distribution, metabolism, and excretion (what the body does to the drug), as well as the *pharmacodynamics*, which is the effect of the drug on the nervous, respiratory, cardiovascular, digestive, and other systems (what the drug does to the body), both have to be considered. Complex physiological/mathematical models are experimentally used to identify and measure potential drug-drug interactions.

Most of the CDS systems employ a more practical, empirical method to deal with the pharmacological complexity of drug interactions. Drugs are checked in pairs, drug A + drug B, as the CDS considers the severity and frequency of an interaction occurring between the two based on the clinical documentation at hand (scientific papers, articles, FDA reports, etc.).

If a patient, for example, takes six drugs and the CDS workflows created a list of an additional 20 candidate drugs in its *Temp CDS DRUG* table, then there are 26 drugs from which to create pairs and check for potential interactions. How many pairs should CDS consider?

The formula for the number of combinations of pairs taken from a set of N members is

$$N!/2(N-2)!$$

For those unfamiliar with the factorial sign !, this formula is equivalent to

$$(N^2 - N)/2$$

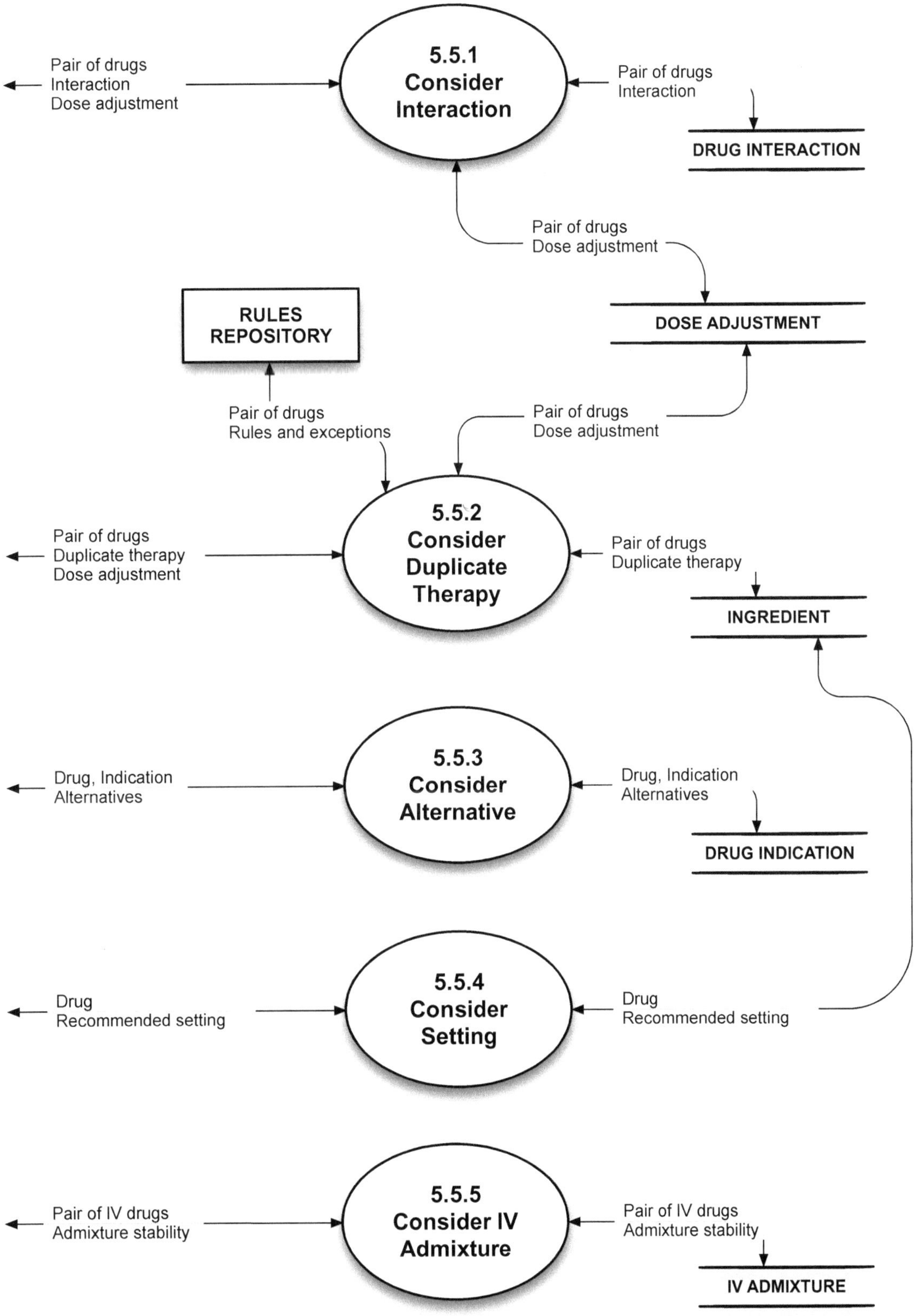

Figure 8.11 DFD 5.5: Consider Patient Drugs.

Thus, for a set of 26 drugs, the CDS has to check 325 pairs of drugs for potential interactions, definitely a job for the computer and not the clinician.

Combinations, for which the order (A then B vs. B then A) is not important, are not to be confused with *permutations*, for which the order is important. Since drug A interacting with drug B is considered in most cases to be the same as drug B interacting with drug A, the CDS uses the combinatorics formula mentioned, for which the order is not important.

The corollary is that there is no need for the CDS to double check on a pair of drugs in both directions: A + B and B + A. A one-sided check is good enough. Alerting the clinician on a potential interaction twice is redundant and may cause alert fatigue.

Each possible pair of drugs is considered by the CDS during process 5.5.1, using information residing in the *DRUG INTERACTION* table, whose structure is discussed in the following. The severity of the interaction, the frequency of occurrence, as well as the quality of the supporting documentation are passed to the *Temp CDS DRUG* table for later summary and display to the clinician. For example, *Erythromycin* and *Ketoconazole* are known to interact, causing a dangerous increase of the *Erythromycin* level in the blood. These two drugs are checked by the CDS as a pair, and a decision related to their potential interaction is passed to process 5.1. If a dose adjustment is needed due to a tentative interaction (and the combination of interacting drugs cannot be avoided), the *DOSE ADJUSTMENT* table will provide process 5.5.1 with the necessary dose adjustment details, such as a recommendation to decrease the dose of *Erythromycin* in the example.

5.5.2 Consider Duplicate Therapy

Duplicate therapy may happen when two or more drugs

- Have the same ingredients: *Senormin* and *Tenormin* both have *Atenolol* as the active ingredient.
- Belong to the same class: *Atenolol* and *Propranolol* are both *beta-blockers.*
- Are prescribed/ordered with the same intent: *Heparin* and *Warfarin* are both *anticoagulants.*

The CDS process 5.5.2 *Consider Duplicate Therapy* checks pairs of drugs and their ingredients to identify potential duplicate therapy in a similar fashion as the previous process checking for interactions.

There are situations for which it is common and appropriate to use more than one drug from the same class; more than one antibiotic may be prescribed to treat infections, and *Insulin* is usually used as a combination of rapid- and intermediate-acting types. Immunosuppressants, anesthetics, and anti-HIV therapies may include more than one member of the same drug category. The CDS discerns these exceptions to duplicate therapy rules using the *RULES REPOSITORY.*

5.5.3 Consider Alternative

The 5.5.3 *Consider Alternative* process identifies pharmacological alternatives to drugs. It considers only pharmacological parameters and aspects: For a specific indication, the CDS identifies the alternative medications that may be used as detailed in the *DRUG INDICATION* table. For example, *Levalbuterol* may be recommended as an alternative to *Albuterol* or *Niacin* as an alternative to *Gemfibrozil*.

This process has nothing to do with the formulary considerations previously mentioned. A *formulary* is a list of drugs preferred by an organization with the sole intent to reduce costs; thus, generics are preferred to brand names, and certain brand names are preferred to others due to contractual agreements with the pharmaceutical suppliers. Process 5.5.3 is concerned only with the clinical aspects of drug alternatives.

5.5.4 Consider Setting

Some drugs should be used only in a certain setting, such as in the hospital and possibly under the supervision of specially trained clinicians. For example, anesthetics such as *Propofol*, or *IV chemotherapeutics* agents should not be used at home and thus should not be prescribed or dispensed by a community pharmacy to individuals. Process 5.5.4 *Consider Setting* checks whether a medication is categorized as one to be used in a hospital only or in a special setting such as the operating room.

5.5.5 Consider IV Admixture

Regarding process 5.5.5 *Consider IV Admixture*, drugs may interact *in vitro*, outside the patient's body, and still be dangerous and potentially harmful. Intravenous solutions and other injectables, such as *Calcium chloride* and *Sodium bicarbonate*, are incompatible since they may precipitate out of solution and become particulate *lime*. The particulate matter poses a significant danger for embolism if injected.

Other solutions, such as *Total Parenteral Nutrition* (*TPN*) mixtures, contain substances that are an excellent growth medium for bacteria and other microorganisms and thus have to be manipulated in a sterile environment, kept under refrigeration, and used within a strict number of hours after dispensation.

The CDS checks drugs for in vitro compatibility and solution stability and offers preparation instructions as well as a recommended time frame the solution may be safely used.

DFD 5.6 Consider Lab

For 5.6 *Consider Lab*, there are three subprocesses related to drugs and laboratory results (Figure 8.12):

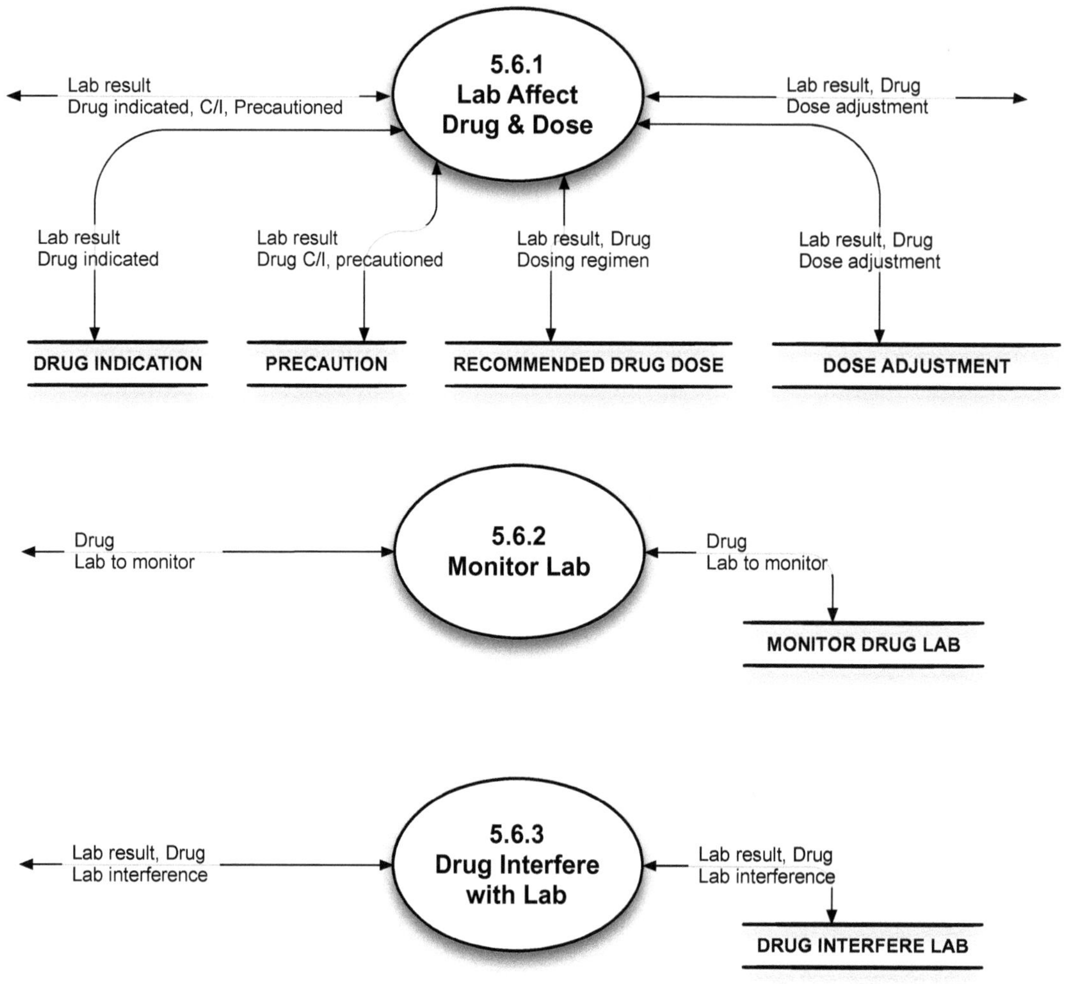

Figure 8.12 DFD 5.6: Consider Lab.

- Lab influencing the drug selected or its dose
- Lab tests that need to be monitored while on a certain drug
- Drugs interfering with lab results

5.6.1 Lab Affect Drug & Dose

A common example of a lab influencing the drug selection, dealt with in 5.6.1 *Lab Affect Drug & Dose*, is the *antibiogram*: the resulting microorganism grown in a laboratory culture and its sensitivity to antibiotics. Urine, sputum, and other bodily fluids and discharges are sent to the microbiology laboratory when an infection is suspected. The resulting *antibiogram* will help a clinician select the optimal antibiotic to treat the patient.

There are drugs that may have a *narrow therapeutic window* (the range between minimal effective dose and the toxic dose is narrow); it is dangerously easy to underdose or overdose the patient. Thus, the drug level in the patient's blood is used to guide treatment: If the level is too low, the clinician will gradually increase the dose; if the drug level is too high, the clinician will decrease the dose or increase the interval between dose administration. *Digoxin, Lithium,* and some *Antiepileptics* are monitored using their levels in the patient's blood. Other drugs need a specific lab test to check their effect on a patient: *PT* or *INR* (International Normalized Ratio) lab results are used to monitor the anti-coagulant effect of *Warfarin*, while *PTT* (Partial Thromboplastin Time) is used to monitor the effect of *Heparin*.

Sometimes, a combination of lab tests is employed to guard against drug toxicity: A patient on *Digoxin* is best treated if the *Digoxin* level as well as the *Potassium* and *Magnesium* blood levels are periodically monitored.

Process 5.6.1 uses lab results and the information in the *DRUG INDICATION* and *RECOMMENDED DRUG DOSE* tables to decide on the optimal drug to prescribe for a specific patient. The *PRECAUTION* may provide information on relevant precautions, and *DOSE ADJUSTMENT* will recommend adjustments such as *"decrease Digoxin dose or administer potassium supplements."*

5.6.2 Monitor Lab

Some medications require laboratory monitoring. The lab tests used for monitoring when on a certain drug are recommended by process 5.6.2 *Monitor Lab* using the table *MONITOR DRUG LAB* discussed in the following material. Receiving a drug name as input, 5.6.2 checks in this table for the lab tests that may need to be monitored as well as the frequency of monitoring. For example, it is best advised for patients on *Griseofulvin* to have a monthly check of their *liver function tests*. The CDS checks if the relevant lab tests have been ordered as a corollary to the drug prescription, and if not, it will recommend the lab orders to the prescribing clinician.

5.6.3 Drug Interfere with Lab

Regarding process 5.6.3 *Drug Interfere with Lab*, drugs may interfere with lab tests, causing false-positive or false-negative results. For example, *Aspirin* may cause a decrease in the *Albumin* level measured. The interference may be caused by physiological process happening in vivo, such as *diuretics* causing enhanced excretion of *Calcium* and thus lowering the resulting blood *Calcium* levels or *Promethazine* increasing the level of *hCG* (human Chorionic Gonadotropin), causing a false-positive pregnancy test result.

The interference may happen in vitro, when a drug interacts with the chemical reagents used by the laboratory, such as *Ascorbic acid* causing a false-negative result for *occult blood* in feces.

Process 5.6.3 checks the list of known drug-lab interferences in the table *DRUG INTERFERE LAB* and passes the relevant information to *Temp CDS DRUG* for further analysis and display to the clinician. Conversely, abnormal lab results that may be caused by drugs a patient is currently taking are flagged accordingly.

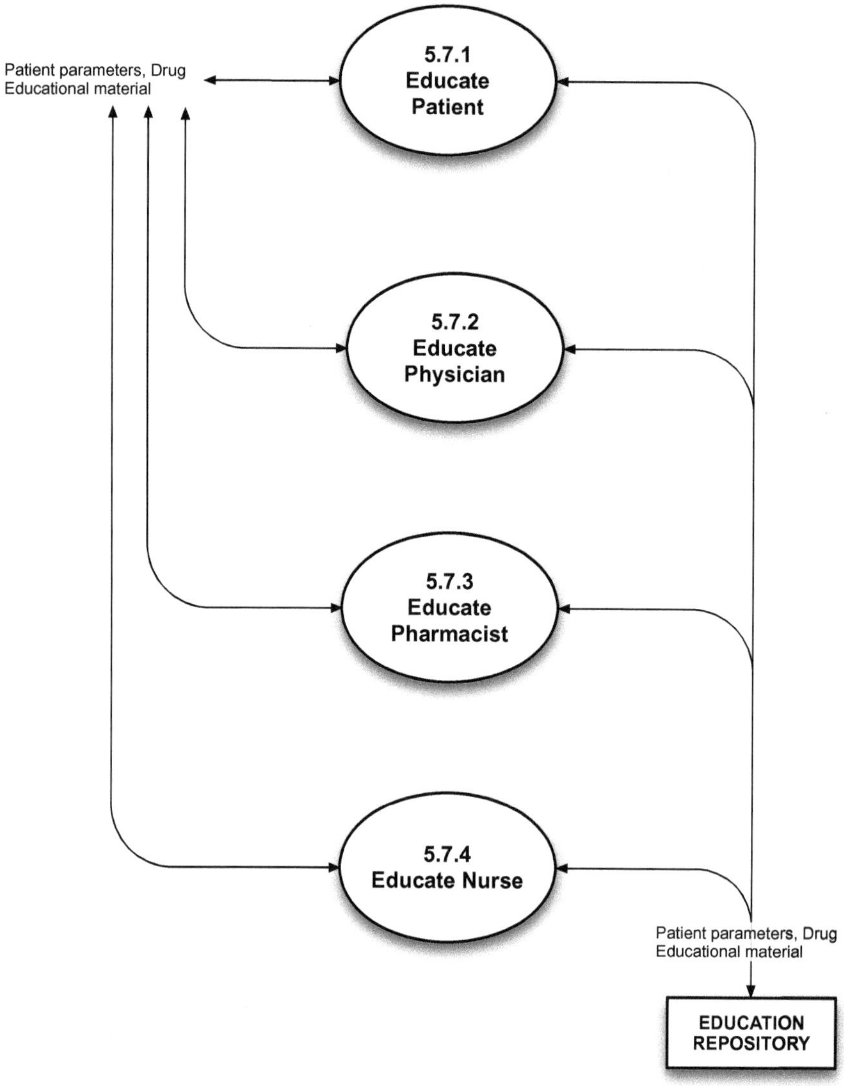

Figure 8.13 DFD 5.7: Educate.

DFD 5.7 Educate

The last CDS process, 5.7 *Educate*, uses patient parameters and extracts educational materials from a repository at different levels of detail to be provided to the patient and next of kin (5.7.1), physician (5.7.2), pharmacist (5.7.3), or nurse (5.7.4) (Figure 8.13).

From the workflows and processes, we focus now on the data structures supporting the CDS.

Data Elements

Analyzing the DFD mentioned, we identify the data element entities in the medications database that may participate in the CDS workflow.

Temp CDS Drug

The temporary table *Temp CDS DRUG* is created for each individual patient, one per CDS session; once it has reached the end of its life cycle, it is either deleted or stored for further analysis and algorithm tweaking by CDS maintenance personnel. It is specific to a single patient's characteristics, and it accumulates rows as processes 5.x contribute information to it in the form of decisions. As mentioned, the decisions are tiered according to severity, frequency of occurrence, and quality of supporting documentation. As a minimum, this table would have the following fields:

Drug: *Erythromycin*
Process: 5.5.1 *Consider Interaction*
Details: *Erythromycin + Ketoconazole*
Tiered decision: *Do NOT use or reduce erythromycin dose*
Severity: *High*
Frequency: *Common*
Quality of supporting documentation: *Good*

Temp CDS Dose

Similar to the *Temp CDS DRUG*, the *Temp CDS DOSE* table accumulates CDS decisions on dose adjustment:

Drug: *Gentamicin*
Process: 5.6.1 *Lab Affect Drug & Dose*
Details: *Gentamicin + Creatinine clearance less than 20 ml/min*
Decision: *Increase intervals between administrations*
Decision details: *Administer once every 48 hours*
Severity: *High*
Frequency: *Common*
Quality of supporting documentation: *Excellent*

Precaution

The *PRECAUTION* table contains a list of situations that justify precaution related to specific drugs and doses, for example, precautions related to lab results or patient conditions, planned procedure, vital sign, and so on. It is complementary to the *DRUG C/I* table, which is a list of drugs and contraindications that one may consider as "stronger" precautions.

Since a precaution may be related to many drugs and one drug may have many precautions, it is a many-to-many (*M:N*) relationship. As explained in previous chapters, *M:N* relationships require an intersection entity in between the two tables, *RECOMMENDED DRUG DOSE* and *PRECAUTION* (Figure 8.14).

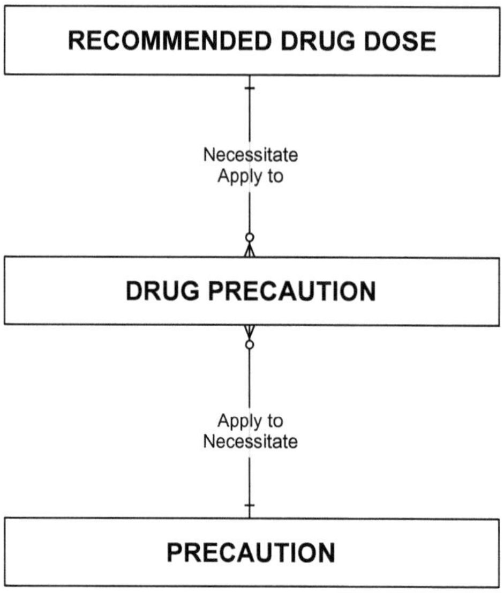

Figure 8.14 Drug precaution intersection entity.

The *DRUG PRECAUTION* intersection table may have the following attributes:

Drug name: *Atenolol*
Precaution type: *Vital sign*
Situation: *Bradycardia*
Situation details: *Heart rate below 60/minute*
Recommendation: *Consider alternative antihypertensive agent*
Severity: *High*
Frequency: *Common*
Quality of supporting documentation: *Excellent*

Drug Interaction

The *DRUG INTERACTION* table holds information on pairs of ingredients and details about their in vivo interaction:

Ingredient A: *Warfarin*
Ingredient B: *Aspirin*
Interaction details: *Increased risk of bleeding*
Time to effect: *Immediate*
Recommendation: *Limit aspirin intake to 100 mg per day, monitor INR*
Severity: *High*
Frequency: *Common*
Quality of supporting documentation: *Excellent*

IV Admixture

Similar to in vivo interactions, the *IV ADMIXTURE* table details in vitro interactions for pairs of ingredients, the solution stability, and other relevant handling requirements:

Ingredient A: *Calcium chloride*
Ingredient B: *Sodium bicarbonate*
Solution stability: *NOT stable: lime precipitate*
Time to effect: *Immediate*
Solution expires after: *not applicable*
Need to refrigerate: *not applicable*
Recommendation: *do NOT mix, do NOT administer through same IV line*
Severity: *High*
Frequency: *Common*
Quality of supporting documentation: *Excellent*

Dose Adjustment

The *DOSE ADJUSTMENT* table contains information related to drug, process step, and recommendations on specific dose adjustments:

Drug: *Tobramycin*
Process: 5.6 *Consider lab*
Details: *Creatinine clearance < 20 ml/min*
Recommendation: *Increase intervals between administrations*
Recommendation details: *Administer every 48 hours and monitor blood level*

Monitor Drug Lab

Regarding the *MONITOR DRUG LAB* table, one ingredient may require several lab tests to be monitored. For example, *Griseofulvin* necessitates several liver function tests, such as *Bilirubin, SGOT (Serum Glutamic-Oxaloacetic Transaminase), SGPT (Serum Glutamic Pyruvic Transaminase), Alkaline phosphatase,* to be performed on a monthly basis. One lab test (such as *SGOT*) may be used to monitor many drugs, such as the effect *statins* may have on hepatic function. Thus, an *M:N* relationship exists between two entities: *INGREDIENT* and *LAB* (Figure 8.15).

The table *MONITOR DRUG LAB* is the intersection table resolving the *M:N* relationship between *INGREDIENT* and *LAB* tables. This table may include information about the frequency lab tests need to be performed when monitoring a drug effect.

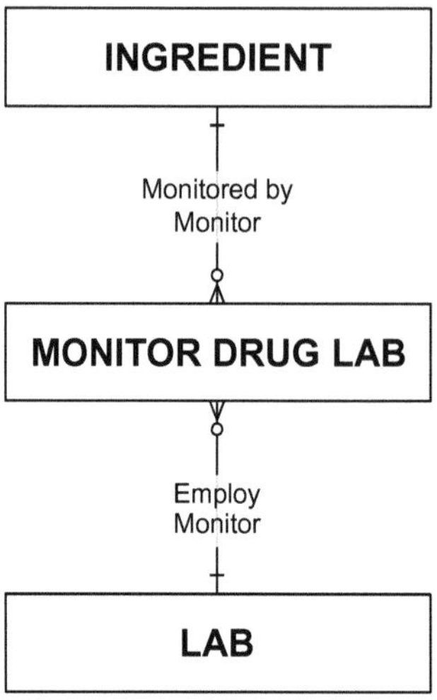

Figure 8.15 Lab to monitor when taking drug.

Drug Interfere Lab

In a similar fashion, a drug or its ingredients may interfere with many lab results, and one lab result may be falsely positive or falsely negative due to many interfering drugs. This is another *M:N* relationship that necessitates an intersection entity, the *DRUG INTERFERE LAB* table (Figure 8.16).

The *DRUG INTERFERE LAB* table includes information on the type of interference (false positive or false negative) as well as an explanation whether the interference is caused by an in vivo or in vitro process, as discussed.

The medications ERD is updated for the needs of the CDS workflow (Figure 8.17). Note that due to space issues, the *DOSE ADJUSTMENT* table is not shown in Figure 8.17.

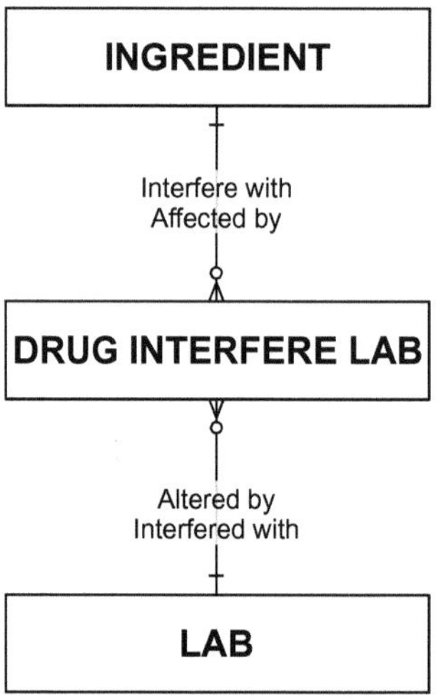

Figure 8.16 Drug interference with lab results.

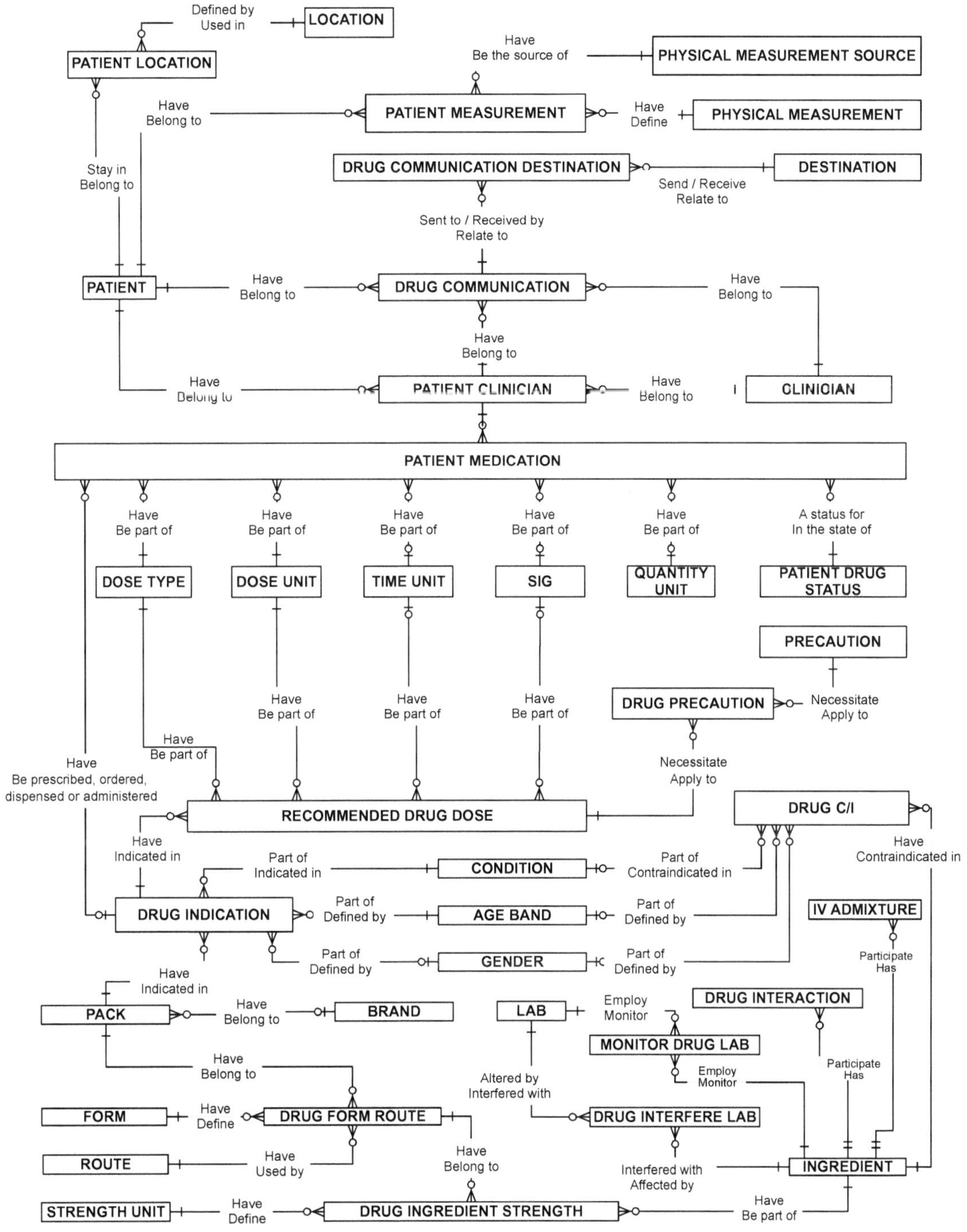

Figure 8.17 ERD updated for CDS.

Barriers to CDS Adoption

The main obstacles to widespread adoption of CDS are related to (20):

- Current systems' limitations, capabilities, usability, integration with EHR, and access to best practice guidelines and EBM
- High costs of purchase and maintenance

- Difficulties developing, implementing, and using CDS as well as demonstrating value and return on investment
- Lack of standards for data representation and dose calculations
- Perception of increased liability if CDS recommendations are not followed by clinicians

Recommendations

As any other project with a big scope, a CDS should be initiated with small tactical goals in mind. The organization's "baby steps" should focus the CDS on a small number of drugs that dominate the list of medications involved in errors:

- *Insulin*
- *Anticoagulants*
- *Narcotic analgesics*
- *Potassium*
- *Chemotherapy*
- *Nephrotoxic agents*

In addition, special attention should be devoted to dosing regimens for the extremes of age or weight: *neonates, pediatrics,* and the *elderly.*

Once a CDS is shown to be actively promoting safety related to the drugs mentioned, the scope can be enlarged to include other drugs, such as expensive medications: *H2 blockers, patient-controlled analgesia, latest antimicrobials,* and so on.

CDS and Genomics: Personalized Medicine

The recent advancements in genetic testing and especially genomic sequencing bring us closer to the much-touted personalized medicine, in which an individual's health is optimized through the use of the patient's own genomic data. The two main research areas as far as pharmacology is concerned are genetics applied to:

- *Pharmacokinetics*: what the body does to the drug
- *Pharmacodynamics*: what the drug does to the body

Genetic variances between individuals were identified in several instances as the principal cause of adverse reactions or nonresponsiveness to treatment with certain drugs.

The goal of pharmacogenomics is to identify through genetic testing the patients who will respond, or conversely may not respond, as expected to a specific therapeutic agent and to predict an individual response to therapy. Examples of genetic tests that have recently moved from experimental medicine to (almost) routine clinical practice include (21,22):

Use of two gene essays (*CYP2C9* and *VKORC1*) in the dose adjustment of *Warfarin*, a widely used anticoagulant with a narrow therapeutic index.

Trastuzumab (*Herceptin*), a monoclonal antibody to HER2 receptor, used to treat HER2-positive breast cancer. *Herceptin* is indicated only for the 10–25% of patients who express the *HER2/neu* gene.

Identifying HIV infections resistant to specific *antiretroviral* therapy.

Identifying patients with low activity of the *TPMT* gene (as being at increased risk for toxicity) or, conversely, high activity of this gene (as potentially nonresponsive patients) when treated with *6-Mercaptopurine* or *Azathioprine* for leukemia.

Identifying patients with *CYP2C19* variants as slow metabolizers of *Tacrolimus*, *Cyclosporine*, and *Omeprazole*, thus necessitating a concomitant dose adjustment.

Reduction of *Irinotecan* dose in patients with a specific *UGT1A1* gene variation.

Contraindication of the chemotherapeutic *Capecitabine* in patients with *DPD* deficiency.

Currently, the FDA requires the integration of pharmacogenomic data into the following product labels (23):

- *Herceptin* for metastatic breast cancer
- *Atomoxetine* for attention deficit disorder
- *Thiopurines* for acute lymphoblastic leukemia
- *Cetuximab* for metastatic colorectal cancer
- *Irinotecan* for colorectal cancer

The list of drugs that can be genetically best fitted to a patient is expected to grow exponentially in the near future, with more specific and sensitive, less-expensive genetic tests reaching market maturity. A CDS will need to integrate this critical knowledge in the decision-making process as an indispensable step toward a safer, personalized practice of medicine.

Summary

The CDS as detailed in this chapter may have more than 20 processes, algorithm steps, with each one potentially looping through many medications and conditions. It is therefore unrealistic for a clinician to perform the CDS manually and stepwise; it is a time-consuming, boring, and repetitive process that cries out for automation. A computer can process in less than 1 second information that may take hours and days for a (very stubborn) human being to grasp.

A tentative algorithm automating the CDS was presented that can summarize hundreds of discrete decisions resulting from the multiple processes mentioned, prioritize and tier them according to hospital rules and clinician preferences, and then display the CDS in an unobtrusive manner for further consideration.

The CDS has the potential to improve quality and contain the rising costs of healthcare. The solution to the currently low CDS adoption may include the need to implement a tiered CDS initially for dangerous drugs only, elimination of predictable alerts, and allowing both hospitals and clinicians to configure the CDS algorithm.

References and Further Reading

1. http://healthit.ahrq.gov/images/sep08cdswebconference/textonly/slide4.html (accessed May 14, 2011).
2. Sim, I., P. Gorman, R.A. Greenes, et al. 2001. Clinical decision support systems for the practice of evidence-based medicine. *Journal of the American Medical Informatics Association* 8: 527–534.
3. Kawamoto, K., C.A. Houlihan, E.A. Balas, et al. 2005. Improving clinical practice using clinical decision support systems: A systematic review of trials to identify features critical to success. *BMJ* doi:10.1136/bmj.38398.500764.8F: 1–8 (March). http://www.bmj.com/content/330/7494/765.full.pdf+html.
4. Glasziou, P., and B. Haynes. 2005. The paths from research to improved health outcomes. *ACP Journal Club* 142: A8–A10.
5. Kohn, L., J. Corrigan, and M. Donaldson. 2000. *To err is human: Building a safer health system*. Committee on Quality of Health Care in America. Institute of Medicine. Washington, DC: National Academies Press.
6. Osheroff, J.A., J.M. Teich, B.F. Middleton, et al. 2006. The roadmap for clinical action on clinical decision support. *AMIA* (June). https://www.amia.org/files/cdsroadmap.pdf.
7. Kuperman, G.J., A. Bobb, T. Payne, et al. 2007. Medication-related clinical decision support in computerized provider order entry system: A review. *Journal of the American Medical Informatics Association* 14: 29–49.
8. Abarca, J., D.C. Malone, E.P. Armstrong, et al. 2004. Concordance of severity ratings provided in four drug interaction compendia. *Journal of the American Pharmacists Association* 44: 136–141.
9. Paterno, M.D., S.M. Maviglia, P.N. Gorman, et al. 2009. Tiering drug-drug interactions alerts by severity increases compliance rates. *Journal of the American Medical Informatics Association* 16: 40–46.
10. Graham, R., M. Mancher, D.M. Wolman, et al. 2011. *Clinical practice guidelines we can trust*. Washington, DC: National Academies Press.
11. Van Ornum, M. 2009. *Electronic prescribing: A safety and implementation guide*. Sudbury, MA: Jones and Bartlett.
12. Lund, A. M. 1997. Expert ratings of usability maxims. A study of the heuristics design experts consider important for good design. *Ergonomics in Design* 5: 15–20.
13. Payne, T.H., W.P. Nichol, P. Hoey, et al. 2002. Characteristics and override rates of order checks in a practitioner order entry system. *Proceedings of AMIA Symposium* 602–606.
14. Van der Sijs, H., J. Aarts, A. Vulto, et al. 2006. Overriding drug safety alerts in computerized physician order entry. *Journal of the American Medical Informatics Association* 13: 138–147.

15. Hsieh, T.C., G.J. Kuperman, T. Jaggi, et al. 2004. Characteristics and consequences of drug allergy alert overrides in a computerized physician order entry system. *Journal of the American Medical Informatics Association* 11: 482–491.
16. Spina, J.R., P.A. Glassman, P. Belperio, et al. 2005. Clinical relevance of automated drug alerts from the perspective of medical providers. *American Journal of Medical Quality* 20: 7–14.
17. Bates, D.W., D.J. Cullen, N. Laird, et al. 1995. Incidence of adverse drug events and potential adverse drug events. Implications for prevention. ADE Prevention Study Group. *Journal of the American Medical Association* 274: 29–34.
18. Kaushal, R., D.W. Bates, C. Landrigan, et al. 2001. Medication errors and adverse drug events in pediatric inpatients. *Journal of the American Medical Association* 285: 2114–2120.
19. Classen, D.C., S.L. Pestotnik, R.S. Evans, et al. 1991. Computerized surveillance of adverse drug events in hospital patients. *Journal of the American Medical Association* 266: 2847–2851.
20. Anderson, P.O., S.M. McGuinness, and P.E. Bourne. 2010. *Pharmacy informatics.* Boca Raton, FL: CRC Press.
21. Grossman, I., and Goldstein, D.B. 2009. Pharmacogenetics and pharmacogenomics. In *Genomic and personalized medicine*, ed. H.F. Willard and G.S. Ginsburg, 321–334. New York: Academic Press, Elsevier.
22. Kawamoto, K., and D.F. Lobach. 2009. Clinical decision support in genomic and personalized medicine. In *Genomic and personalized medicine*, ed. H.F. Willard and G.S. Ginsburg, 242–251. New York: Academic Press, Elsevier.
23. Spraggs, C.F., B.T. Koshy, M.R. Edbrooke, et al. 2009. Role of pharmacogenomics in drug development. In *Genomic and personalized medicine*, ed. H.F. Willard and G.S. Ginsburg, 345. New York: Academic Press, Elsevier.
24. Berner, E.S. 2006. *Clinical decision support systems: Theory and practice.* New York: Springer.
25. Mendonca, E.A., R. Jenders, and Y.A. Lussier. 2003. Clinical decision support systems. In *Healthcare information systems*, ed. K. Beaver, 221–235. Boca Raton, FL: CRC Press.
26. Osheroff, J.A. (editor). 2009. *Improving medications use and outcomes with clinical decision support: A step-by-step guide.* HIMSS.
27. Osheroff, J.A., Pifer, E.A., Teich, J.M. et al. 2005. *Improving outcomes with clinical decision support: An implementer's guide.* Chicago: HIMSS.
28. Shortliffe, E.H., and L.E. Perreault. 2001. *Computer applications in health care and biomedicine.* New York: Springer.
29. Wright, A., S. Phansalkar, M. Bloomrosen, et al. 2010. Best practices in clinical decision support: The case of preventive case reminders. *Applied Clinical Informatics* 1: 331–345.

Review Questions

1. What are the consequences of false-positive and false-negative alarms?
2. Consider two populations: healthy and sick. When new laboratory test results are charted, the distributions of the two populations are noted to overlap. How can we minimize the new test's false-positive and false-negative rates?
3. A patient is taking *warfarin* (an anticoagulant) at home. During admission to the hospital, a physician orders heparin, also an anticoagulant. Which CDS processes are expected to be involved in the decision-making process and alert the physician about the potential danger of the combination?
4. A patient is on eight different medications. The physician considers adding two antibiotics to treat his pneumonia. How many pairs of medications should the drug-drug interaction process consider?
5. A patient is prescribed two types of *insulin*: regular and slow acting. Which process decides whether the combination is appropriate or should be considered duplicate therapy?
6. A patient with end-stage renal disease needs an antibiotic known to be excreted by the kidney. Which CDS processes should advise the physician prescribing the drug?
7. Green leafy vegetables are known to reduce the effectiveness of the anticoagulant *warfarin*. Which process would be responsible to inform a physician about this interaction?
8. *Acitretin* is potentially harmful to the fetus if taken by the mother. The CDS used by the physician fires a critical alert when the doctor tries to prescribe this drug to a 16-year-old male patient. What should be configured differently in the CDS setup?
9. *Erythromycin* and *ketoconazole* may interact, with the result being a dangerously high blood level of erythromycin. What are the possible CDS recommendations related to this interaction? What would be considered excessive and thus the CDS should not contemplate?
10. A female patient on *promethazine* has her positive pregnancy test flagged as suspect and potentially false positive. Which CDS process decided to alert about this possible wrong result?

Chapter 9

Report

Data is like salt water, you can drown in it but you can't drink it without filtering it. We are drowning in data while dying of thirst for information.

Coleman Reid

The characteristics of the system input were introduced in the chapter on the user interface. What the system should do with the information it controls was presented in the chapter on clinical decision support (CDS). The electronic health record (EHR) output, stakeholders involved, types of reports, and related methodology issues are the subjects of this chapter.

We differentiate between single and multiple patients, clinical versus administrative and real-time versus retrospective reports. Manual, automatic, and hybrid reports, as well as claims and registry-based methods of reporting, are detailed with the special challenges of using an EHR for the purpose of quality and meaningful use reporting.

Outcome, process, and structure are the main dimensions that will help us categorize quality. The calculation of quality measures using a numerator, denominator, and exclusion criteria is explained, followed by the need for a data warehouse (DW) for reporting purposes.

From clinical reporting in general, using dataflow and entity relationship diagrams, we focus on the specifics related to the *Medications* system reports.

Motivation

The reasons for participating in a reporting effort are as follows:

- Commitment to continuous improvement of the practice of medicine
- Track patients with specific conditions for research and quality benchmarking purposes
- Identify hazards or benefits, which may be apparent only in large datasets
- Fulfill certification requirements of professional associations
- Financial incentives/penalties

The last point merits some details: The *Physician Quality Reporting Initiative* (*PQRI*) was established in 2006 as part of the *Tax Relief and Healthcare Act* (1). What started as a voluntary bonus paid to eligible clinicians reporting on several quality measures to *Medicare* will become mandatory in 2015 under the recently passed *Patient Protection and Affordable Care Act* (*PPACA*) (2) and the slightly different name *Physician Quality Reporting System* (*PQRS*).

Clinicians paid under the *Medicare Physician Fee Schedule* (*MPFS*) who reported successfully on quality measures related to their specialty were entitled to a 2% bonus in 2009 and 2010; 1% bonus in 2011; and 0.5% each year for 2012–2014. Starting in 2015, there will be no more federal carrots: Successfully reporting to *Medicare* will not entitle eligible clinicians to any incentives. The stick is that clinicians who will not report on quality measures will be penalized 1.5% of their *MPFS* in 2015 and then 2% in 2016 and each year afterward.

Under the *American Recovery and Reinvestment Act* (*ARRA*) of 2009, incentives are provided for eligible professionals demonstrating "meaningful use" (*MU*) of certified healthcare information technology (HIT). These *MU* incentives are in addition to those mentioned for *PQRS*. Early adopters will benefit from up to $44,000 over 5 years under the *Medicare EHR Incentive Program* and $63,750 over 6 years under the *Medicaid EHR Incentive Program*, while eligible hospitals may benefit from millions of dollars if they report on a number of measures stipulated in the *ARRA* (3). As with *PQRS*, beginning 2015, those who do not report on their meaningful use of HIT will face negative payment adjustments in their future *Medicare* (but not *Medicaid*) reimbursements.

Types of Reports

A report may be related to a single patient, such as a discharge summary letter, or may be based on multiple patients, such as a report on the mortality of patients undergoing a specific surgical procedure. Some reports are mandated by law, such as the surgical procedure report, while other reports are voluntary at the moment, although accompanied by the federal carrot and stick, such as the reporting on the meaningful use of HIT. Reports may be of a clinical nature or purely administrative, and some are not directly related to patients at all, such as inventory reports (Figure 9.1).

Reports on a single patient can be straightforward, as the system queries multiple repositories, databases, and tables with the patient's unique identifier and collects discrete data elements and parses them into a given XML (eXtensible Markup Language) format detailed in the next chapter. These reports may be the patient's list of medications, discharge or visit summary, immunizations to be transmitted to a state registry, lab values to be reported to a syndromic surveillance public health agency, or a copy of the patient's personal health information.

Measuring Healthcare Quality

Reporting on multiple patients is a different kettle of fish. These reports are more complex than those on single patients since they try to measure the quality of healthcare, a vast, convoluted and thorny subject.

In his seminal paper (4), Donabedian identified three main aspects of healthcare quality: outcome, process, and structure, with the following emphasis: *"Outcomes, by and large, remain the ultimate validation of the effectiveness and quality of medical care."*

- *Outcome measures* reflect the impact of healthcare on a patient's status, for example, the rate of hospital-acquired infection after *coronary artery bypass graft surgery*. Other commonly used outcome measures are *mortality*, *length of stay* (*LOS*), *costs*, *readmission rate*, and *patient satisfaction*. Outcomes may seem like the "gold standard" of quality in healthcare, but these measures are influenced by multiple factors, many not under the provider's control. For example, if a surgeon operates on much sicker patients than his or her peers, there should be no surprise if this surgeon's reported mortality is higher than that of others. A need to quantify the term *much sicker* becomes necessary. Thus, *risk adjustment*—mathematical methods to correct and compensate for different characteristics of the populations measured—is needed to compare "apples to apples." Poor outcomes may result from the wrong treatment being perfectly carried out or the right treatment being poorly applied. Donabedian and others mentioned that the most important outcomes may be also the most difficult to measure and have cautioned against picking *mortality* as the quintessential (and the easiest-to-quantify) surrogate for other measures, such as *Functional disability*, which may prove more difficult to quantify.
- *Process measures* evaluate how healthcare is delivered: the actions, interactions, steps, procedures, activities of a healthcare-providing entity compared to the best practices, existing guidelines, and accepted standards of care. For example, the percentage of diabetic patients who have their eyes checked regularly. According to the U.S. Department of Health and Human Services

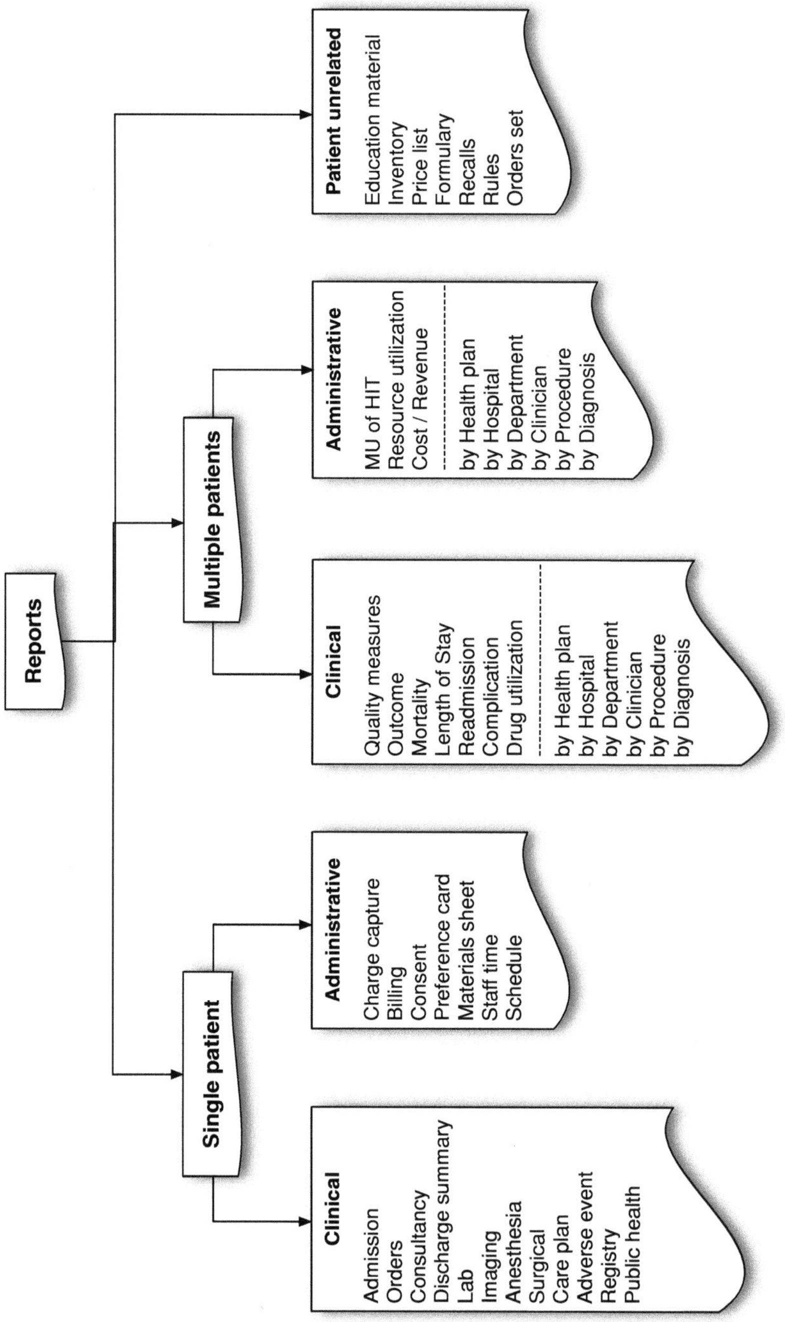

Figure 9.1 Types of reports.

Agency for Healthcare Research and Quality (AHRQ), the majority of the quality measures used for public reporting are process measures. Common sense may lead one to believe that good processes will foster good outcomes, but unfortunately that is not always the case.

- *Structure measures* assess the capacity, systems, organization, and capabilities to provide healthcare, for example, the percentage of physicians using computerized physician order entry (CPOE) or the percentage of patients with their medications list electronically documented. Although never proven to be the correct strategy, many tried to improve processes while restructuring healthcare, diligently expecting to improve outcomes. Oxman et al. summarized such a potential fallacy in an entertaining article, titled: *"A Surrealistic Mega-analysis of Redisorganization Theories"* (5).

Goals for the Healthcare System

An influential analytical framework, published by the *Institute of Medicine (IOM),* sets six goals for the healthcare system (6):

1. *Safe*: *Primum non nocere*, which is Latin for *"First do not harm,"* avoiding harm to patients from the care intended to help them
2. *Effective*: provide healthcare to all who could benefit, avoiding underuse; refrain from providing care to those not likely to benefit, avoiding misuse
3. *Patient centered*: provide care that considers individual needs, preferences, and values.
4. *Timely*: reduce delays and waits, which may prove harmful to the patient
5. *Efficient*: minimize waste of equipment, energy, and ideas
6. *Equitable*: provide equal quality of care to patients from different backgrounds (socioeconomic, gender, ethnicity, etc.)

Dimensions of Quality Measures

Quality measures may be organized around multiple dimensions:

- Goal: *safety, effectiveness, efficiency*
- Type of measure: *outcome, process, structure*
- Disease/condition: *diabetes, congestive heart failure*
- Settings: *physician, hospital, health plan*
- Data source: *claims, registry, EHR*

Evaluating Quality Measures

How good is a measure, and how suitable is it for one's practice? The *National Quality Forum* (*NQF*) and *AHRQ* have established a number of criteria to consider when evaluating a quality measure (7–9):

- *Importance*: The topic measured is of relevance to stakeholders; it is clinically important, having a high prevalence/incidence in the population and a high burden of illness, or it can be quantified/stratified and there is a potential for improvement.
- *Scientific soundness*: There is explicit evidence to support the measure's rationale; the results should be reliable, comprehensible, reproducible, and valid over time. Measures should be standardized and audited at the national level and adjusted for age, gender, ethnicity, income status, and the like.
- *Feasibility*: The data supporting the measure are timely, accessible, and available, and the cost of data abstraction and collection is justified by the potential for quality improvement.

Organizations Involved in Quality Reports

Starting in the mid-1980s with hospital-specific reports on mortality and continuing with more granular aspects of healthcare, today there are hundreds of organizations involved with reporting on medical quality. According to the *AHRQ*, Table 9.1 is a partial list of organizations active in promoting quality of care (10).

The government, employers, insurance companies, professional organizations, and consumer groups encourage and support these reporting efforts, which cover most of the healthcare spectrum: physicians, nurses, pharmacists, hospitals, health plans, states, and so on.

PQRS and MU Measures

The *Center for Medicare and Medicaid Services* (*CMS*) maintains exhaustive lists of measures and links to relevant sites clarifying the reporting needs of clinicians and hospitals (11). For example, the *2011 Physician Quality Reporting Measures List* has 240 measures, such as the following:

> *Asthma—Pharmacologic Therapy*: Percentage of patients aged 5 through 50 years with a diagnosis of mild, moderate, or severe persistent asthma who were prescribed either the preferred long-term control medication (inhaled corticosteroid) or an acceptable alternative treatment.

Table 9.1 Organizations Involved with Quality Measures

Government and state	Centers for Medicare and Medicaid Services (CMS)
	Child Health Care Quality Toolbox
	Hospital Quality Initiative
	National Quality Measures Clearinghouse
	National Academy for State Health Policy
	National Association for Health Data Organizations
National quality organizations	Joint Commission on Accreditation of Healthcare Organizations
	National Quality Forum (NQF)
	Child and Adolescent Health Measurement Initiative
	National Committee for Quality Assurance (NCQA)
	Foundation for Accountability Archives
	Hospital Quality Alliance
	Ambulatory Care Quality Alliance
	Pharmacy Quality Alliance
Quality improvement initiatives	Institute for Clinical Systems Improvement
	Institute for Healthcare Improvement
	National Initiative for Children's Healthcare Quality
Plan and provider associations	American Hospital Association
	American Medical Group Association
	American Board of Medical Specialties
	America's Health and Insurance Plans
Employer/purchaser	National Business Coalition on Health
	Leapfrog Group
	Bridges to Excellence
Foundations and consortia	Robert Wood Johnson Foundation
	Commonwealth Fund
	Yale School of Public Health
	California Healthcare Foundation
	RAND Corporation

Or, a different quality measure:

> *Aspirin at Arrival for Acute Myocardial Infarction (AMI)*: Percentage of patients, regardless of age, with an emergency department discharge diagnosis of AMI who had documentation of receiving aspirin within 24 hours before emergency department arrival or during emergency department stay.

MU reports are less numerous than *PQRS* measures but are similar in structure (12). For example,

> *Maintain active medications list*: More than 80% of patients have at least one entry recorded as structured data.

Or, a different *MU* measure:

> *Computer provider order entry (CPOE) for medication orders*: More than 30% of patients with at least one medication in their medication list have at least one medication ordered through CPOE.

Some *PQRS* measures are also part of the *MU* reports (13). For example, the following is *PQRI* quality measurement 5, *NQF* 0081, and it is also one of the measures to consider for reporting in the *MU* final rule:

> *Heart Failure: Angiotensin-Converting Enzyme (ACE) Inhibitor or Angiotensin Receptor Blocker (ARB) Therapy for Left Ventricular Systolic Dysfunction (LVSD)*
>
> Description: Percentage of patients aged 18 years and older with a diagnosis of heart failure and LVSD who were prescribed ACE inhibitor or ARB therapy.

There is an overlap between requirements of the two reporting systems—*PQRS* and *MU*—which makes an already complex issue a bit confusing. Let us take a closer look at the structure of quality measures.

Anatomy of a Quality Measure

A quality measure has the following characteristics (14–16):

- Name: *Discharge on antithrombotic therapy*
- Description: *Percentage of patients aged 18 years and older with a diagnosis of ischemic stroke or transient ischemic attack (TIA) who were prescribed antiplatelet therapy at discharge*

- Identifiers: *NQF 0435, Joint Commission STK-2, PQRS 32, HITSP* (HIT Standards Panel) *1.3.6.1.4.1.33895.1.2.16001.1.1*
- Author: *Joint Commission*
- Verifying authority: *NQF*
- Set of related measures: *Stroke and stroke rehabilitation*
- Measure type: *Process*
- Effective date: *October 1, 2009, to March 31, 2012*
- Rationale: *The effectiveness of antithrombotic/antiplatelet agents in reducing stroke mortality, stroke-related morbidity and recurrence rates has been studied in several large clinical trials. While the use of these agents … Data at this time suggest … and so on.*
- References: *Adams HP et al. Guidelines for the Early Management of Adults with Ischemic Stroke … 2007;38:1655–1711, and so on.*
- Calculation: *Proportion, ratio, performance rate*

Most measures require a ratio (*percentage*) to be reported. This ratio is compared to a threshold, which may be an absolute number or it may drive a comparison with others. The need for calculating a ratio translates into a numerator, such as *prescribed antithrombotic therapy at discharge*, and a denominator, such as *patients aged 18 years and older with a diagnosis of ischemic stroke or transient ischemic attack (TIA)*:

- Denominator: *Patients aged ≥ 18 years on date of encounter AND Diagnosis for ischemic stroke or transient ischemic attack (ICD-9-CM, International Classification of Diseases, Ninth Revision, Clinical Modification): 433.01, 433.11, 433.21, 433.31, 433.81, 433.91, 434.01, 434.11, and so on AND Patient encounter during the reporting period (CPT, Current Procedural Terminology): 99221, 99222, 99223, 99238, and so on.*
- Numerator: *Patients from the denominator who have discharge orders including one of the following antithrombotic agents: heparin, fondaparinux, argatroban, dalteparin, enoxaparin, warfarin, lepirudin, aspirin and aspirin combinations with buffering agents, dipyridamole, clopidogrel, tinzaparin, hydroxymethylglutaryl-CoA reductase inhibitors, and so on.*

These are inclusion criteria: what goes into the calculation. Both numerator and denominator may have exclusion criteria: what should not be counted, for example,

- Denominator exclusions: *Patient has expired OR had a length of stay over 120 days OR has left against medical advice OR is on a clinical trial OR is scheduled for an elective carotid intervention OR there is a valid reason for not prescribing antithrombotic therapy and so on.*
- Numerator exclusions: *Prescription for antithrombotic therapy was cancelled, discontinued, put on hold … OR Heparin lock, heparin flush and so on*

Let us calculate an example:

Denominator: *120 patients over 18 years of age with the diagnoses of ischemic stroke during 1 year.*
Excluded from denominator: *5 patients have transferred to other facilities, 5 have died, and 10 had valid medical reason not to prescribe them antithrombotic therapy at discharge.* 20 patients are excluded, and the denominator becomes 120 – 20 = 100 patients.
Numerator: *80 patients from the final denominator group have been prescribed antithrombotic therapy at discharge.*
Excluded from numerator: *3 patients with Heparin lock and 1 patient with antithrombotic therapy on hold.* Numerator is 80 – 4 = 76 patients.
The quality measure performance rate: *Numerator/Denominator* = 76/100 = 76% *performance rate.* This rate can be compared to a specific threshold set by one of the organizations mentioned, or it can be used to compare performance to other institutions, health plans, providers, and the like.

Reporting Methods

There are several methods for reporting on quality/MU measures: medical records, claims based, registry, EHR, and hybrid.

Medical records are a natural data source used for reporting. They are also the most time- and resource-consuming method of creating a report: A person manually pulls the information needed for the numerator and denominator calculation from the old-fashioned paper medical record. One can imagine the tediousness of such a method and its potential for errors, especially with the abundance and complexity of the exclusion criteria. Still, for some organizations the manual abstraction of information from paper-based medical records for reporting purposes is the painful norm.

The *claims-based* method of reporting uses the *CPT* codes, specifically the *G codes* or the *Category II* subset for performance measurement (*0001F-7025F*). *CPT* codes are similar to other structured coding systems, such as *ICD-9*, explained in the next chapter, except *CPT* codes are used to communicate services rendered (not diagnosis) and are an integral part of the billing process for medical services in the United States. For example (17):

G8027: Heart failure patient with left ventricular systolic dysfunction (LVSD) documented to be on either angiotensin-converting enzyme-inhibitor or angiotensin-receptor blocker (ACE-I or ARB) therapy

Or

CPT Cat II code 4009F: Angiotensin-converting enzyme (ACE) inhibitor or angiotensin receptor blocker (ARB) therapy prescribed.

For exclusion criteria, a different *G* code is used or the same *CPT II* code with a modifier attached, for example:

G8029: Clinician documented that heart failure patient was not an eligible candidate for either angiotensin-converting enzyme inhibitor or angiotensin receptor blocker (ACE-I or ARB) therapy measure

Or

CPT Cat II code 4009F WITH modifier 1P, 2P, or 3P: Angiotensin-converting enzyme (ACE) inhibitor or angiotensin receptor blocker (ARB) therapy excluded

The reasons for excluding a patient from a measure calculation using a *CPT II* code may be

1P—Medical reason (contraindication, drug interaction)
2P—Patient reason (religious, socioeconomic)
3P—System reason (insurance, coverage)
8P—Unspecified reason

The claims-based method is still tedious for reporting on quality measures as it creates an additional burden to an already-complex billing process. The special codes mentioned for reporting using a claims mechanism may in turn delay *CMS* payments to the providers, which would not like the idea of not being paid for services rendered just for the sake of "quality reporting."

A *Registry* is a collection of medical data related to patients with a specific diagnosis or procedure. Usually, a registry is less complex than an EHR, focusing on a subset of the population and containing only the data elements pertinent to one disease or one procedure. Registries are used for keeping track of medical devices such as implants and for reporting purposes such as for pay-for-performance programs or quality measures. *CMS* has published specific requirements for registry submission of quality measures during 2011 (18). When compared to claims-based reporting, using a registry has several advantages (19):

- A registry employs data validation mechanisms that may significantly reduce reporting errors.
- It allows physicians to report retrospectively in a single transaction at the end of the reporting period, reducing the administrative overhead.
- A registry provides physicians with tools to view the data reported prior to submission as well as their own performance during the reporting period, allowing comparison to their peers.
- The registry method separates the reporting process from claims processing, eliminating payment delays for services rendered due to quality reporting issues.

At first glance, the *EHR* seems like the perfect source for the data elements needed for reporting purposes, but the data from the EHR may suffer from the following issues (20):

- Data originating from a number of modules or disparate vendors may reside in several databases and thus need to be consolidated from multiple sources.
- Data may exist as free text since they have not gone through the coding process yet. Free text is notoriously difficult to use for reporting purposes, unless a natural language-processing tool is employed.
- A concept may have more than one code. For example, there are multiple terminologies for medications (as explained in the chapter on interoperability), which may translate into difficulties with mapping the correct concept to the right code.
- One code may have more than one meaning. See the next chapter for details on synonymy and ambiguity.
- Information may be contradictory, for example, one physician prescribes a drug and another clinician puts it on hold, leaving the decision on whether the drug is active to a manual disambiguation process.
- Exclusion criteria are complex and cannot be easily translated into machine language.
- Partial information may still exist on paper.

If only one data element is missing for the automated extraction and calculation, the EHR-based reporting process necessitates reverting to manual abstraction. This may become an unsustainable data collection process; pulling a paper chart from medical records and searching for one data element is almost as expensive and time consuming as searching for many parameters.

Recognizing the need for both EHR and paper medical records as data sources for reports, we can design a pragmatic method that optimizes the process of data extraction: a *hybrid* method that automatically pulls the data elements from the EHR and points to those elements that are still on paper records or need further disambiguation for the manual abstractor to clarify.

The process of data extraction from multiple sources and its transformation/cleansing and loading into a Data Warehouse (DW) is detailed below.

Medications Reports

Approximately a fifth of the 240 *PQRS* measures *CMS* has published for 2011 are related to medications (21); immunization-/vaccination-related reports not included. On top of the *CMS* quality measures, there is a need for additional medication-specific reports (22,23):

- Alerts triggered by CPOE or electronic prescription applications. The type and frequency of the alerts can be used to adjust the application settings (see Chapter 8).

- Alerts overridden.
- Alerts that resulted in order or prescription change.
- Formulary adherence.
- Preferred therapeutic agents.
- Patience adherence and compliance with therapy.
- Pharmacist intervention: type and frequency. Reimbursement schemes need proof of pharmacist intervention for certain situations or medications.
- Medication use evaluation reports on high-volume, high-risk or high-cost drugs.
- *Drug Enforcement Administration* (*DEA*) reports on narcotics.
- Research on safety and effectiveness of medications.
- Patients on medications that the *Food and Drug Administration* (*FDA*) has recalled.

For *MU* (stage 1), the following reports on measures are related to medications:

- Percentage of patients with at least one medication entered using CPOE
- Percentage of prescriptions generated and transmitted electronically
- Percentage of patients with their active medications list maintained electronically
- Percentage of patients with their allergies list maintained electronically
- Percentage of transitions of care for which medications reconciliation was performed
- Implementation of one CDS rule as well as the drug-drug and drug-allergy checks
- Implementation of a formulary check system
- Provide patients with clinical summaries and timely access to their clinical information

Most reports related to medications will need the drug name, ingredients, class, route, strength, dose, and timing of prescription/administration—discrete data elements discussed in previous chapters.

Data Warehouse

Why can't we use the source databases populated by the EHR for our reporting needs? There are several reasons why a DW is necessary:

- Information resides in multiple source databases and needs to be consolidated for reporting purposes.
- A patient in one system needs to be matched to the same patient in other systems. This issue is more complicated in the United States, where there is no *unique patient identifier* (the *Social Security Number, SSN,* is not a good

candidate for a patient unique identifier due to reasons that are out of scope of this discussion). Thus, the need for complex algorithms using surrogate identifiers and combinations thereof (such as *name* and *SSN* and *gender* and *age* and *zip code,* etc.) to match patients in different systems.

- Reports are based on complex queries, which need to filter, aggregate, and order data elements from disparate tables. These queries require multiple *join* operations on database tables, which in turn may have a negative impact on the EHR performance, slowing the transactional, source systems.
- Storing more than 2–3 years worth of information in the source systems may slow the EHR; thus, data need to be purged periodically into less-active data archives. A DW stores information spanning much longer time frames, typically 10–20 years, for analytical purposes.
- The discrete data elements may be available in one source and not the other; see discussion in this chapter on partial information residing in paper medical records.
- The same data element may exist in more than one source; thus, inconsistency issues need to be resolved.
- Conversion of units may be necessary for reports: A medication dose defined as *drops/minute* in one system needs to be consolidated with a dose range measured in *milligrams per day* from other systems (see discussion on units conversion in Chapter 6).
- Definitions of terms may depend on the source: In one system, *myocardial infarct* is defined as a chronic condition, while in another system, it may be defined as an acute disease.
- The EHR source databases are optimized for transactions—*Select, Insert, Update,* and *Delete*—operations performed concurrently by multiple users. To be efficient, a transactional database must capture the transactions without redundancy.
- The EHR databases are accessed by multiple users at the same time; a locking mechanism must be enforced so one user will not be able to inadvertently override the same fact concomitantly introduced by other users.
- An EHR source database is thus normalized, has minimal redundancy, and utilizes a locking mechanism to prevent inconsistencies. Unfortunately, these requirements create a data store that is not optimized for reporting since a complex query would necessitate joining multiple normalized tables—an operation that is time consuming and burdensome for the processor.
- A DW, on the other hand, is "read only"; it does not need to employ a locking mechanism since users do not update it. The DW is updated periodically by only one process— *Extract-Transform-Load* (*ETL*), explained further in this section.
- A DW provides access to cleansed and consolidated information in a format that is optimized for complex queries.

In his book, Rainardi provided a comprehensive definition of a DW (24):

> *A data warehouse is a system that retrieves and consolidates data periodically from the source systems into a dimensional or normalized data store. It usually keeps years of history and it is queried for business intelligence or other analytical activities. It is usually updated in batches, not every time a transaction happens in the source system.*

Let us take a look at a DW system (Figure 9.2). The main process in a DW system is the *ETL*. The *ETL* pulls data elements from the source systems on the left side of Figure 9.2—claims, registry, EHR or paper medical records—usually on a daily basis, and using a set of rules from the *RULES REPOSITORY*, *ETL* performs a cleansing/transformation process. *ETL* identifies the "dirty" data and then has three options:

- If data is correct according to the *RULES REPOSITORY*, it is loaded into the DW.
- If data is incorrect according to the *RULES REPOSITORY* or it needs editing, mapping, consolidation, or aggregation—activities *ETL* can automatically perform—it is transformed and loaded into the DW.
- If data is incorrect and it is beyond the capabilities of *ETL* to cleanse, repair, or transform, *ETL* rejects it, and the data is not loaded into the DW.

All *ETL* activities and the sources' data quality are reported to the *Correction Engine*. The *Correction Engine* stores the data quality reports in the *Metadata repository* and uses this information to send correcting suggestions to the *DW Administrator*, so data quality on the source side can be incrementally improved.

A hybrid reporting method uses the *ETL* to identify the locations of data elements known to exist only on paper records. The locations are sent to the human abstractor, who uses an application to manually extract and send back to *ETL* these data elements. Data in need of disambiguation is also clarified by the *Manual abstraction* process and then sent to *ETL*. *ETL* performs an audit on the data quality of these manually extracted elements, and if found to be correct according to the *RULES REPOSITORY*, it loads them into the DW.

The important aspect of the DW system described in Figure 9.2 is the fact that users do not update or interact directly with the DW only *ETL* feeds the DW. From a user's perspective, the DW is read only: to be used for reports, spreadsheets, dashboard, scorecards, ad hoc queries, and other analytical activities on the right side of the diagram.

The DW may store information in *normalized, dimensional,* or *cube* formats:

1. *Normalized database:* This type of data store should be familiar to the reader by now. Information is stored in multiple tables connected by relationships, with minimal redundancy according to a set of strict normalization rules (see Chapter 1 and Figure 8.17). The schema of a normalized relational

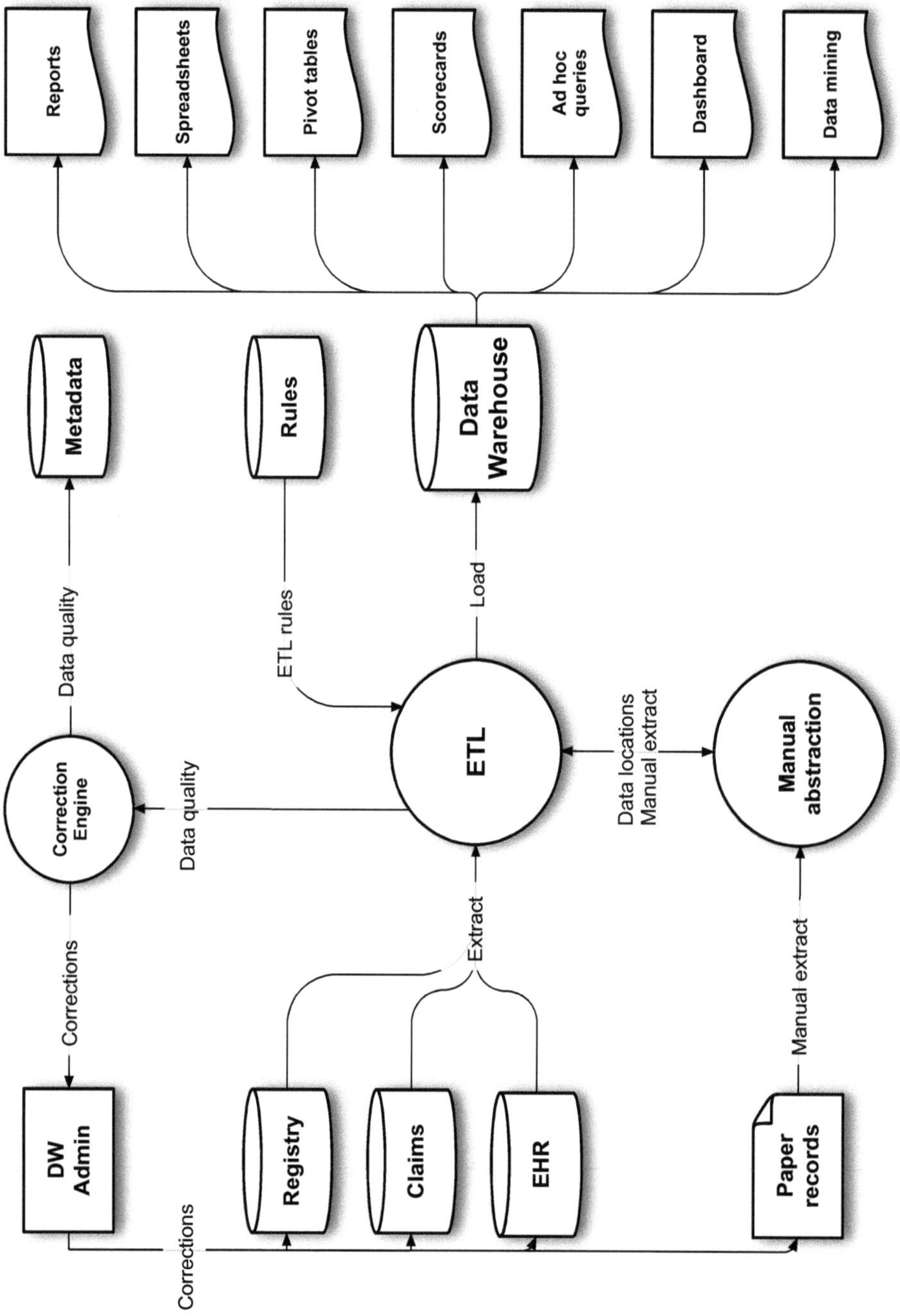

Figure 9.2 Data warehouse system.

database is the *Entity-Relationship Diagram* (*ERD*), which has entities as tables and their attributes as fields.

2. *A dimensional data store* contains a collection of *data marts* (24,25). A dimensional *data mart* is a group of related *fact* tables with their corresponding *dimensions* denormalized in a *star*, *snowflake*, or *galaxy* schema. The fields in the *fact* table are measures such as aggregates or calculated values of the *dimensions*. The data arrangement (schema) for a dimensional data store may resemble that of Figure 9.3.

A *star* schema has only one level of dimensions, and it does not have any subdimensions. For example, a *fact* table would contain the *percentage of patients aged 18 years and older with a diagnosis of heart failure and LVSD (LVEF [left ventricular ejection fraction] < 40%) who were prescribed an ACE inhibitor or ARB therapy*. The *dimensions* would store the various parameters for the *fact* table, such as a list of the *ACE* or *ARB inhibitors*, the *ICD-9* codes for *heart failure*, the *LVEF* results, and so on.

A *snowflake* schema (Figure 9.4) is similar to the star schema, but dimensions may have subdimensions. A *Medications* table is a dimension with the subdimension of

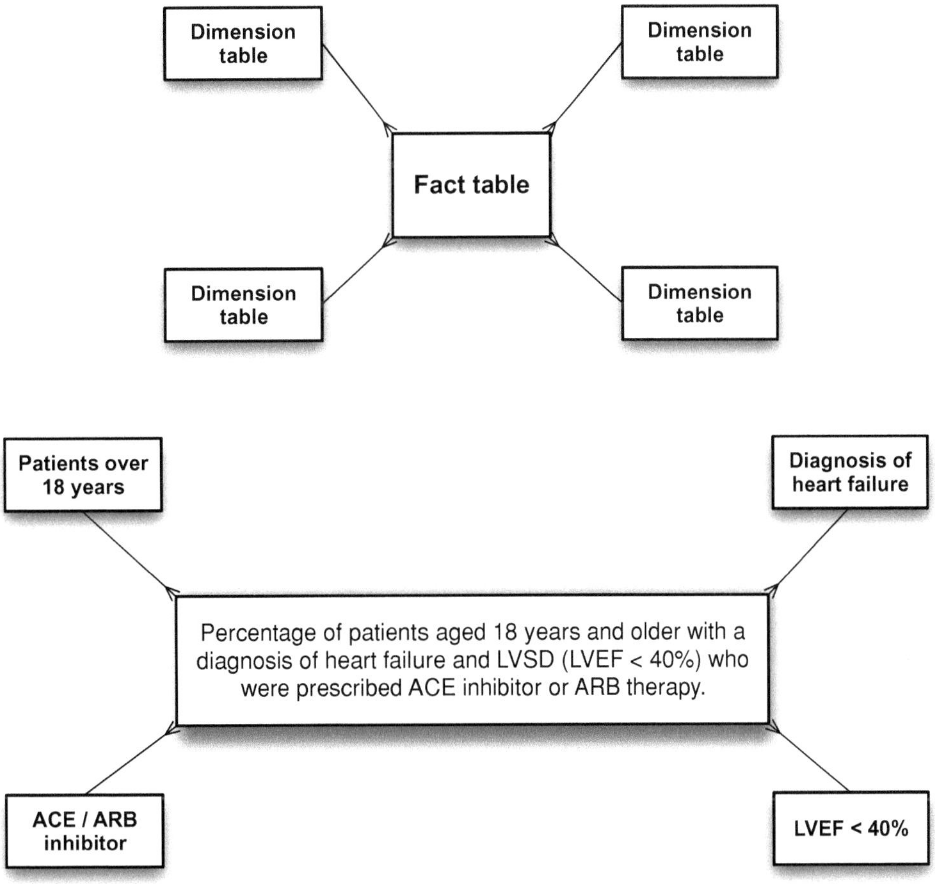

Figure 9.3 Star schema.

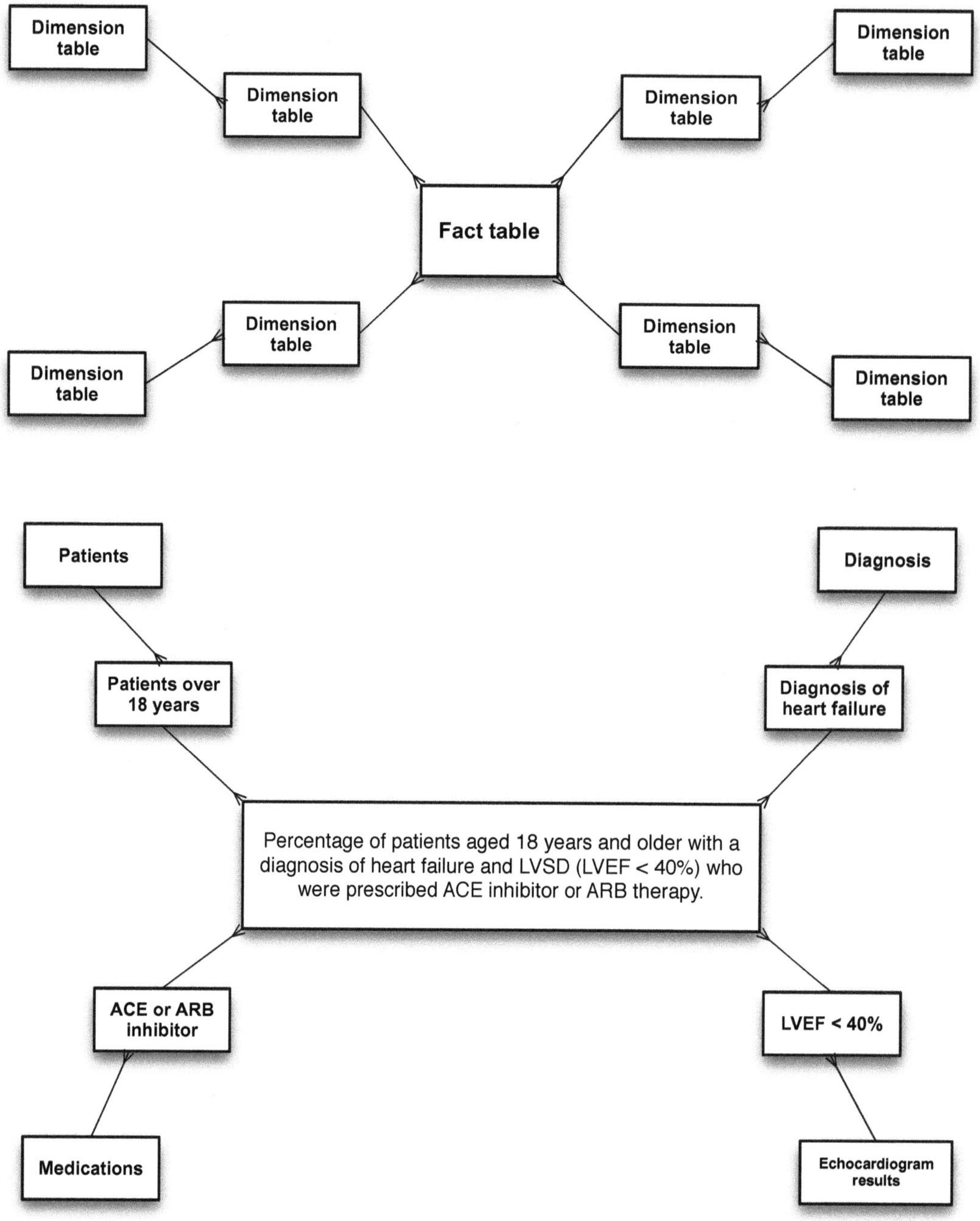

Figure 9.4 Snowflake schema.

ACE or ARB inhibitors as a subset table; similarly, the *Diagnosis* dimension will have as a subdimension a list of the codes for *heart failure* and the like.

In a *galaxy* or *fact constellation* schema (Figure 9.5), common *dimensions* surround two or more related *fact* tables. For example, the *Diagnosis* dimension is used for defining the patients with *heart failure* as well as the patients who are contraindicated *ACE or ARB inhibitors* therapy; it is a common *dimension* for two *fact* tables.

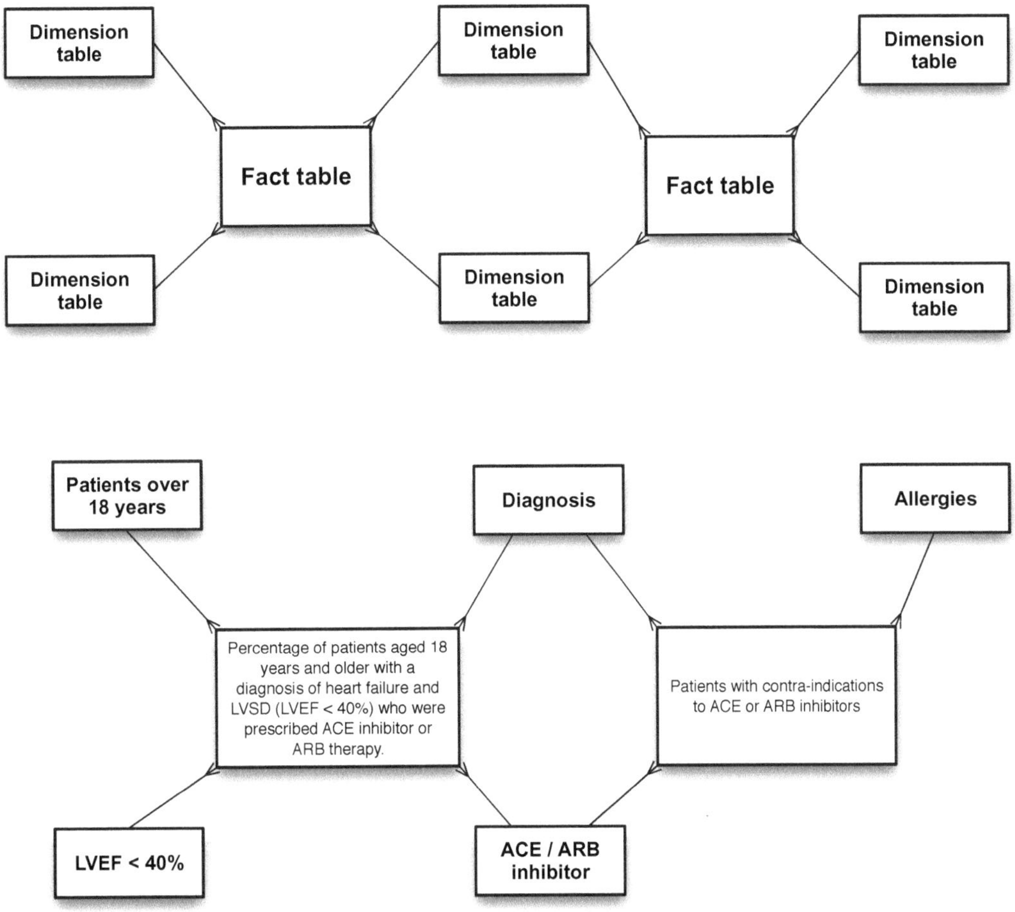

Figure 9.5 Galaxy schema.

3. *A multidimensional database/cube* is a form of data store in which data are located in cells, and the position of each cell is defined by a number of dimensions. When there are three dimensions, the *multidimensional database* is called a *cube*; with more than three dimensions the data store is called a *hypercube* (Figure 9.6).

Each cell in the cube depicted in Figure 9.6 contains the calculated result (*83.2%*) for one quality measure (*NQF 0435—Percentage of patients aged 18 years and older with a diagnosis of ischemic stroke or transient ischemic attack [TIA] who were prescribed antiplatelet therapy at discharge)* for one physician (*Dr. S.*) per month (*Feb 2012*).

Data Mining

Data mining is the process of automatically discovering new and useful patterns in large databases. It is different from information retrieval, such as finding a record in a data store or searching the Web for a specific page. Data

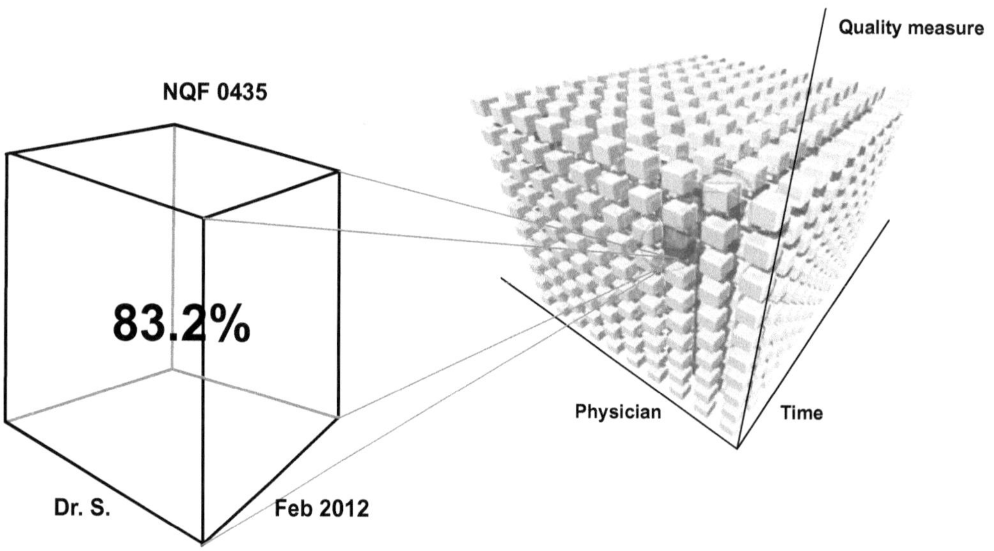

Figure 9.6 Multidimensional cube.

mining tasks are generally divided into two major categories: descriptive and prescriptive (26):

Descriptive tasks have the objectives to derive patterns such as correlations, trends, clusters, and anomalies that summarize the underlying data. Some medications' side effects have a very low frequency; thus, large databases need to be consolidated and mined to find meaningful knowledge about these rare, negative effects. For example, the correlation between the long-term use of *Rofecoxib* (*Vioxx*) and *cardiovascular* events was discovered only after the drug was on the market for 5 years and some 80 million people had been treated with this drug (27). A rare event needs a large database to be mined to find it.

Prescriptive tasks try to predict future values based on data existing in the data stores. The parameter to be predicted is known as the *dependent* variable, and the attributes used for making the prediction are known as the *independent* variables.

If the *dependent* variable is discrete (such as *mortality*), then a *classification* model is used. If it is continuous (such as *cost*), then a *regression* technique is employed. The vast subject of data mining is well covered by Reference 26.

CDS versus Reports

Both *CDS* (see Chapter 8) and *Report* workflows produce an output based on a complex query process. The differences between the two reporting systems are summarized in Table 9.2.

A real-time, online CDS may help improve the quality measures, one patient at a time, while an off-line reporting system employs a higher-level strategy, overarching multiple individuals, processes, and outcomes in order to improve healthcare.

Table 9.2 CDS versus Quality Measures

CDS	*Quality Measures*
Real time/online	Retrospective/off line
Single patient	Multiple patients
Patient is in the vicinity of a clinician	Patient is not in the clinician's vicinity
Rules triggered may lead to clinical action	Post factum analysis
May prevent errors in real time	Cannot prevent errors in real time
Data elements may be free text/ unstructured	Data elements are usually coded/ structured
May impact EHR performance	No impact on EHR performance
No need for Extraction → Transfer → Load	ETL is necessary
Based on production, live systems	Based on data warehouse, data marts
May eventually improve quality measures	Used for long-term improvement programs

Processes

DFD 6 Report Workflow

Let us isolate the DFD 6 *Report* workflow from DFD 0 (Figure 9.7). All actors—patients, providers, payers, and regulatory agencies—may request a report on a single identified person or on multiple de-identified persons, such as the summary of a quality measure report. There are several preconditions that must be satisfied before a system can proceed with the *Report* workflow:

- User is uniquely identified.
- User is authorized in lieu of reporting.
- Access and permission levels are defined granularly (by both patient and clinician).
- Report is uniquely identified.
- All system's activities are logged.

The *CLINICAL DATA WAREHOUSE* in Figure 9.7 is the collection of data marts discussed: dimensional or normalized databases, multidimensional cubes and hypercubes with cleansed, integrated information from various EHR and health information exchange (HIE) sources.

The system creates a single- or multiple-patient report (Figure 9.8), depending on the security permissions and access rights defined in a repository of *REPORT RULES*.

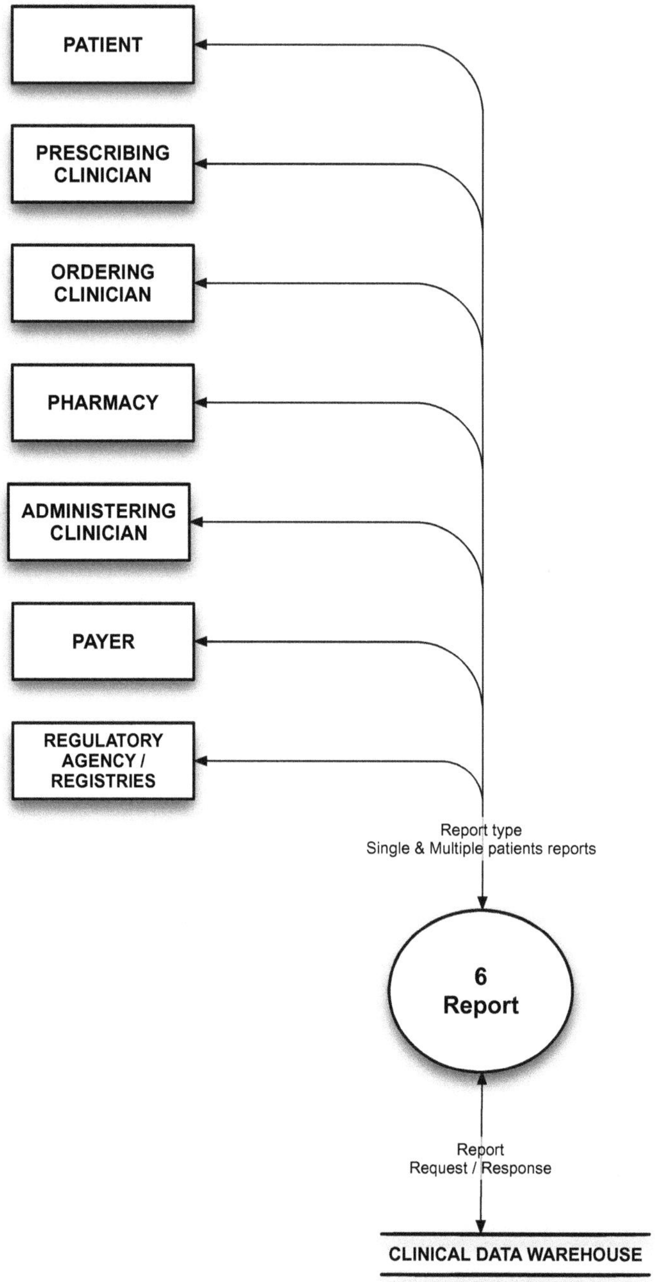

Figure 9.7 Report workflow: Isolated from DFD 0.

DFD 6.1 Report on Single Patient

The user provides the *Patient ID*. The system lists available single-patient reports (summary of emergency department [ED] visit, cardiology catheterization lab report, last-year lab values, etc.) (Figure 9.9). During process 6.1.1, the user selects a report type from the list of available reports and defines additional parameters. According to the *REPORT RULES, Patient ID & parameters* and *Report ID*, process

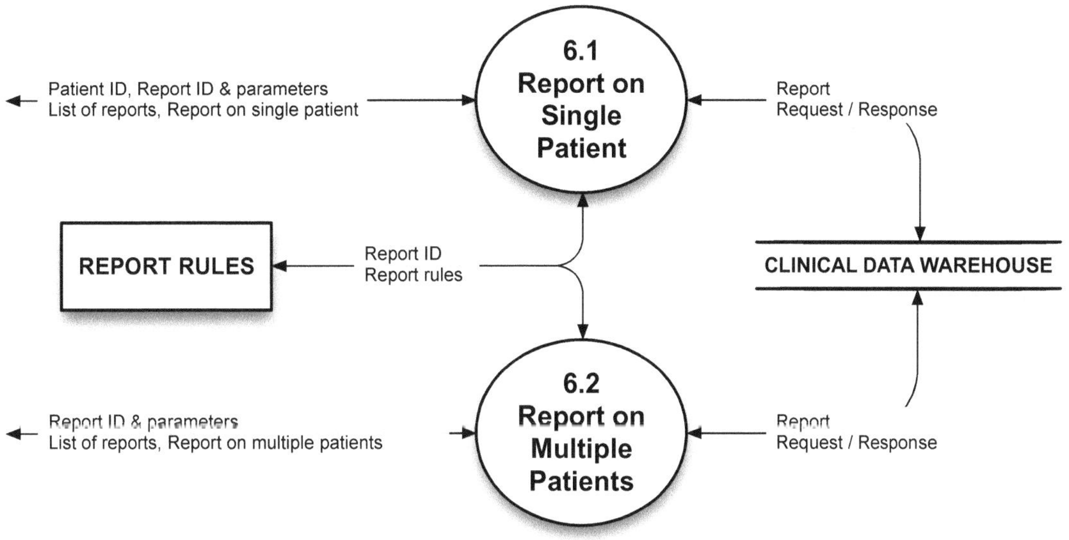

Figure 9.8 DFD 6: Report workflow.

6.1.2 prepares a query. The query most probably includes one or more basic *Structured Query Language* (*SQL*) operations—filter, aggregate, order:

- Filter (*Select, Where*): such as only data pertinent to the given *Patient ID* is selected
- Aggregate (*Sum, Max, Min, Avg, Join, Group by, Having, etc.*): for example, lab values grouped or averaged
- Order (*Order by*): values listed in chronological order

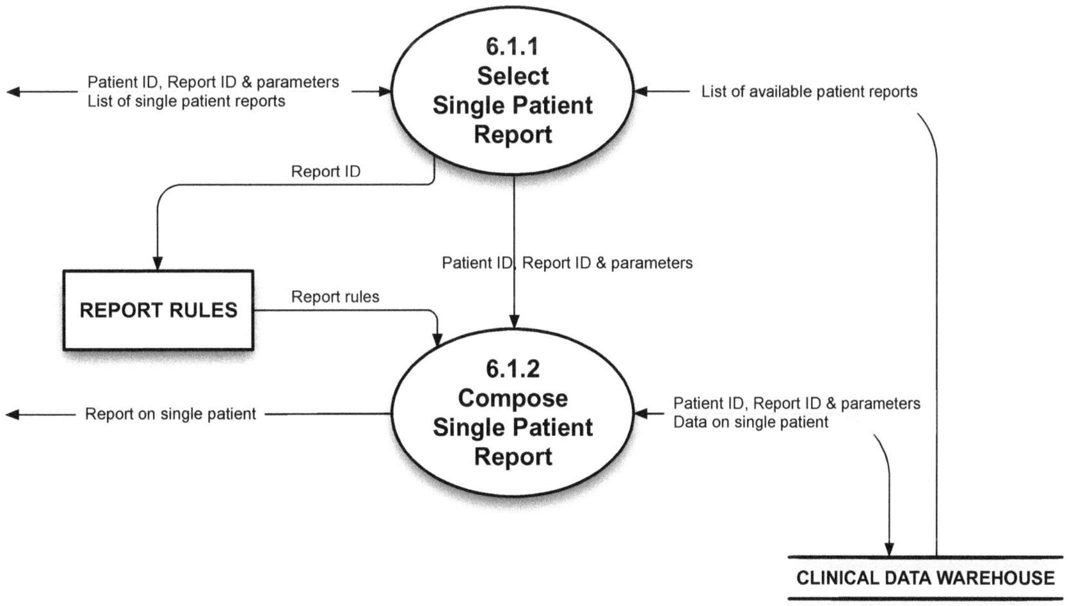

Figure 9.9 DFD 6.1: Report on Single Patient.

Process 6.1.2 pulls data from the warehouse according to specific rules and query parameters and using basic SQL operations prepares a single-patient report.

DFD 6.2 Report on Multiple Patients

During process 6.2.1 the user selects the type of report/quality measure defines additional parameters such as the time frame of the report and the group level (physician, department, hospital, etc.) (Figure 9.10). Process 6.2.2, according to the *REPORT RULES*, pulls the necessary data elements, the inclusion and exclusion criteria to calculate the denominator. Next, process 6.2.3 pulls the data elements for the numerator calculation. Once the numerator and denominator have been acquired, the system can proceed with the actual measure calculation as the rate or percentage number.

Process 6.2.4 identifies the data elements known to exist only on paper and points their locations to the manual abstractor. (The application used by the abstractor to input these manually extracted data elements is not shown in Figure 9.10.)

Finally, process 6.2.5 prepares the report in XML format (28) and sends it to the user.

Note that *REPORT RULES* is a different entity than the *CLINICAL DATA WAREHOUSE*; the separation between the data used for reporting (*DW*) and the logic used for preparing a report (*RULES*) allows rules to change, new ones to be added, or others to be retired without disturbing the underlying data structures. The same *REPORT RULES* engine can be installed at different institutions, and the complex logic contained can be reused on different DWs with minimal configuration.

Data Elements

The data elements expected to participate in the *Report* workflow are (28)

- Type of report: *single/multiple patients*
- Patient ID (only for single-patient reports)
- Report name/report unique identifier
- Other report identifiers: *NQF, Joint commission, PQRS, HITSP*
- Grouping level: *clinician, department, hospital,* and so on
- Description
- Author
- Set of related measures
- Measure type
- Report time frame
- Rationale
- References
- Report rules

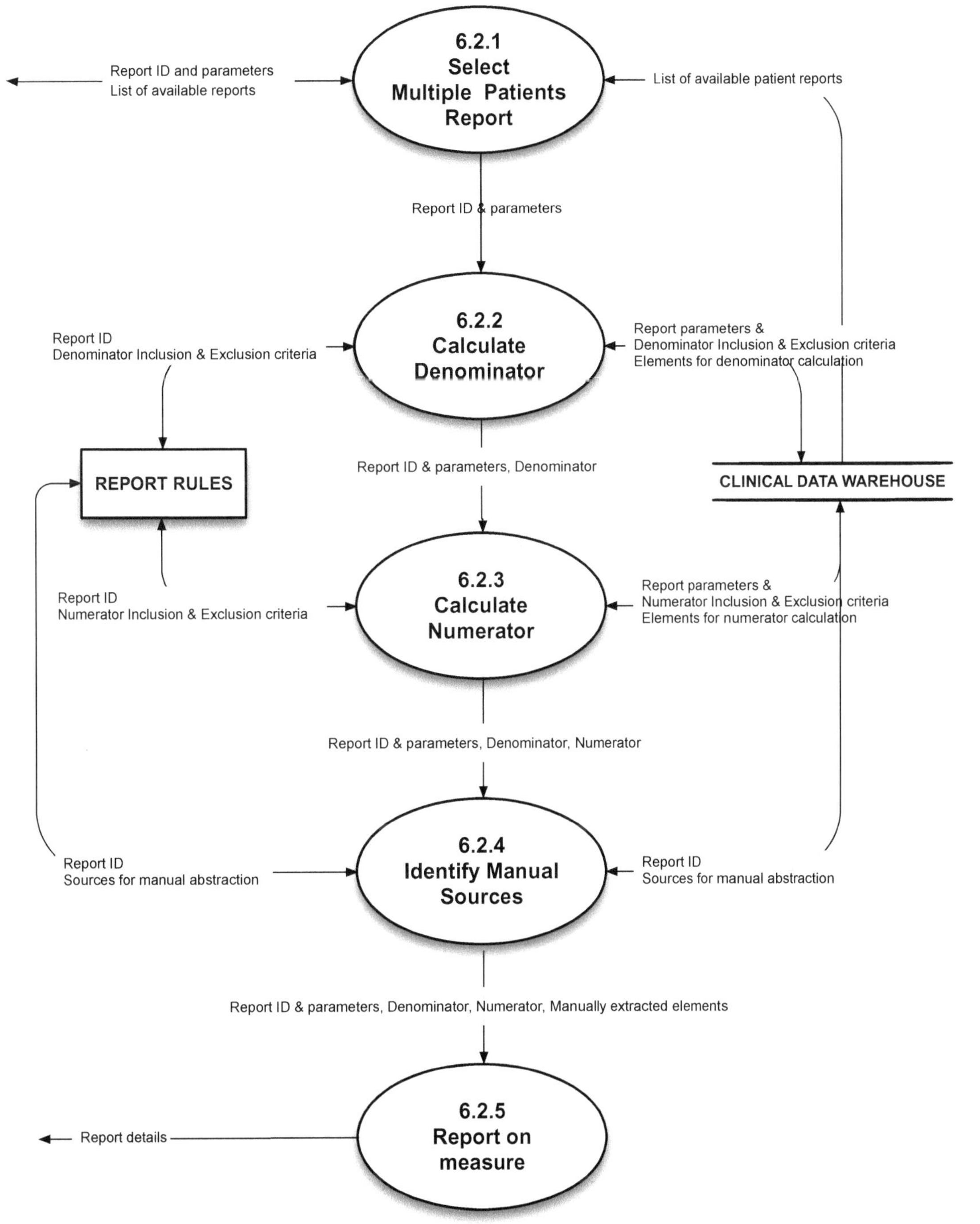

Figure 9.10 DFD 6.2: Report on Multiple Patients.

- Denominator: inclusion/exclusion criteria and data elements
- Numerator: inclusion/exclusion criteria and data elements
- Manually abstracted elements: locations and values
- Measure calculation
- User and date and time stamp of report creation
- Format used and version

Summary

This chapter detailed the types of health-related reports, the rationale for quality measures, the organizations involved with quality reporting, *PQRS* and *MU* report characteristics, as well as several principles used in the *Report* workflow.

The anatomy of a quality measure was used to identify the main data elements—denominator, numerator—and the logic employed in the calculation of a quality measure, using the inclusion and exclusion criteria.

Several methods of reporting, such as claims based, registry and EHR, were introduced together with the pragmatic need for a hybrid method by which both automated and manual extraction processes are used for reporting from the currently mixed paper/computer sources. The need for a DW was presented as well as the basic principles of the DW and mining.

The chapter ended with the processes and data to support the *Report* workflow.

References

1. https://www.cms.gov/PQRS/Downloads/PQRITaxReliefHealthCareAct.pdf (accessed May 28, 2011).
2. https://www.cms.gov/LegislativeUpdate/downloads/PPACA.pdf (accessed May 28, 2011).
3. https://www.cms.gov/EHRIncentivePrograms/ (accessed May 28, 2011).
4. Donabedian, A. 1966. Evaluating the quality of medical care. *Milbank Memorial Fund Quarterly* 44: 166–203.
5. Oxman, A.D., D.L. Sackett, I. Chalmers, et al. 2005. A surrealistic mega-analysis of redisorganization theories. *Journal of the Royal Society of Medicine* 98: 563–568.
6. Institute of Medicine. 2001. *Crossing the quality chasm: A new health system for the 21st century.* Washington, DC: National Academy Press.
7. http://www.qualityforum.org/uploadedFiles/Quality_Forum/Measuring_Performance/tbEvalCriteria2008-08-28Final.pdf (accessed June 1, 2011).
8. http://www.qualitymeasures.ahrq.gov/selecting-and-using/validity.aspx (accessed June 2, 2011).
9. https://www.talkingquality.ahrq.gov/content/create/select.aspx (accessed June 2, 2011).
10. http://www.ahrq.gov/ (accessed May 28, 2011).
11. https://www.cms.gov/PQRS/ (accessed June 3, 2011).
12. Blumenthal, D., and M. Tavenner. 2010. The "meaningful use" regulation for electronic health records. *The New England Journal of Medicine—Health Policy and Reform* 10.1056/NEJMp1006114. http://healthpolicyandreform.nejm.org/?p=3732 (accessed June 4, 2011).
13. Overview of clinical quality measures reporting in the Centers for Medicare & Medicaid Services (CMS) Final Rule on Meaningful Use. http://www.cms.org/uploads/MUClinicalQualityFinal.pdf (accessed June 4, 2011).
14. TN906 Quality measures technical note Version 1.1. http://www.hitsp.org/ConstructSet_Details.aspx?&PrefixAlpha=5&PrefixNumeric=906 (accessed June 1, 2011).

15. CMS 2011 Physician quality reporting measure specifications manual. http://www.cms.gov/PQRS/15_MeasuresCodes.asp (accessed June 1, 2011).
16. Specifications manual for national hospital inpatient quality measures. Version 3.3a Section 2.8 (STK). http://qualitynet.org/dcs/ContentServer?c=Page&pagename=QnetPublic%2FPage%2FQnetTier4&cid=1228760129036 (accessed June 1, 2011).
17. http://www.cms.gov/PVRP/Downloads/PVRPCoreStarterSetSpecificationsAndInstruction.pdf (accessed June 21, 2011).
18. https://www.cms.gov/PQRS/Downloads/2011RegistrySubmissionRequirements_12-08-2010_Final.pdf (accessed June 22, 2011).
19. http://wiki.oemr.org/wiki/File:PQRI_Simplified_2009.pdf (accessed June 22, 2011).
20. http://healthit.hhs.gov/portal/server.pt/gateway/PTARGS_0_10741_954697_0_0_18/corley-testimony-05-19-11.pdf (accessed June 22, 2011).
21. https://www.cms.gov/PQRS/Downloads/2011_PhysQualRptg_MeasuresList_033111.pdf (accessed July 3, 2011).
22. Van Ornum, M. 2009. *Electronic prescribing—A safety and implementation guide.* Sudbury, MA: Jones and Bartlett.
23. Anderson, P.O., S.M. McGuinness, and P.E. Bourne. 2010. *Pharmacy informatics.* Boca Raton, FL: CRC Press.
24. Rainardi, V. 2008. *Building a data warehouse with examples in SQL server.* New York: Apress.
25. Kimball, R., L. Reeves, M. Ross, et al. 2008. *The data warehouse lifecycle toolkit. Expert methods for designing, developing and deploying data warehouses.* New York: Wiley.
26. Tan, P., M. Steinbach, and V. Kumar. 2006. *Introduction to data mining.* Boston: Addison-Wesley.
27. http://en.wikipedia.org/wiki/Rofecoxib (accessed June 30, 2011).
28. http://wiki.oemr.org/wiki/PQRI_XML_Output_for_CQM (accessed July 3, 2011).

Review Questions

1. What are the main types of health-related reports?
2. What are the financial incentives for reporting under the PQRS and MU initiatives?
3. What are the three main aspects of healthcare quality according to the Donabedian model?
4. What are the IOM-declared goals for the U.S. healthcare system?
5. Consider the following quality measure: *PQRS # 170: Coronary Artery Bypass Graft (CABG): Beta-Blockers Administered at Discharge Percentage of patients aged 18 years and older undergoing isolated CABG surgery who were discharged on beta-*blockers. Using the following data calculate the quality measure: Total number of patients aged 18 years and older during 2011 = 13,200. Of these, 2,883 were diagnosed with heart disease; 594 patients had CABG surgery: 22 patients also had mitral valve repair, 15 also had tricuspid valve repair, and 5 patients had CABG and mitral and tricuspid valve repair during the same surgery. Eight patients were transferred to another facility, 13 had died, and 31 had a contraindication to beta-blockers. There were 380 patients on beta-blockers at admission, and 420 patients received this treatment at discharge.
6. List the current quality reporting methods from the most to the least efficient.
7. Why is a data warehouse needed for reporting?
8. What are the three main schemas for dimensional data stores?
9. What are the main types of SQL operations used for reporting purposes?
10. Assume there are 5,000 physicians in an organization, which reports on 240 quality measures using a three-dimensional cube as its data mart. How many cells would accumulate in this cube after 1 year if the quality measures are stored once per month per physician?

Chapter 10

Interoperability Standards and Vocabularies

The wonderful thing about standards is that there are so many of them to choose from.

Andrew Tanenbaum

The last chapter presents the interoperability challenges facing the electronic health record (EHR) in general and the medications system specifically. How are the actors, the disparate components presented in the previous chapters of this book, supposed to play with others in the proverbial healthcare information technology sandbox? The short answer is by adhering to standards and vocabularies.

Interoperability, standards, terminologies, and classifications are defined. From the organizations active in standardization in medicine, we identify the main standards and vocabularies in HIT. The chapter is biased toward the U.S. market, in lieu of the recently published *Meaningful Use (MU)* part of the American Recovery and Reinvestment Act.

Next, we zoom in on the standards and vocabularies relevant to the domain of this book as we identify the interoperability needs: workflows and data structures related to the *Medications* system.

In between, we briefly touch, and only from an information technology perspective, on philosophical issues of language, context, meaning, ambiguity, and so on.

Interoperability

The *raison d'etre*, the whole reason, for IT existence is to enable information sharing. This basic fact is sometimes forgotten in discussions about the higher, noble roles of Healthcare Information Technology (HIT): improving healthcare quality and containing the rising costs of providing care. But without information sharing, higher goals cannot be achieved.

As defined by the *Institute of Electrical and Electronics Engineers (IEEE)*, interoperability is *"the ability of two or more systems or components to exchange information and to use the information that has been exchanged"* (1).

A meaningful information exchange between two entities necessitates that both sender and receiver (be it human or machine) agree on a set of communication rules:

- The grammar of the language used
- The meaning of the words—the vocabulary, terminology assumed to be in effect

Interoperability has several capability levels as defined by Tolk et al. (2) in the *Levels of Conceptual Interoperability Model (LCIM)* (Table 10.1). Engineers initially published *LCIM* for the modeling and simulation domain (3), but *LCIM* can be applied to other communication channels as well, such as humans or machines exchanging information:

No Interoperability → Technical → Lexical and Syntactic → Semantic → Context → Dynamic → Conceptual

The *LCIM* shows that in order to achieve the level of semantic (language) interoperability, the communicating entities need to achieve both lexical (terms) and syntactical (grammar) interoperability as prerequisites. For the highest level of conceptual interoperability (message meaning), the context, modeling assumptions, and constraints as well as the dynamics of workflow and processes must be agreed on by the participating entities.

Open Systems Interconnection Model

How are machines communicating? How are different hardware configurations, with disparate operating systems and software applications, interoperating? How does information hop over numerous and various machines, wired and wireless media? How does the Web work? The answer is by adhering to standards. The *Open Systems Interconnection (OSI)* model, developed by the *International Organization for Standardization (ISO)* divides a communication system into layers (4) (Table 10.2). Each layer has specific responsibilities,

Table 10.1 Levels of Conceptual Interoperability Model between Two Entities

Level	*Description*	*Assumption on Common*	*Information Exchanged*
6. Conceptual	Entities are aware of each others' information, processes, context, and modeling assumptions	Conceptual model	Assumptions, constraints, logic
5. Dynamic	Entities may change in time the production/consumption of information based on changed meaning or context	Execution model	Effect
4. Pragmatic	Entities are aware of context and meaning	Workflow model	Context
3. Semantic	Entities exchange discrete, semantically parsed terms	Reference model	Content, meaning of words
2. Syntactic	Entities exchange data in the right format, order but without meaning	Data structure	Format, structure
1. Technical	Entities have the technical infrastructure to exchange data	Communication protocol	Bits, bytes, symbols
0. No interoperability	—	—	—

operates according to known standards and protocols, and provides services to the layer above it while receiving services from the layer below it (Figure 10.1).

Encapsulation is one of the key concepts of the *OSI* model: At each layer, the message element *adds* one envelope. By the time a message reaches the physical layer of the sending entity, it has seven envelopes. Each envelope defines the standard and protocol used by a specific layer as applied to the message within.

At the receiving end, starting at the physical level and going up to the application level, each *OSI* layer *removes* one envelope from the message. Once the message has reached the application layer of the receiving end, all its envelopes have been removed; the message content is supposed to be the same as it was at the sending end.

Each layer interacts only with the layer above and the one below. Error correction is a process that may happen at several layers, while security is implemented at all levels. At each layer, machines adhere to protocols and standards, which are numerous and evolving. This arrangement enables interoperability for all parties adhering to the *OSI* model.

Table 10.2 OSI Model

Layer	*Function*	*Examples*
7. Application	Identify communication partners, determine resource availability, and synchronize communication	HTTP, FTP, SMTP, SNMP
6. Presentation	Data representation, encryption and decryption, convert machine-dependent data to machine-independent data	ASCII, MIDI, MPEG
5. Session	Establish, manage, terminate, and restart the connections between the local and remote applications	TCP, RPC
4. Transport	End-to-end, error-free connections and reliability, flow control	TCP, UDP
3. Network	Path determination, route, relay, fragmentation, reassembly, and logical addressing	IP, ICMP
2. Data link	Streams bits, physical addressing, detect and correct physical layer errors	Ethernet, PPP, Token Ring,
1. Physical	Define the electrical and physical specs of devices (copper, optical cable), layout of pins, voltages, repeaters, hubs, adapters	RS232, full duplex, T1, E1, 10BaseT

Note: See the acronyms list at the back of the book for definitions of the abbreviations in the Examples column.

The application layer number 7 is the only domain-specific aspect of *OSI* model; it interacts with other applications directly, without intermediaries. The scope of the seventh layer of the *OSI* model includes only the basic technical communication needs of an application: communication partners, resources, and synchronization. Semantics and meanings of a message are not within the scope of the *OSI* model. The *OSI* model covers the technical aspects of interoperability but leaves the issue of semantics for further clarifications.

Medicine, as other domains, is in the process of stabilizing the existing structure of standards and vocabularies by deciding on rules and terminologies to be used in specific workflows. The following HIT-related standards and terminologies are applicable at the *OSI* seventh application layer, with the goal of enabling semantic interoperability between communicating parties. The other six layers of the *OSI* model are being taken care of by other, not necessarily medical, standards organizations. Layer 7 in the *OSI* model is the source of the name *HL7—Health Level 7*, a large standardization organization presented in this chapter.

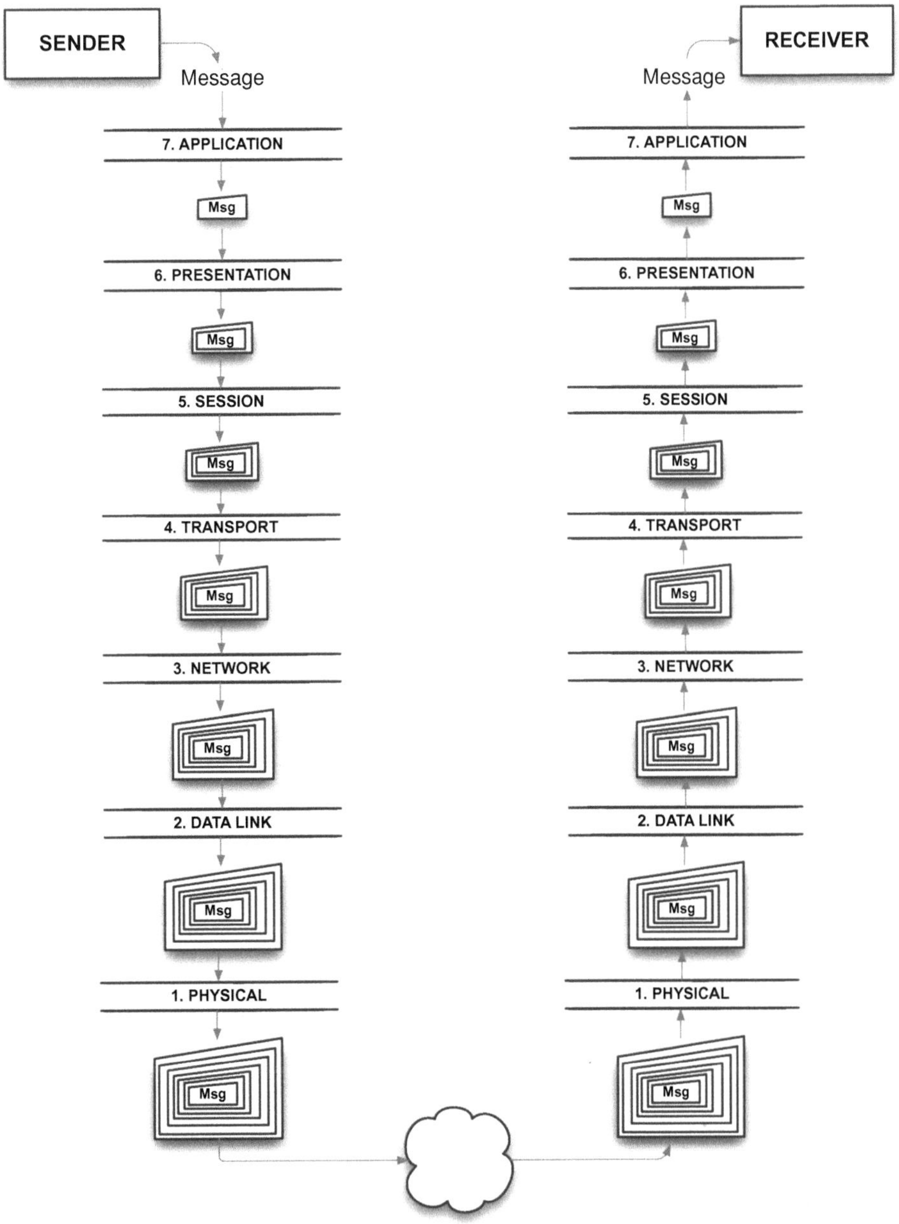

Figure 10.1 OSI model at work.

Language and Ontologies

On one side of the spectrum, there are human languages; rich in expressivity but from a computer perspective suffering from both synonymy and ambiguity.

- *Synonymy*: one concept, many terms
- *Ambiguity*: one term, many concepts

Machines find it difficult to solve synonymy and ambiguity problems unless presented with additional data and logic to solve the human language puzzles.

On the other side, the computer languages are well structured (*SQL* stands for *Structured Query Language*) but lack in expressivity. In between these extremes, we may find semistructured languages such as *eXtensible Markup Language* (*XML*).

The meaning of a concept may depend on the context, and it may change, becoming unstable/dynamic. A concept in a terminology is a slowly changing dimension; it is not expected to change frequently, like values in a banking transaction system. Still, a concept may evolve in time. What is needed is a formal, relatively stable representation of knowledge so it may describe a domain—an *ontology* (5):

- Concepts
- Relationships
- Attributes
- Events
- Rules
- Axioms
- Classifications

An *ontology* is a compendium of terminologies that machines may be able to use and reason about concepts in a domain.

Problems with ontologies, standards, and vocabularies are to be expected when there is more than one. For example, when two terminologies coexist, each with its own perspective on a medical issue, then the relationships between matching concepts may be "many to many":

Terminology A: One item → Terminology B: Many matching items
Terminology B: One item → Terminology A: Many matching items

This situation in turn necessitates a mapping layer between two terminologies, either a simple intersection entity or a complex disambiguation mechanism. Layers and mechanisms usually translate into higher system complexity and increased costs.

Interfaces

Three entities, *A, B,* and *C,* are communicating, two nodes at a time (sender/receiver). How many interfaces are needed? The answer is three interfaces: *A-B, A-C, B-C.* If there are six entities communicating (*A, B, C, D, E, F*), then 15 interfaces are required: *A-B, A-C, A-D, A-E, A-F, B-C, B-D, B-E, B-F, C-D, C-E, C-F, D-E, D-F, E-F.*

The formula defining the number of interfaces between *N* communicating entities comes from combinatorics (6):

$$N!/2(N-2)!$$

Or, if you do not like the factorial sign (!) as perhaps too loud, the formula is equivalent to

$$(N^2 - N)/2$$

Imagine one facility that may need 100 entities to communicate; this equates to 4,950 interfaces. At a cost of around $20,000 a pop, the bill for interfaces may easily surpass US$100 million.

One study by Walker et al., focusing on the lack of semantic interoperability standards in healthcare in the United States, estimated $77.8 billion per year would be saved if such a standard would be implemented (7).

Rocket Science Standards

The importance of adhering to only one standard for units of measure was revealed September 23, 1999, when the *Mars Climate Orbiter* crashed on the red planet due to an error in the interface specifications between two software modules. While the output sent by one module was in *metric* units of *Newton-second* (*N-s*), the input received by the other module was thought erroneously as *English* units: *pound-second* (*lbf-s*). A semantic interoperability issue causing a spectacular, totally unexpected, early demise of the Mars space probe.

U.S. Government and MU

A standard helps disparate components interoperate safely while it decreases the costs of achieving interoperability: The more a standard is used, the larger the market for it will be and the more cost effective it becomes to adhere to the standard—a multiplier effect.

Not surprisingly, the *Office of the National Coordinator* (*ONC*) for HIT stated in the U.S. federal strategic plan (8): *"Interoperability: To effectively exchange health information, health IT systems and products must use consistent, specific data and technical standards."*

To promote EHR adoption, the U.S. federal government will pay $27–32 billion over the next years to hospitals and physicians who are demonstrating *MU* of HIT. The monetary incentive for users of certified EHR modules is expected to foster innovation in the U.S. HIT industry while increasing the users' adherence to a much-needed *lingua franca* for the medical domain—a list of standards and vocabularies. Gains from eventually adhering to standards are expected to be higher than the costs of implementing the American Recovery and Reinvestment Act (ARRA) *Health Information Technology for Economic and Clinical Health* (*HITECH*) act. The complete final rule on standards was published by the US Department of Health and Human Services (2010) as part of the ARRA HITECH (code 45 part 170) with specific recommendations on standards and vocabularies to be used by the industry (9).

The *Health Information Technology Standards Panel* (*HITSP*) is responsible for standards harmonization. On his blog, Dr. John Halamka, chair of HITSP, shares an excellent analysis and summary of the current U.S. HIT standards (10).

Qualities of a Modern Clinical Terminology

An influential paper by Cimino in 1998, "Desiderata for Controlled Medical Vocabularies in the Twenty-First Century," summarizes the common requirements for a modern clinical terminology (11):

- *Content*: Scope and quality are necessary to satisfy domain coverage (concepts) and human readable terms (definitions). Multiple languages need to be supported. Content is dynamic and expected to change.
- *Concept orientation*: One concept may have multiple synonyms, but it has only one meaning (nonambiguity). One term may have multiple meanings if each relates to a different concept (homonymy).
- *Persistence*: Once created, a term must maintain its meanings. A term may become redundant, incorrect, or retired.
- *Unique identifiers*: This should come as no surprise for the reader of this book. Each concept in an IT system must be uniquely identified, and the primary key (PK) identifier should have no meaning (PK is not an attribute of the concept).
- *Multidimensional classification*: Clinical concepts are best organized/classified in a polyhierarchy. One concept may have more than one parent concept. In Chapter 2, we briefly introduced a polyhierarchy classification system with multiple parents-children relationships, as related to the classification of drugs; a similar requirement exists for other medical terminologies.
- *Identification <> Classification*: The means to identify a concept should be independent of the means to classify it. The concept's definition and classification are represented by its attributes and not by the unique identifier.
- *No NEC*: The use of vague terms such as *Not Elsewhere Classified* (*NEC*) should be minimized or eliminated if possible. *NEC* is too ambiguous to be used effectively in any terminology. It becomes the default garbage can for a rapidly growing class of *NEC* instances on each and every branch of the hierarchy where *NEC* is permitted.
- *Granularity*: Different levels of granularity should be allowed for different types of users: clinicians, administrators, billers, and so on.
- *Graceful evolution*: The whole structure of a terminology should allow changes to concepts, relationships, and meanings during time. A living terminology is an evolving, dynamic terminology.
- *Machine usable context*: Since the meaning of a concept depends on the context, the context should be machine understandable.

The creators of the *Systematized Nomenclature of Medicine Clinical Terms* (*SNOMED CT*) have added three requirements to those presented (6):

- *Understandable*: Definitions should be concise and understood by clinicians.
- *Reproducible*: Conceptual meaning should be independent of the technology used.
- *Usable*: If there is no use in healthcare, it should not be part of the terminology.

With these goals in mind, let us take a look at the organizations involved in HIT standardization.

Healthcare Standards Organizations

The *ISO* defines a *standard* as a published document, a set of rules established by consensus, approved by a recognized body to be used consistently as good principles, practices, or guidelines. A standard controls how people develop and manage materials, products, services, technologies, processes, and systems. The *ISO* has more than 160 members, one from each country. The *American National Standards Institute* (*ANSI*) is the U.S. representative member in the *ISO*.

Either under the umbrella of one of the mentioned bodies or acting independently—some as nonprofits—the organizations in Table 10.3 are actively involved in standardization efforts in medicine in general and HIT specifically. The list is

Table 10.3 Healthcare Standardization Organizations

Organization	*Standard/ Vocabulary*	*Domain*
ADA	CDT	Dental terminology
AMA	CPT	Diagnostic and surgical procedures terminology
ANSI	ASC X.12	Billing, claims-processing standard
APA	DSM	Psychiatry terminology
ASTM	CCR	Clinical documents XML standard
CDC	CVX	Immunization vocabulary
CDISC	SDTM	Clinical, pharmaceutical trials
CMS	PQRI DRG	Quality reporting specs Diagnosis-related vocabulary/classification
Department of HHS	HIPAA	Privacy and security regulations

(continued)

Table 10.3 Healthcare Standardization Organizations (Continued)

Organization	*Standard/ Vocabulary*	*Domain*
FDA	NDC UNII	Drugs vocabulary Ingredients/chemical substances vocabulary
HITSP	Harmonization	Standards and vocabularies
HL7	v2.x, v3	Interoperability standard Clinical and administrative data messaging standard Reports requirements: public health, biosurveillance
HL7 RIM	v3	Information model
HL7	CCD, CDA	Clinical documents XML standard
ICD	v9, v10, v11	Diseases vocabulary/classification
IEEE/ISO	11073	Communications standard for monitors Bedside devices
IHTSDO	SNOMED CT 2009	Clinical terminology
IHE	XDS	Cross-enterprise document-sharing profile
NANDA		Nursing terminology
NCPDP	v8.1, v10.6	Pharmacy communications standard
NEMA	DICOM	Imaging communications standard
NIST	FIPS 197, 180	Security standards
NLM	UMLS	Ontology—a compendium of multiple vocabularies
	RxNorm	Medications vocabulary
	MIM	Genetic disorders vocabulary
OpenEHR		EHR architecture
Regenstrief I.	LOINC	Laboratory test names vocabulary
	UCUM	Units of measurement vocabulary
VHA	NDF RT	Drug classification vocabulary

Note: See the acronyms list at the back of the book for definitions of the abbreviations.

not meant to be all inclusive, and it is biased toward the U.S. market. The numerous abbreviations/acronyms are defined in the Appendix.

The more a standard is used, the more appeal it has for others: HL7 is the standard used in most U.S. hospitals for messaging medical discrete data elements. Adhering to the HL7 messaging standard makes sense; instead of working with many messaging standards, one has to deal only with HL7. It usually results in significant cost savings: Instead of developing a number of interfaces, one can connect to a single HL7 engine that will do the rest.

On the other hand, there are domains, such as drug bar coding, with no standard defined, many sponsors, and no clear recommendations as of 2011. The market currently has some 11 different types of bar codes for drugs (see Chapter 6, "Administer/eMAR"). The proliferation of bar code standards for medications makes it difficult for any specific one to become a de facto leader and may cause issues with bar code interoperability and meaningful communications with other EHR components: *eRx* (electronic prescription), *CPOE* (computerized physician order entry), *eMAR* (administer/electronic Medication Administration Record), and so on.

EHR Standards and Vocabularies

As previously explained, an EHR needs to communicate with many external entities, actors, and repositories. Imagine the EHR on one side of the communication channel and the rest of the world on the other. In between, there must be a layer to enable interoperability. This layer (part of the OSI model) uses standards and terminologies so it can map concepts from the "world" to the EHR and vice versa (Figure 10.2). The interoperability layer from Figure 10.2 is presented in Table 10.4 in a tabular format and with a bit more detail.

The vocabulary, classification, or terminology actually used depends on the workflow/functional domain (Diagnosis, Lab, Medications) and specialty (Pediatrics, Surgery, Psychiatry).

Most of the vocabularies have some *classification* properties even if not explicitly stated. A dictionary classifying a word as noun or verb is a good example; even if the main goal of the dictionary/vocabulary is to define concepts, it has some classification aspects as well: A word is classified as a noun, verb, adjective, and so on.

The distinction between a vocabulary and a classification system in medicine may become blurred. When a domain uses more than one vocabulary or classification system, for example, *ICD9* (*International Classification of Diseases, Ninth Revision*) and *SNOMED* CT, for documenting the concept of diagnosis, one has to assume the need for a mechanism to map between the two terminologies as well as a disambiguation mechanism.

From a high-level review of the vast domain of healthcare standardization, we zoom in now on the standards and vocabularies relevant to the medications domain.

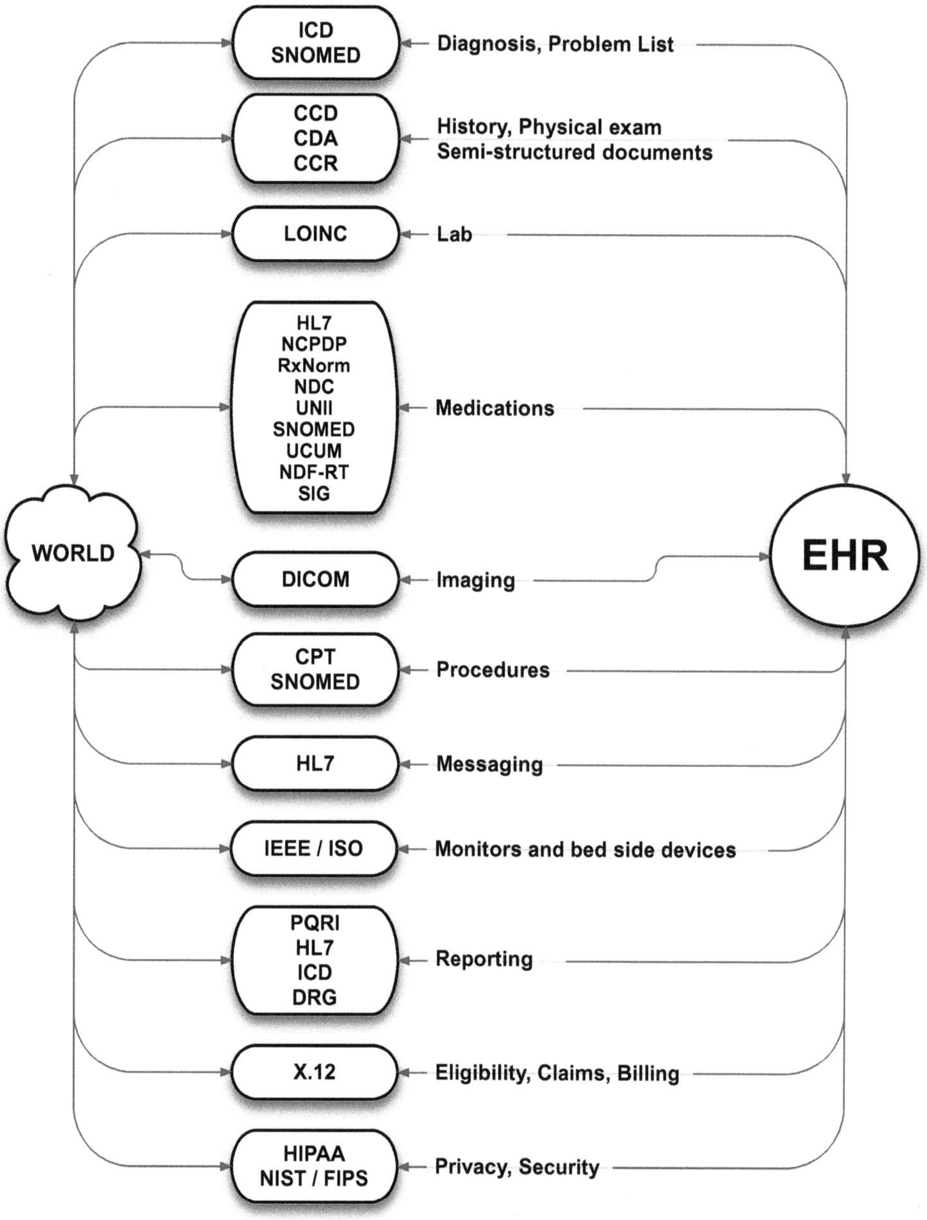

Figure 10.2 Interoperability layer between EHR and the "world."

Medications Standards and Vocabularies

There are two main organizations that publish standards related to the medications domain: *HL7* and the *National Council for Prescription Drug Programs (NCPDP)*. Both are built around the concept of request/response between two parties (as presented in previous chapters):

- *HL7* is a general healthcare-related messaging standard. As such, it has segments and types of messages (such as *OMP, OPR, RDE*) and events (such as

Table 10.4 Clinical Domains: Standards and Vocabularies

Domain	*Standard/Vocabulary*
Diagnosis	ICD9 CM
Problem list	SNOMED D axis
History, physical exam, and other XML communications	HL7 CCD HITSP C32 CDA R2 CCR E2369
Lab	LOINC 2.27 UCUM
Medications	HL7 NCPDP RxNorm NDC UNII SNOMED C axis UCUM USP NDF-RT SIG
Imaging	DICOM
Procedures terminology	CPT 4 SNOMED P axis
Nursing terminology	NANDA NIC NMDS
Communications and messaging	HL7 v2.x, v3 IHE XDS
Immunization	CDC CVX
Monitors/bedside devices	IEEE/ISO 11073
Psychiatry	DSM-4 ICD-10
Dental	ADA CDT
Genetics	GO MIM

(*continued*)

Table 10.4 Clinical Domains: Standards and Vocabularies (Continued)

Domain	*Standard/Vocabulary*
Reporting	PQRI 2009 HL7 2.3.1, 2.5.1 ICD DRG
Eligibility, billing, claims	ANSI ASC X.12
Privacy and security	HIPAA NIST FIPS 140, 180, 197

Note: See the acronyms list at the back of the book for definitions of the abbreviations.

EO09, EO12, EO26) specifically tailored for communications about medications with the pharmacy.

- *NCPDP* is the standard for electronic prescriptions communications (*eRx*).
- Terminologies related to the medications domain are presented in Table 10.5.

Multiple classifications are a painful reality in the medical world in general and the medications domain specifically. The *National Drug File Reference Terminology* (*NDF-RT*), for example, is the preferred drug classification system, but at the same time, there are other competing systems presented and used as vocabularies, such as *RxNorm*, *SNOMED CT*, or the *National Drug Code* (*NDC*).

Table 10.5 Medications Domain Vocabularies

Medications Domain	*Vocabulary*
Drugs	RxNorm (preferred) NDC SNOMED CT USP
Ingredients/chemicals	UNII (preferred) SNOMED C axis USP
Units of measure	UCUM
Classification	NDF RT (preferred) SNOMED CT
SIG	Structured SIG

Note: See the acronyms list at the back of the book for definitions of the abbreviations.

Much overlap can be expected between coexisting terminologies. This in turn may cause ambiguity, redundancy, and inconsistency.

In addition, there is a "need to get the whole picture"; the user may need *Pharmacology*, *CDS* (clinical decision support), and *Formulary*—all at the same time, within one medications system, and consequently the legitimate requirements to integrate several terminologies or use a third-party application for this purpose. The next section describes the interoperability issues and solutions from the medications domain perspective.

Processes

DFD 7 Update/Sync Workflow

In the initial dataflow diagram, DFD 0 (Figure 2.3), workflow number 7 *Update/Sync* is responsible for all communications the *Medications* system may have with external actors as far as information update and synchronization are concerned. Note the separate workflow—number 6 *Report*—is in charge of data aggregation, filtering, query, and reporting; since they have been discussed in another chapter, reporting issues are not part of workflow number 7 scope.

In the same manner we have isolated workflows from DFD 0 in previous chapters, we isolate (copy and paste) workflow number 7 *Update/Sync* (Figure 10.3).

Workflow number 7 *Update/Sync* has two main generic actors: *HEALTH INFORMATION EXCHANGE* (*HIE/HUB*) and *DRUG KNOWLEDGE AUTHORITY*.

HIE/Regional Health Information Organization/HUB

The *HIE/Regional Health Information Organization* (*RHIO*)/*HUB* actor produces or consumes clinical information related to patients. It acts as either an aggregator

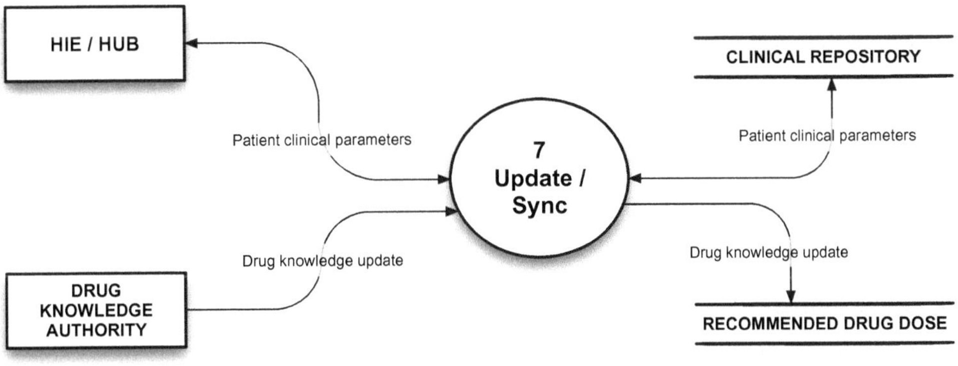

Figure 10.3 Update/Sync workflow: Isolated from DFD 0.

of patient clinical data from multiple sources or a hub facilitating patient-related information exchange. There are several types of *RHIO* facilitating *HIE*:

- *Centralized:* participants send their data to a central repository
- *Federated:* data stays with participant and *RHIO* maintains only the pointers to data locations
- *Hybrid:* mixture of both centralized and federated models

There are pros and cons for each type, most in the realm of ownership, access control, scalability, complexity, governance, and so on.

The dataflow between *HIE/RHIO/HUB* and the medications system is bidirectional. On the *Medications* side, the clinical data received from the *HIE* is stored in the *CLINICAL REPOSITORY*, which has been analyzed throughout this book. The *Medications* system may pull information from the *HIE* or may send data, push it to the *HIE* periodically, at specific triggers or following some rules.

The goal for *RHIO* is to enable information sharing, but first the obstacle of mistrust between various parties needs to be solved. The U.S. government invests heavily in building the *Nationwide Health Information Network* (*NHIN*), an interoperable infrastructure to facilitate standards-based, secure, nationwide, communication between patients, providers, and others. *NHIN* is planned to become a "net of networks," the backbone for other *HIE* organizations to interoperate.

The *Data Use and Reciprocal Support Agreement* (*DURSA*) template is used to define *NHIN* relationships with other *HIE* organizations (12). *NHIN* recommends the same *DURSA* template as a starting point for defining the relationships of a *RHIO* with other entities. *DURSA* may accelerate the process of *HIE* adoption and utilization by clarifying issues such as

- *RHIO* basic operation model: centralized, federated or hybrid?
- Who participates in *HIE*? What are the terms and conditions?
- Which standards and vocabularies are used?
- How is patient uniquely identified over multiple entities?
- How are providers identified?
- What data may be exchanged?
- How is data routed?
- What relationships exist with other *HIE* organizations?
- How is communication initiated? What are the triggers?
- Control over access to personal health information? Special circumstances?
- Sender/receiver responsibilities
- *HIE* performance and capabilities expectations
- Dispute resolution
- Privacy and security
- Other *HIE* governance issues

The *DURSA* framework has the potential to build trust quickly between parties while defining mutually accepted *HIE* policies, procedures, standards, and terminologies.

Drug Knowledge Authority

The generic *DRUG KNOWLEDGE AUTHORITY* external actor provides reliable and timely information on drugs. It is not the name of an existing entity or vendor, just the placeholder for one. The dataflow is unidirectional: from the "source of truth" toward the *Medications* system. Compared to *HIE*, the *DRUG KNOWLEDGE AUTHORITY* actor is usually less problematic in terms of trust, ownership, complexity, and other political issues. Still, the standards and vocabularies used in the information update need to be defined and mutually accepted by the *Medications* system and the *DRUG KNOWLEDGE AUTHORITY*.

Periodically, either the *DRUG KNOWLEDGE AUTHORITY* pushes an update to the *Medications* system or the *Medications* system pulls the update from this actor. The data stores affected by these updates are the tables defined in the entity-relationship diagram (ERD) around the entity *RECOMMENDED DRUG DOSE* and *CDS*-related repositories. Let us examine the *Update/Sync* workflow (Figure 10.4).

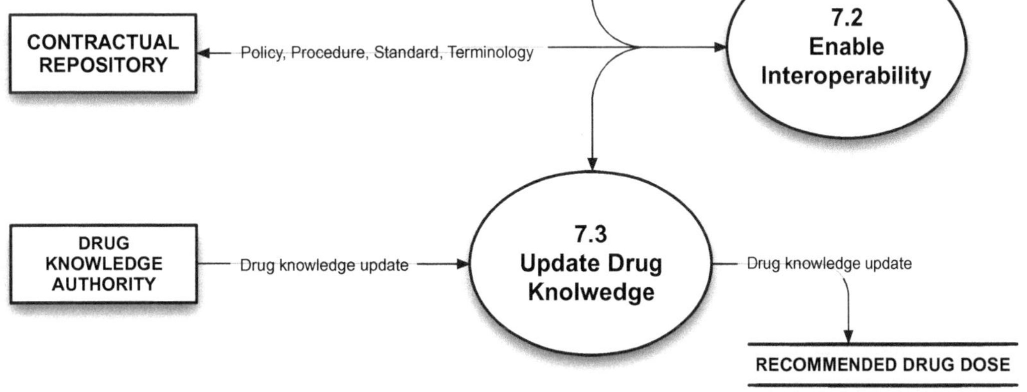

Figure 10.4 DFD 7: Update/Sync workflow.

What becomes evident as missing from the initial DFD 0 is the need for a *CONTRACTUAL REPOSITORY* where the following information resides and can be used by machines in setting a communication channel:

- Policies: who may participate in the *HIE*, what information type may be exchanged
- Procedures: sender/receiver, data elements exchanged, push versus pull, timing, triggers, and rules
- Standards
- Terminologies

Process 7.2 *Enable Interoperability* is responsible for providing this information to the other 7.x processes: 7.1 *Exchange Patient Data* and 7.3 *Update Drug Knowledge*.

Consider the following scenario:

1. A patient is discharged from the hospital.
2. This triggers a rule, which sends a message to 7.1 to prepare and distribute a discharge summary to the primary care physician and patient.
3. Process 7.1 sends a query to 7.2.
4. Process 7.2 checks the *CONTRACTUAL REPOSITORY* for the relevant *RHIO*, policies, participants, communication mechanism, standards and terminologies in effect, and so on and communicates results to 7.1.
5. Process 7.1 retrieves information from the *CLINICAL REPOSITORY* (diagnosis, problem list, labs, etc.) and from the *PATIENT MEDICATION* table (relevant drugs and doses) and creates a semistructured discharge summary based on the *Continuity of Care Document* (*CCD*), one of the three recommended XML standards).
6. The XML-based *CCD* summary is sent to the *RHIO/HIE/HUB* for storage (in a centralized model) and is either pushed to or pulled by the primary care physician (into the EHR) or by the patient (into a *PHR* [*Personal Health Record*] application).

Data Elements

At the EHR level, there are several data types:

- Structured, discrete data elements: most adhering to a version of the *HL7* messaging standard.
- Semistructured documents: XML flavors—*Continuity of Care Record* (*CCR*), *Clinical Document Architecture* (*CDA*), or *CCD*
- Imaging: *DICOM* (*Digital Imaging and Communications in Medicine*)

- Unstructured documents: text, e-mail or SMS (Short Message Service) content—no HIT standard
- Other media: voice, video—no HIT standard

From the *Medications* system perspective, there are two main data structures: patient related and drug related. Using the ERD we have built in previous chapters, we can identify the actual entities, the patient- and drug-related tables in the *Medications* system data model (Figures 10.5 and 10.6, respectively).

When process 7.1 exchanges information with *CLINICAL REPOSITORY* (DFD 7), the patient-related tables in the *Medications* system, which may be involved, are on the upper part of the ERD. These tables may participate in the *HIE* with a *RHIO*:

→ Receiving data from the *RHIO* and storing it in the PATIENT (upper) part of the ERD
← Sending data from PATIENT-related tables to the RHIO

The *DRUG KNOWLEDGE AUTHORITY* through process 7.3 (Figure 10.4) may update more than the *RECOMMENDED DRUG DOSE* table; it may actually access all the tables identified on the lower DRUG part of the ERD.

Data types exchanged during communications between the *Medications* system and external entities are expected to be either discrete data elements or semistructured/unstructured documents.

Discrete Data Elements

The discrete data elements usually arrive as messages or are parsed from semistructured documents. For automation sake, machines need to decide

- Whether message content can be used as is
- Which mapping layer to use (translation dictionary)
- Disambiguation mechanism

Candidates for discrete data elements are as follows:

- Patient identifier
- Weight/height
- Diagnosis
- Procedure
- Drug
- Ingredient
- Form
- Strength
- Dose
- Unit of measure

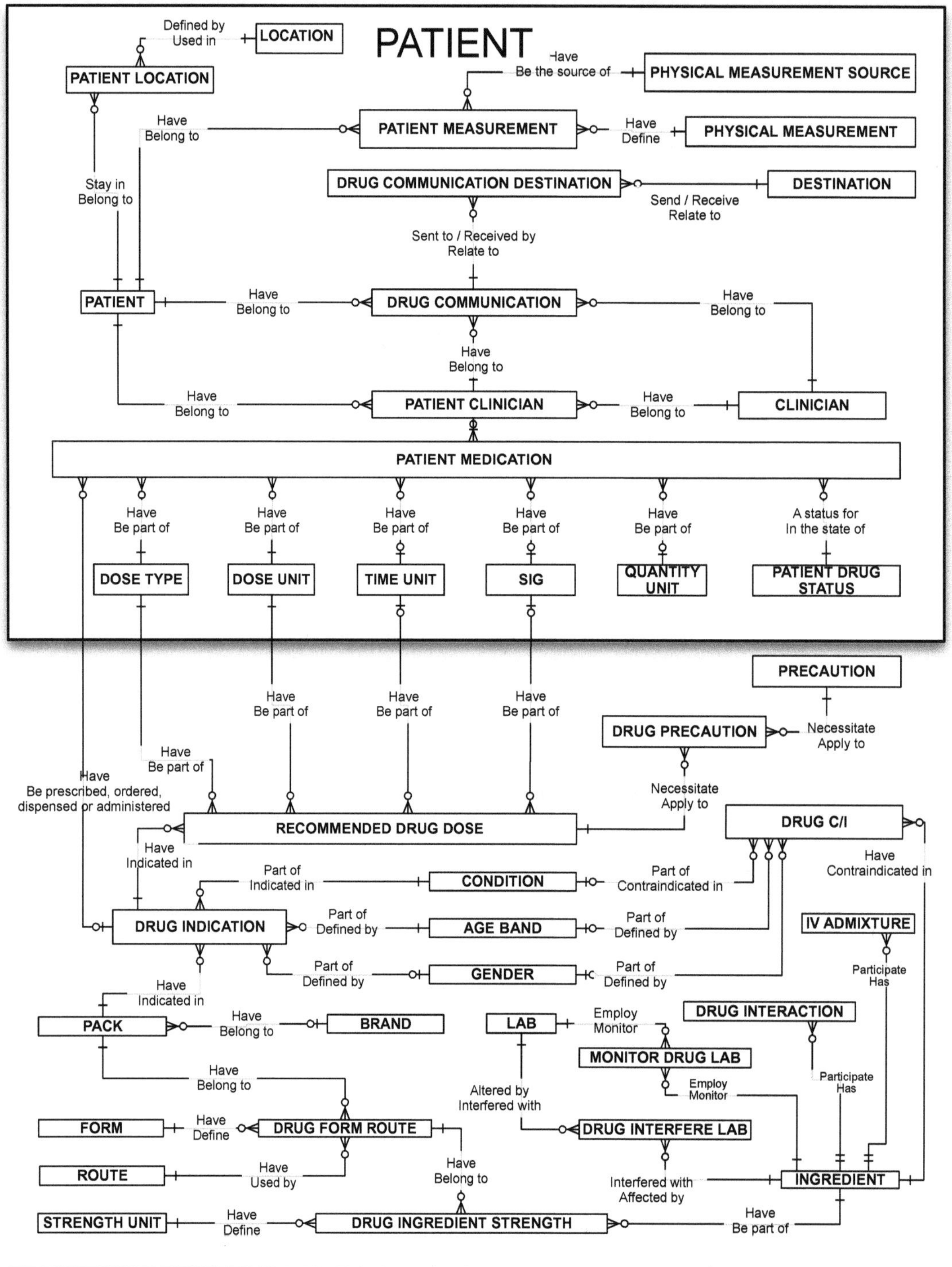

Figure 10.5 Conceptual model: Patient-related tables.

Semistructured Documents

Some of the communications in medicine cannot be atomized to discrete elements (yet?) and need to remain for the time being as the dreaded "free text." Dreaded, since without a *Natural Language Processing* (*NLP*) in place,

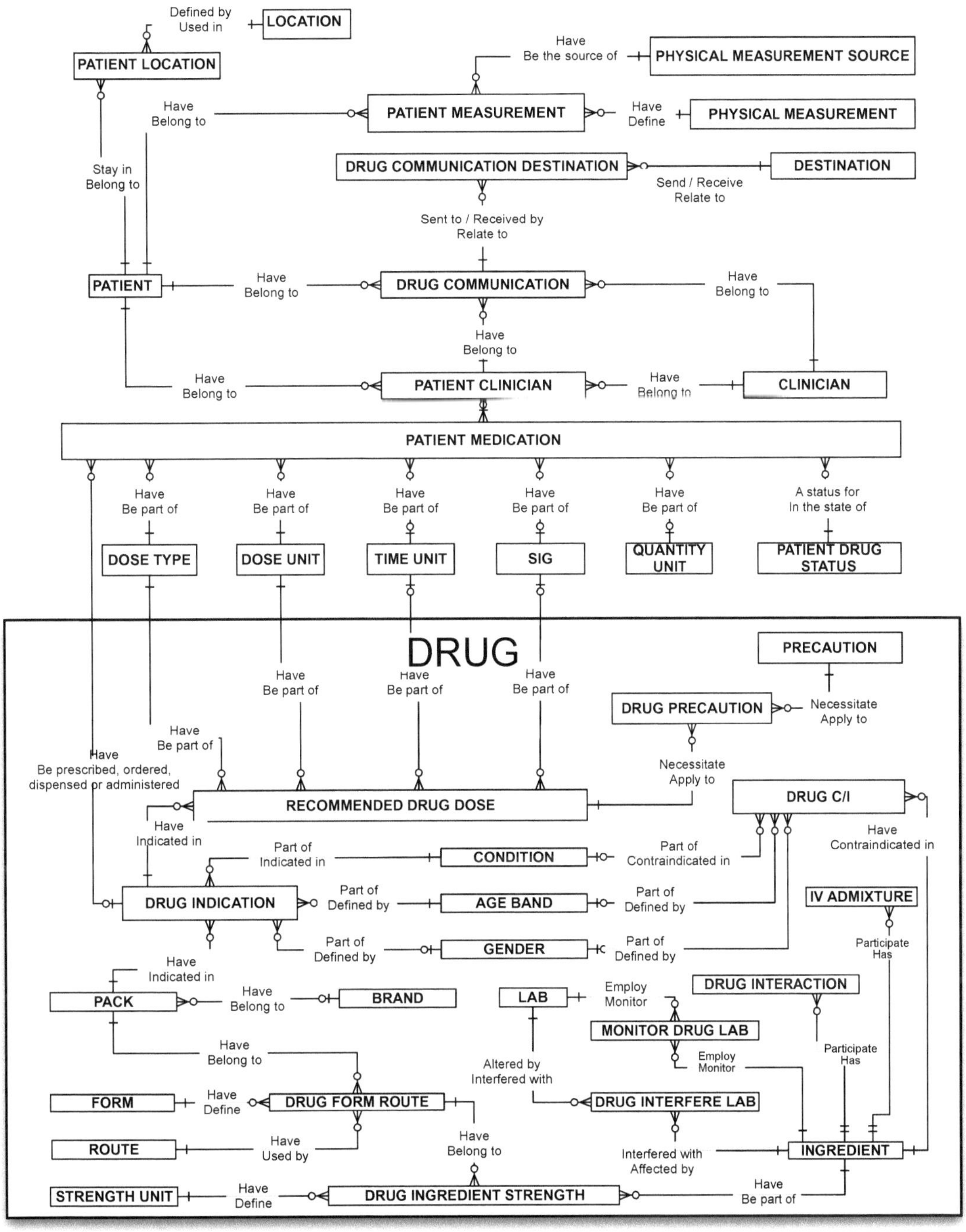

Figure 10.6 Conceptual model: Drug-related tables.

unstructured free text cannot be used by machines for reasoning; it cannot participate directly in machine logic.

Using XML, systems can parse without losing meaning or context such information as a structured *Diagnosis* from the semistructured:

- *History and Physical exam*
- *Patient discharge summary*

- *Notes*
- *Consultations*

Semistructured XML documents can help tame the notorious "free-text" beast. Created with a simple set of markup symbols, XML documents are both human and machine readable. A system adhering to *CCD, CDA,* or *CCR* (XML standards) can communicate meaningfully with others using semistructured documents.

Summary

The last chapter of the book presented the rationale for standardization and defined interoperability, vocabulary, and classification as well as a short version of the *OSI* model. The qualities of a modern medical ontology were introduced.

From a high-level view of the standardization organizations, we focused on the standards and terminologies in the U.S. HIT market, using the lens of the ARRA HITECH meaningful use rule. Next, we zoomed in on standards and vocabularies related specifically to the *Medications* system.

The synchronization and update processes, as sketched in DFD 7, revealed some of the intricate and delicate issues of interoperability (*HIE, RHIO, NHIN*).

Last, we grouped the data structures within the *Medications* system ERD as they may be influenced by interoperability into patient- and drug-related tables.

References

1. http://en.wikipedia.org/wiki/Interoperability (accessed March 14, 2011).
2. Tolk, A., C. Turnitsa, and S. Diallo. 2008. Implied ontological representation within the levels of conceptual interoperability model. *Intelligent Decision Technologies* 2: 3–19.
3. Wang, W., A. Tolk, and W. Wang. 2009. The levels of conceptual interoperability model: Applying systems engineering principles to M&S. *SpringSim'09 Proceedings.*
4. http://en.wikipedia.org/wiki/OSI_model (accessed March 14, 2011).
5. http://en.wikipedia.org/wiki/Ontology_(information_science) (accessed March 14, 2011).
6. Benson, T. 2010. *Principles of health interoperability HL7 and SNOMED.* New York: Springer.
7. Walker, J., E. Pan, D. Johnston, et al. 2005. The value of health care information exchange and interoperability. Health Affairs. http://content.healthaffairs.org/content/suppl/2005/02/07/hlthaff.w5.10.DC1 (accessed March 14, 2011).
8. The ONC coordinated federal health information technology strategic plan 2008–2012. http://dhhs.nv.gov/HOLD/HIT/docs/ONC20082012HITStrategicPlanSummary.pdf (accessed March 14, 2011).

9. Electronic code of federal regulations—Health information technology standards, implementations specifications, certification criteria and certification programs. http://ecfr.gpoaccess.gov/cgi/t/text/text-idx?c=ecfr&tpl=/ecfrbrowse/Title45/45cfr170_main_02.tpl (accessed March 16, 2011).
10. Summary of the ARRA HITECH rule on standards by John Halamka MD. http://geekdoctor.blogspot.com/2010/07/analysis-of-final-standards-rule.html (accessed March 16, 2011).
11. Cimino, J. 1998. Desiderata for controlled medical vocabularies in the twenty-first century. *Methods of Information in Medicine* 37: 394–403.
12. NHIN Cooperative Workgroup. Data Use and Reciprocal Support Agreement (DURSA). 2009. http://healthit.hhs.gov/ (accessed March 19, 2011).

Review Questions

1. What levels of interoperability, according to LCIM, must be reached before achieving conceptual interoperability?
2. Follow a message through the layers of the OSI model from sender to receiver.
3. What are the dimensions of an ontology?
4. Assume each entity needs to communicate with all the others in the following list: Emergency Department, Surgery, Internal Medicine, Pediatrics, Lab, Pathology, Imaging, Operating Room, Postanesthesia Care Unit, Intensive Care Unit, Neonatal Intensive Care Unit, Cardiology, Endocrinology, Psychiatry, Dietary Services, Physiotherapy, Billing, Medical Records. How many interfaces are needed?
5. List the qualities of a modern medical terminology according to Cimino.
6. In the following figure, draw the lines between clinical domains and the respective standards and vocabularies as applied in the United States.

CLINICAL DOMAINS

Diagnosis
Problem List
History & Physical exam
Lab
Medications
Imaging
Procedures
Communications
Summaries
Bed side devices
Reporting
Eligibility, claims
Privacy & Security

STANDARD/VOCABULARY

SNOMED
HL7
ICD
XML: CCD, CDA, CCR
LOINC
NCPDP
RxNorm
NDC
UNII
UCUM
NDF-RT
DICOM
CPT 4
IEEE/ISO
PQRI
DRG
ASC X.12
HIPAA
NIST/FIPS

7. Reporting to immunization registries and other quality programs is an expected part of HIE. Explain why reporting issues are not detailed in DFD 7.
8. List the issues HIE participants need to agree on when joining a RHIO.
9. List the tables in the *Medications* system ERD that may be involved in HIE in the following scenario: New patient data arrive from RHIO just as a patient is being admitted to the hospital: patient name, address, gender, date of birth, height, weight, medications taken at home, drugs and doses just ordered by the admitting physician.
10. List the tables in the ERD that a *Drug Knowledge Authority* may update periodically.

Appendix

Acronyms List

ADA	American Dental Association
ADC	Automated Dispensing Cabinet
AHIC	American Health Information Community
AHRQ	Agency for Healthcare Research and Quality
AMA	American Medical Association
ANSI	American National Standards Institute
APA	American Psychiatric Association
ARRA	American Recovery and Reinvestment Act
ASC	Accredited Standards Committee
ASCII	American Standard Code for Information Interchange
ASP	Application Service Provider
ASTM	American Society for Testing and Materials
BCMA	Bar Code Medication Administration
CAH	Critical Access Hospital
CAQH	Council for Affordable Quality Healthcare
CCD	Continuity of Care Document
CCHIT	Certification Commission for Health Information Technology
CCR	Continuity of Care Record
CDA	Clinical Document Architecture
CDC	Centers for Disease Control and Prevention
CDISC	Clinical Data Interchange Standards Consortium
CDS	Clinical Decision Support

CDT	Current Dental Terminology
CMS	Centers for Medicare & Medicaid Services
CPOE	Computerized Physician (Provider) Order Entry
CPT	Current Procedural Terminology
CVX	Codes for Vaccine Administered
DAW	Dispense As Written
DD	Data Dictionary
DEA	Drug Enforcement Administration
DFD	Data Flow Diagram
DICOM	Digital Imaging and Communications in Medicine
DRG	Diagnosis Related Group
DSM	Diagnostic and Statistical Manual of mental disorders
DURSA	Data Use and Reciprocal Support Agreement
EBM	Evidence Based Medicine
ED	Emergency Department
EHR	Electronic Health Record
ERD	Entity Relationship Diagram
eRx	Electronic prescription
FDA	Food & Drug Administration
FIPS	Federal Information Processing Standards
FK	Foreign Key
FTP	File Transfer Protocol
GIPSE	Geocoded Interoperable Population Summary Exchange
GO	Gene Ontology
HHS	Department of Health and Human Services
HIE	Health Information Exchange
HIPA	AHealth Insurance Portability and Accountability Act of 1996
HIT	Health Information Technology
HITECH	Health Information Technology for Economic and Clinical Health
HITPCHIT	Policy Committee
HITSCHIT	Standards Committee
HITSPHIT	Standards Panel
HL7	Health Level Seven

HL7 RIM	Health Level Seven Reference Information Model
HTTP	Hypertext Transfer Protocol
ICD	International Classification of Diseases
ICMP	Internet Control Message Protocol
IEEE	Institute of Electrical & Electronics Engineers
IHE	Integrating the Healthcare Enterprise
IHTSDO	International Health Terminology Standards Development Organization
IP	Internet Protocol
ISMP	Institute for Safe Medication Practices
ISO	International Organization for Standardization
JCAHO	Joint Commission on Accreditation of Healthcare Organizations
LCIM	Levels of Conceptual Interoperability Model
LOINC	Logical Observation Identifiers Names and Codes
MIDI	Musical Industry Device Interface
MIM	Mendelian Inheritance in Man
MPEG	Moving Picture Experts Group
MPFS	Medicare Physician Fee Schedule
MRN	Medical Record Number
MAR	Medication Administration Record
NANDA	North American Nursing Diagnosis
NCPDP	National Council for Prescription Drug Programs
NCQA	National Committee for Quality Assurance
NCVHS	National Committee on Vital and Health Statistics
NDC	National Drug Code
NDF RT	National Drug File Reference Terminology
NEMA	National Electrical Manufacturers Association
NHIN	Nationwide Health Information Network
NIST	National Institute of Standards & Technology
NLM	National Library of Medicine
NPI	National Provider Identifier
NQF	National Quality Forum
OASIS	Organization for the Advancement of Structured Information Standards
ONCHIT	Office of the National Coordinator for Healthcare Information Technology

OSI	Open Systems Interconnection
OTC	Over The Counter
PBM	Pharmacy Benefit Management
PHR	Personal Health Record
PHSA	Public Health Service Act
PK	Primary Key
PPACA	Patient Protection and Affordable Care Act
PPP	Point to Point Protocol
PQRI	Physician Quality Reporting Initiative
PQRS	Physician Quality Reporting System
QRDA	Quality Reporting Document Architecture
REST	Representational State Transfer
RHIO	Regional Health Information Organization
RPC	Remote Procedure Call
RS 232	Recommended Standard 232
Rx	Prescription
SDOs	Standards Development Organizations
SDTM	Study Data Tabulation Model
SMTP	Simple Mail Transfer Protocol
SNMP	Simple Network Management Protocol
SNOMED CT	Systematized Nomenclature of Medicine Clinical Terms
SOAP	Simple Object Access Protocol
SSA	Structured Systems Analysis
STD	State transition Diagram
TCP	Transmission Control Protocol
UCUM	Unified Code for Units of Measure
UDP	User Datagram Protocol
UID	Unique Identifier
UMLS	Unified Medical Language System
UNII	Unique Ingredient Identifier

UPC	Universal Product Code
USPUS	Pharmacopeia
VHA	Veterans Health Administration
XDS	Cross-enterprise Document Sharing
XML	eXtensible Markup Language

Answers to Review Questions

Chapter 1

1. Narrative tends to become too long a document.
   Text cannot be easily partitioned.
   Text can easily get ambiguous.
   Narrative is difficult to maintain in a consistent state.
   Text tends to become redundant.
2. Graphical
   Top-down
   Partitioned analysis
   Minimal redundancy
   Predicts system behavior
   Transparent
   Easy to learn and use
   Technology independent
3. There is a data flow between external entities, outside the system boundaries.
4. There is a data flow between two data stores, without a process in between.
5. An external entity has a direct read/write access to a data store without any process in between.
6. Process 9.2 *CDS* provides information to process 9.1 without a data store. Where does it get this information?
7. *Clinical Repository* is a "black hole"
   There are only input data flows and no output. The opposite situation—"spontaneous generation"—data stores should be avoided as well.
8. Even though the processes are numbered 1.1, 1.2, 1.3, and 1.4, the numbers do not reflect the order of processing; the numbering is used for leveling the DFDs and not for defining actual processing order. Thus, order cannot be defined by a DFD.
9. Errors in the figure accompanying Question 9:
   PATIENTS is plural; it should be *PATIENT* as singular
   Dr. Livingstone, as nice as he or she may be, cannot be the name of an entity.
   THING is too ambiguous to name an entity.
   TYLENOL and *ACETAMINOPHEN* are instances of entities and cannot be used for entity names.
   "May treat," "Take," and *"Has"* are only one direction of the relationships. The name of a relationship should state both directions.
   There is a relationship without name.
   There is a relationship *"To be determined."*

There is an incorrect relationship (just a line, without any symbols at the ends) between *PHARMACY* and *TYLENOL*.

"Always" (or *"must"* or *"may"*) should not be part of the relationship name.

The forms at the ends of the line define the cardinality and optionality of a relationship.

10.

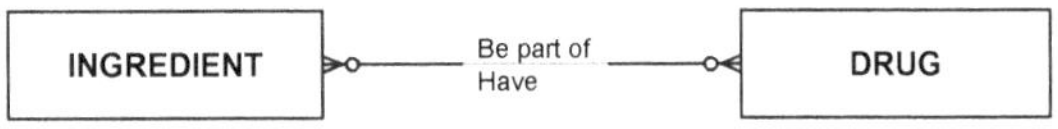

Conceptual level

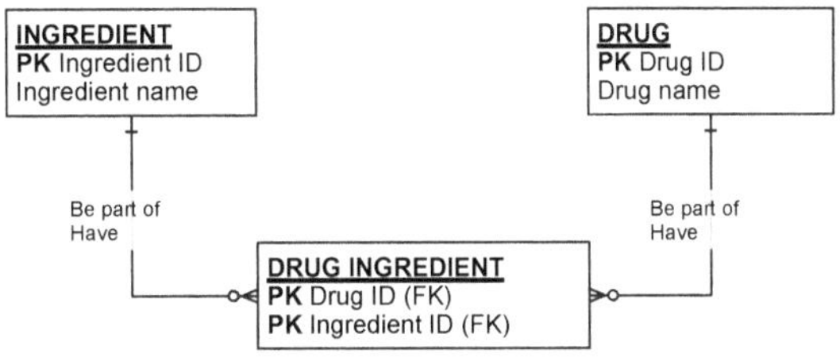

Logical level

(Intersection entity has a compound PK)

Chapter 2

1. The figure with Question 1 depicts a context diagram since there is only one process: 0 Medications System.
   There should be no data stores on a context diagram; thus, the error is the two data stores illustrated: *RECOMMENDED DRUG DOSE* and *CLINICAL REPOSITORY*.
2. The relationship between *PATIENT* and *PATIENT LAB RESULT* is *M:N*. One instance of *PATIENT LAB RESULT* can belong to only one patient and not to many.
3. The relationship between *BRAND* and *PACK* is wrong. It should be the opposite of what is depicted: One *BRAND* may have many *PACKs*.
   The relationship of *ROUTE* should not be with *BRAND*.
   Same error with *FORM*: The relationship should not be with *BRAND*.
4. Drugs are indicated in relation to *PACK* and not to *INGREDIENT*. Drugs are contraindicated according to *INGREDIENT* and not to *PACK*.
5. The concepts of *TIME*, *DOSE*, *QUANTITY*, and *STRENGTH* are not entities, but attributes. The concepts that should be depicted in the model are

   TIME UNIT
   DOSE UNIT
   QUANTITY UNIT

STRENGTH UNIT

6. A drug indication must have a condition defined; thus, the relationship is mandatory. Contraindication may or may not have a condition defined since ingredients may be contraindicated in relation to age band or gender without any condition; thus, the relationship is optional.
7. A drug indication may not have a gender defined.
   Same situation with contraindications: It may not have the gender as part of the definition. Thus, the relationships are optional and not mandatory.
8. Both *RECOMMENDED DRUG DOSE* and *PATIENT MEDICATION* may not have the frequency and *SIG* defined; thus, the relationships are optional and not mandatory.
   PATIENT MEDICATION may not have duration and quantity defined, so these relationships are optional as well.
9. The child's weight = 22 pounds = 10 kg.
   Each dose may be up to 10 kg × 12 mg/kg/dose = 120 mg/dose.
   120 mg/dose/100 mg/ml = 1.2 ml/dose
   60 mg/kg/day × 10 kg = 600 mg/day
   600 mg/day/100 mg/ml = 6 ml/day
   30 ml/bottle/1.2 ml/dose = 25 doses/bottle
10. Acetaminophen in the example:
    Form: Solution
    Strength: 100 mg/ml
    Route: PO (by mouth)
    Recommended dose: 10–12 mg/kg/dose
    Frequency: 5 doses/day
    Maximal dose: 60 mg/kg/day
    Dispensed original package: 30-ml/bottle

Chapter 3

1. Prescribing clinician accesses *PATIENT MEDICATION* directly, without a process in between.
2. The process needs to access a data source to retrieve the information provided to the *PRESCRIBING CLINICIAN*. The data source is not illustrated on the diagram.
3. The data stores *PATIENT MEDICATION* and *RECOMMENDED DRUG DOSE* have been interchanged.

4. The data stores related to external actors such as *PBM* should not be detailed in the DFD of the system; it is beyond the system boundaries. (In addition, an actor should not access a data store directly, but only through a process.)
5. Figure 3.2 illustrates DFD 1.
   CDS is workflow number 5, so it should not appear on a lower-level DFD.
6. Figure 3.6 illustrates DFD 1.2: 1.2.1, 1.2.2, and 1.2.3.
   Process 1.5 *Consider Formulary* is a higher-level DFD and thus is not depicted on this diagram.
7. Figure 3.1 is a high-level view of the *Prescribe* workflow.
   Decomposed in DFD 1 (Figure 3.2), it was replaced by an external actor, *NON-DRUG REPOSITORY*, and one data store, *PATIENT MEDICATION*
8. The processes depicted in the figure for Question 8 in which a dose is selected before a drug was chosen and before the patient data were reviewed, as awkward as it may seem, are correctly numbered. The numbering of processes on a DFD does not imply their actual order or sequence of processing by the system.
9. The entity to provide these discrete data elements is the *PATIENT* table.
10. The safer therapeutic alternative is *CDS* responsibility (workflow 5).
    The less-expensive alternative is *PBM* responsibility (via process 1.5 *Consider Formulary*).

Chapter 4

1. Sources of information for drug and dose:
   RECOMMENDED DRUG DOSE
   PATIENT MEDICATION
   ORDER SET REPOSITORY
   Formulary—PHARMACY
   CDS
2. Workflows/processes influencing CPOE drug and dose:
   Select Drug
   Select Dose
   Use Order Set
   CDS
   Consider Formulary
   Reconcile Meds

3. Similar processes in eRx and CPOE:
   Communicate
   Review Patient Data
   Select Drug
   Select Dose
   Consider Formulary
   CDS
   Sign

4. Table 4.1

5. The source of *Formulary* information is the *PHARMACY.*
   The process responsible for retrieving the *Formulary* information is 2.5 *Consider Formulary.*

6. *Order Set* categories:
   Patient details
   Order Set Title (condition, age band, A/D/T)
   Vitals to be monitored
   Respiratory/Oxygen therapy
   IV access and fluids
   Nutrition and Diet
   Labs
   Medications
   Imaging
   Consults
   Nursing care plans
   Blood products
   Patient education

7. Benefits of using *Order Sets*
   Improved quality by reducing variability of care processes (standardization)
   Improved adherence with guidelines, best practices, and evidence-based medicine
   Reduced errors while ordering
   Increased speed of documentation
   Reduced costs while eliminating unnecessary tests or treatments

8. Move the *Priority* attribute (column) from *DRUG COMMUNICATION* to *DRUG COMMUNICATION DESTINATION* as in this figure.

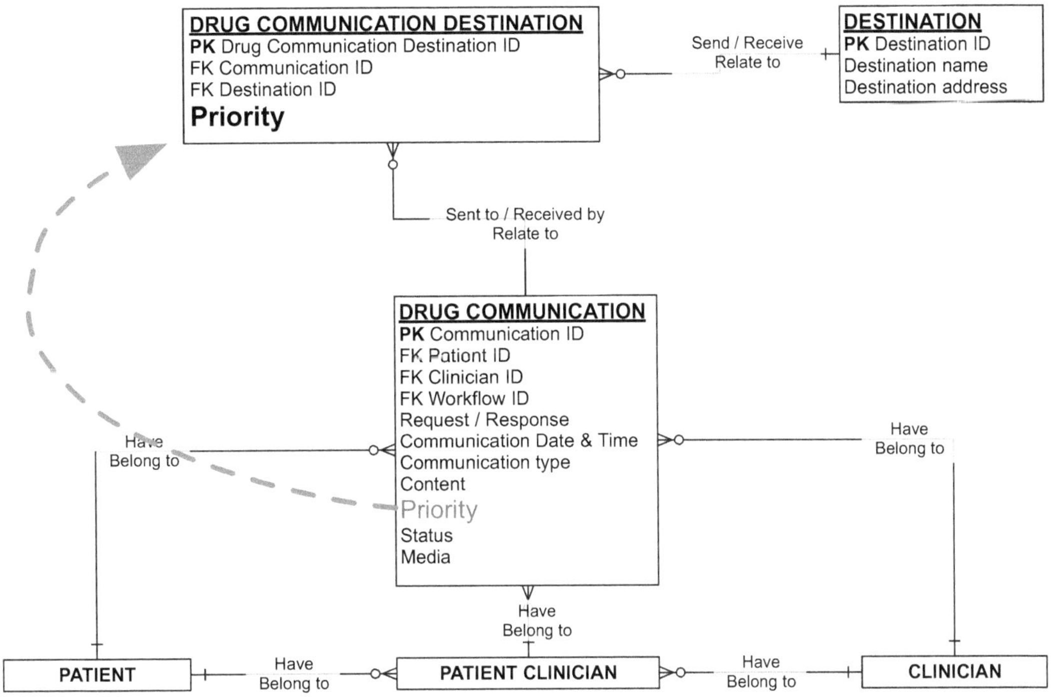

9. 100 patients × 526,000 min/year × 10 vital signs/patient × 1 row/vital sign/min = 526 million rows/year.

10.

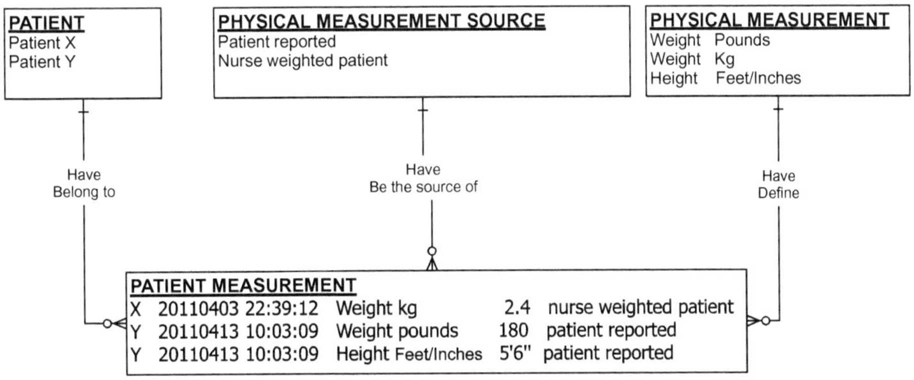

Chapter 5

1.

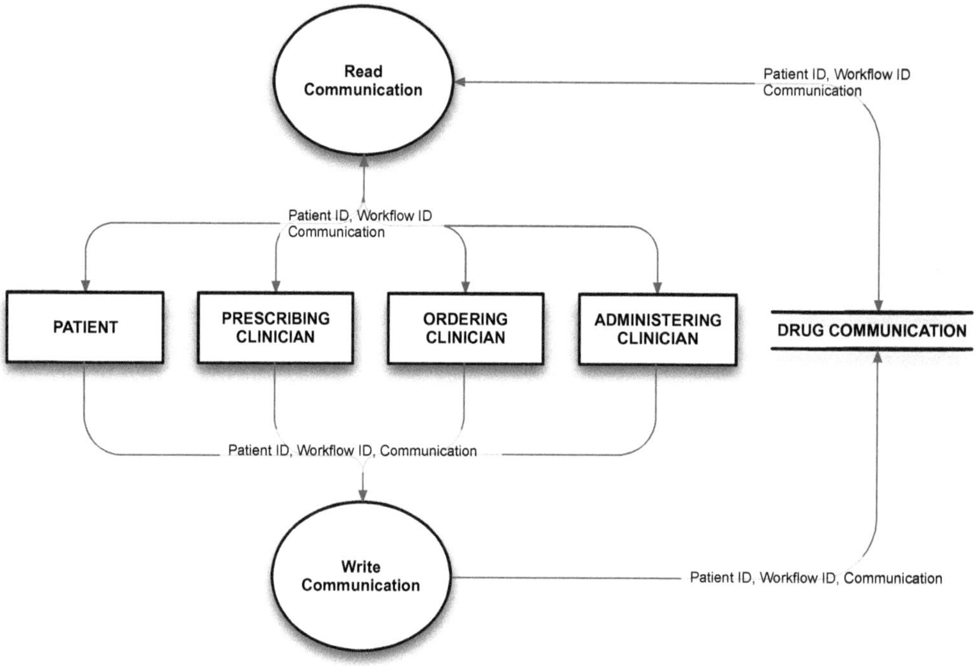

2. *DRUG COMMUNICATION*
3. *NON-DRUG REPOSITORY*
4. *PBM/FORMULARY*
5. Process 3.3 *Select Drug* receives information on a drug availability from *INVENTORY.*
6. Process 3.5 *Consider Formulary* receives information on therapeutic alternatives from *FORMULARY.*
7. Unfortunately, this type of error is very difficult to prevent with the safety mechanisms defined in this chapter. Using an automated video surveillance system may prevent this error.
8. Process 3.5 *Consider Formulary* using information from the *FORMULARY* entity
9. Process 3.2 *Review Patient Data* and *CDS*
10. *DRUG COMMUNICATION*

Chapter 6

1. Scans badge (validates *Clinician*)
   Scans patient's wristband (validates *Patient*)
   Scans unit dose (validates *Drug, Dose, Route*)
   Scans Pump (validates *Drug, Dose, Time*)
   Sign
2. Documenting the self-administration of drugs
   Monitoring drug effects
   CDS
3. Pharmacy communicates with the other actors through a common data store: *DRUG COMMUNICATION* table.
4. This error may be prevented by process 4.4.3 or 4.4.4
5. *Patient weight* = 110 pounds = 50 kg.
   Drug Rate per min = 50 kg × 1.2 mcg/kg/min = 60 mcg/min.
   Drug Rate per hour = 60 mcg/min × 60 min/hour = 3,600 mcg/hour = 3.6 mg/hour.
6. *Drug Rate* = 3 kg × 0.2 mcg/kg/min = 0.6 mcg/min.
   Strength = 1.8 mg/100 ml = 1,800 mcg/100 ml = 18 mcg/ml.
   Volume Rate = 0.6 mcg/min/18 mcg/ml = 0.0333 ml/min.
   Drops Rate = 0. 0333 ml/min × 60 drops/ml = 2 drops/min.
7. *Volume Rate* = 0.8 ml/hour = 0.01333 ml/min. That is less than the minimum possible with a drop-based infusion device, which is 0.0167 ml/min = 1 drop/min.
8. Patient may document the self-administration of prescribed, ordered, OTC, and dispensed drugs.
9. *Volume rate* = 6 drops/min = 0.1 ml/min.
   Drug rate = 0.1 ml/min × 10 mg/ml = 1 mg/min = 1,000 mcg/min.
   Dose = 1,000 mcg/minute/100 kg = 10 mcg/kg/min.
10. It is impossible to "concentrate" a drug solution into a higher strength by eliminating the solution's fluids. The only way to increase the strength of a solution is to add more of the drug.

Chapter 7

1. Product A is slightly more effective, and product B is significantly more efficient.
2. The medications list is separated from the prescription pad, so the clinician cannot edit the existing drugs list.
   Display does not show other parameters, such as diagnosis, problems, and lab.

Low-density display is being translated into a high number of clicks in order to view many screens.
It does not support the clinician's cognitive processes.

3. It is not clear which value is abnormal.
   There is only a small * sign to denote abnormality and no use of color.
   Range of a test being numeric needs further cognitive processing on the clinician's behalf.
   The magnitude or direction of abnormal values is not displayed. Is this value very high or just a bit low?
   There is no history of previous lab results.
   No trends are displayed over time.
   No comparison can be made between different lab tests.
4. There are lab tests (Potassium) interspersed between the Glucose lab values
   Similarly, there is a KCl order in between Insulin orders.
   Cause and effect and titration to effect phenomena cannot be discerned since Glucose lab values cannot be visually compared to the Insulin treatment.
   Magnitude of change is not displayed.
   The lab time axis goes from past at the top to present at the bottom.
   The Orders time axis is in the opposite direction: from present at the top to past at the bottom.
5. (a) 30 screens
   (b) 2 screens
6. Instead of repeating information on seven lines, use only one line with drop-down menus for route, drug name, strength, strength unit, dose, dose unit, maximal dose, and maximal dose unit.
7. The application is not correlating the route (Eye) with the form (Rectal suppository). If the two were correlated, the form would constrain the routes and vice versa.
8. The application is not correlating the drug, form, and route with the correct strengths available.
9. Hide beyond hyperlinks the address and contact details and reveal instead the allergies, weight, and height in a central visible location.
10. CDS alerts and formulary recommendations are too dominant.
    The font size and color are alarming.
    Multiple alerts are open at the same time, covering other information.
    The clinician workflow is interrupted until the system is satisfied.
    The effort necessary to overcome the alerts is disproportionate to the severity and significance of the alerts.
    The clinician has no control over and no clear escape from the situation displayed.
    Thus, the user is not expected "to be in a good mood when done."

Chapter 8

1. False-positive alarms cause alert fatigue, which may render the CDS useless. False-negative alarms induce a false state of security and may cause a medication error.
2. If the two population distributions are overlapping, one cannot decrease both the false-positive and the false-negative test results.
3. Drug-drug interaction and duplicate therapy (same intent).
4. There are 8 + 2 = 10 drugs in the set. Thus, there would be $(10^2 - 10)/2 = 45$ pairs to consider.
5. The duplicate therapy process using the list of rules exceptions in *RULES REPOSITORY* should be able to conclude the combination of regular and slow-acting insulin is common and appropriate.
6. Filter drug.
   Adjust dose.
   Consider patient condition.
   Consider lab processes.
7. Consider drug lifestyle.
8. CDS algorithm should bypass pregnancy considerations for male patients. Information on the dangers a drug may pose to women of childbearing age should be part of the education materials, not the CDS false-positive alerts.
9. Do not use *erythromycin* and *ketoconazole* or use the combination but decrease *erythromycin* dose or increase intervals between *erythromycin* doses.
   CDS should not consider both *erythromycin* + *ketoconazole* and *ketoconazole* + *erythromycin*. That would be excessive.
10. Process 5.6.3 Drug Interfere with Lab.

Chapter 9

1. Single versus multiple patients
   Clinical versus administrative
   Patient-unrelated reports
2. PQRS reporting entitles physicians to a yearly bonus until the end of 2014. Those opting for no PQRS reporting may suffer annual negative adjustments of their Medicare Physician Fee Schedule.
   Reporting on MU entitles a physician or a hospital to financial incentives through Medicare or Medicaid.

3. Outcomes, processes, and structure
4. Safe
   Effective
   Efficient
   Timely
   Patient centered
   Equitable
5. Initial denominator = 594 patients.
   Not all patients had *"isolated" CABG,* and 42 patients need to be excluded because they had CABG plus additional valve(s) repair: 22 mitral valve, 15 tricuspid valve, and 5 patients had both valves repaired = 42.
   Additional patients excluded from denominator: 8 transferred, 13 died, and 31 had contraindications to beta-blocker therapy = 52 patients.
   Denominator - Exclusions = 594 – 42 – 52 = 500 patients.
   420 (patients on beta-blockers at discharge)/500 (denominator) = 84%.
6. EHR based
   Hybrid
   Registry
   Claims based
   Paper medical records
7. Data are extracted from multiple sources, cleansed, transformed, and consolidated in a data warehouse in a format that is optimized for query and analysis and does not have an impact on EHR performance.
8. Star, snowflake, and galaxy/fact constellation schemas
9. Filter, aggregate, and order
10. 5,000 physicians × 240 measures × 12 months = 14.4 million cells/year

Chapter 10

1. Levels of interoperability:
   Technical
   Syntactic
   Semantic
   Pragmatic
   Dynamic
   Conceptual
2. Application → Presentation → Session → Transport → Network → Data link → Physical → Physical → Data link → Network → Transport → Session → Presentation → Application

3. Dimensions of ontology:
   Concepts
   Attributes
   Classification
   Relationships
   Events
   Rules
   Axioms

4. There are 18 entities listed. $(18^2 - 18)/2 = 153$ interfaces.

5. Qualities of a modern medical terminology:
   Content orientation
   Concept orientation
   Persistence
   Unique (meaningless) identifiers
   Multidimensional classification
   No NEC
   Granularity
   Graceful evolution
   Machine-usable context

6.

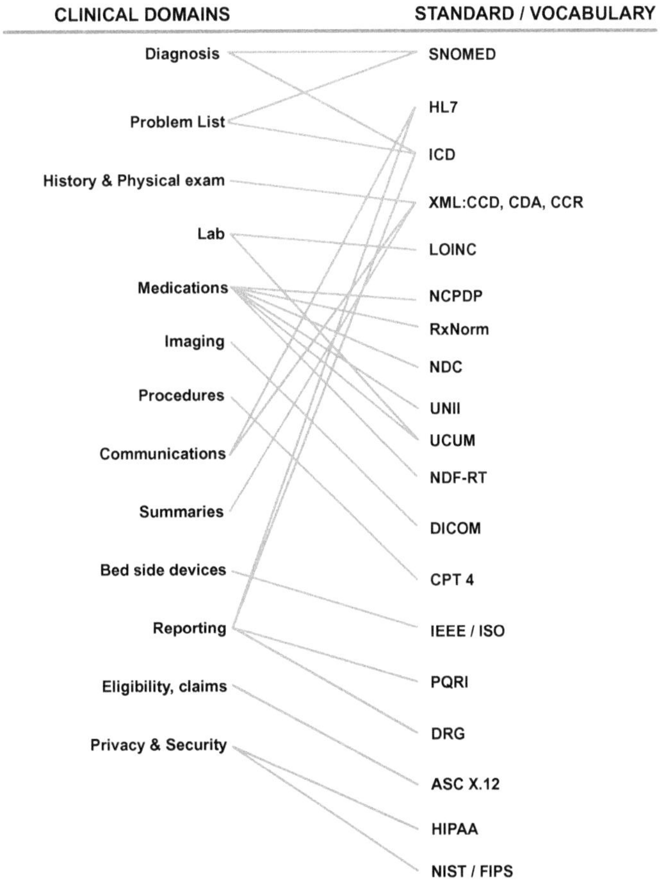

7. Reporting is the responsibility of workflow number 6 and is detailed in its own chapter with DFD 6.

8. RHIO issues:
   Centralized, federated, or hybrid model
   Who participates?
   Standards and vocabularies
   Data exchanged
   How are data routed?
   How is patient uniquely identified?
   How are providers identified?
   Who controls access to data?
   How is communication initiated? What are the triggers?
   HIE performance
   Sender/receiver responsibilities
   Dispute resolution
   Privacy and security

9. Patient-related tables in ERD, potentially involved in HIE:
   PATIENT
   PATIENT LOCATION
   PATIENT MEASUREMENT
   DRUG COMMUNICATION
   DRUG COMMUNICATION DESTINATION
   CLINICIAN
   PATIENT CLINICIAN
   PATIENT DRUG STATUS
   PATIENT MEDICATION

10. Drug-related tables in ERD, potentially involved in an update:
    RECOMMENDED DRUG DOSE
    DOSE TYPE
    DOSE UNIT
    TIME UNIT
    SIG
    PRECAUTION
    DRUG PRECAUTION
    DRUG INDICATION
    DRUG C/I
    CONDITION
    AGE BAND
    GENDER
    PACK
    BRAND
    FORM

ROUTE
INGREDIENT
DRUG FORM ROUTE
STRENGTH UNIT
DRUG INGREDIENT STRENGTH
LAB DRUG INTERACTION
MONITOR DRUG LAB
IV ADMIXTURE
DRUG INTERFERE LAB

Index

A

D

N

O

P

Q

R

S

X

Z